Advanced Numerical Modelling of Wave Structure Interactions

Editors

David M Kelly
Centre for Environment, Fisheries and Aquaculture Science (CEFAS)
Pakefield Road, Lowestoft, NR33 0HT, UK

Aggelos Dimakopoulos
HR Wallingford, Wallingford OX10 8BA, UK

Pablo Higuera
Department of Civil and Environmental Engineering
Faculty of Engineering, University of Auckland
Auckland, 1010, New Zealand

T0174403

CRC Press
Taylor & Francis Group
Boca Raton London New York

CRC Press is an imprint of the
Taylor & Francis Group, an **informa** business

A SCIENCE PUBLISHERS BOOK

CRC Press
Taylor & Francis Group
6000 Broken Sound Parkway NW, Suite 300
Boca Raton, FL 33487-2742

© 2021 by Taylor & Francis Group, LLC
CRC Press is an imprint of Taylor & Francis Group, an Informa business

No claim to original U.S. Government works

Version Date: 20200918

ISBN 13: 978-0-815-35997-5 (hbk)
ISBN 13: 978-0-367-61938-1 (pbk)
ISBN 13: 978-1-351-11954-2 (ebk)

This book contains information obtained from authentic and highly regarded sources. Reasonable efforts have been made to publish reliable data and information, but the author and publisher cannot assume responsibility for the validity of all materials or the consequences of their use. The authors and publishers have attempted to trace the copyright holders of all material reproduced in this publication and apologize to copyright holders if permission to publish in this form has not been obtained. If any copyright material has not been acknowledged please write and let us know so we may rectify in any future reprint.

Except as permitted under U.S. Copyright Law, no part of this book may be reprinted, reproduced, transmitted, or utilized in any form by any electronic, mechanical, or other means, now known or hereafter invented, including photocopying, microfilming, and recording, or in any information storage or retrieval system, without written permission from the publishers.

For permission to photocopy or use material electronically from this work, please access www.copyright.com (http://www.copyright.com/) or contact the Copyright Clearance Center, Inc. (CCC), 222 Rosewood Drive, Danvers, MA 01923, 978-750-8400. CCC is a not-for-profit organization that provides licenses and registration for a variety of users. For organizations that have been granted a photocopy license by the CCC, a separate system of payment has been arranged.

Trademark Notice: Product or corporate names may be trademarks or registered trademarks, and are used only for identification and explanation without intent to infringe.

Library of Congress Cataloging-in-Publication Data

Names: Kelly, David M., editor. | Dimakopoulos, Aggelos, editor. |
    Caubilla, Pablo Higuera, editor.
Title: Advanced numerical modelling of wave structure interactions /
    editors, David M Kelly, Aggelos Dimakopoulos, Pablo Higuera Caubilla.
Description: First edition. | Boca Raton, FL : CRC Press/Taylor & Francis
    Group, [2021] | Includes bibliographical references and index. |
    Summary: "Due to the growth in cheap computing power, over the last
    decade the numerical modelling of wave structure interaction has entered
    a new phase. Increasingly sophisticated mathematical models now allow
    for the most realistic modelling of fluid structure interaction ever
    achieved. Advanced Numerical Modelling of Wave Structure Interaction
    contains contributions from a wide range of experts in the field of
    numerical modelling for coastal engineering. Chapters present
    state-of-the-art numerical approaches for aspects of fluid structure
    interaction ranging from wave-breakwater interaction to the effect of
    vegetation on currents. This book aims to serve as a comprehensive guide
    to up-to-date numerical techniques used for a wide variety of fluid
    structure interaction problems within a coastal engineering context"--
    Provided by publisher.
Identifiers: LCCN 2020024251 | ISBN 9780815359975 (hardcover)
Subjects: LCSH: Wave resistance (Hydrodynamics)--Mathematical models. |
    Hydraulic structures--Mathematical models.
Classification: LCC TA654.55 .A38 2021 | DDC 627--dc23
LC record available at https://lccn.loc.gov/2020024251

**Visit the Taylor & Francis Web site at
http://www.taylorandfrancis.com**

**and the CRC Press Web site at
http://www.routledge.com**

Foreword

I am not a numerical modeller myself, but over the last 30–40 years I have been an interested user of numerical models, and of their results, to help illuminate the interactions between waves and coastal structures. It is therefore as a 'customer' for numerical models of wave/structure interactions as essential complements to empirical design tools, and physical modelling, that I welcome this guidance on the developing toolkit of wave/structure numerical models, often termed Computational Fluid Dynamics (CFD).

In any discussion on the development of new tools, I am reminded that in 1847, John Scott Russell wrote to the Institution of Civil Engineers that: "*...it may be considered rather hard by the young engineer that (s)he should be left to be guided entirely by circumstances, without the aid of any one general principle for her/his assistance, ... when (s)he has to decide on a system for best opposing the force of the sea...*" Not only did numerical modelling not exist at the time, but then nor did physical modelling (until the 1930s), or even any useful analytical methods.

Over the last nearly 100 years, physical modelling of water waves and flows at structures, along coasts, and in harbours, has developed substantially, as in more recent years have numerical models, providing answers to questions that might be too complex or too extensive to model in realistic physical modelling facilities.

The authors of this book have focussed on near-structure fluid/structure interactions giving the reader descriptions of some of the latest advances, but with an emphasis on practical use of these advanced tools in solving real-world problems. But as well reviewing what is possible now, the authors have also given pointers as to how the capabilities of these tools will develop in future.

It is anticipated that this book will be of both interest and utility to engineers in consultancy as well as to researchers at MSc or PhD levels.

I very much hope that readers will find this book helpful in understanding what numerical modelling tools are available, how they work, and indeed alert you to the problems to be overcome in applying these advanced methods to real-life problems.

William Allsop
Abingdon and Edinburgh

September 2020

Contents

Introduction

A distinguishing characteristic of anthropogenically engineered coastal environments is the highly complex interaction that exists between solid structures and the waves that impinge upon them. Similar interchanges also occur between waves and natural structures, such as reefs or rock outcrops. The interaction takes place because of the strong influence that the structure shape imposes on the hydrodynamics, which is often complex, necessarily fully three-dimensional and two-way, because of the dynamically induced movement of the structure that can result from the hydrodynamic forcing. In turn, structure motion simultaneously leads to changes in the hydrodynamics processes themselves. The problem of modelling fluid structure interactions accurately has grown in importance over the last two decades with the increasing deployment of various types of marine devices, breakwaters protecting port terminals, land reclamations and biological or natural defense solutions; all of which lie within the coastal zone.

Throughout the 1990s and early 2000s the boundary integral element method (BIEM) was used extensively, and almost exclusively, to simulate non-linear water waves and the interaction between steep waves and structures. Whilst this method has proved to be efficient for the determination of fluid-structure interaction (especially in two spatial dimensions) it is based on potential flow theory, which requires the flow to be both irrotational and inviscid. More recently, within the last 10- to 20-years, Computational Fluid Dynamics (CFD) approaches have been employed to solve the full fluid motion described by the Navier-Stokes, or Reynolds Averaged Navier-Stokes (RANS), equations incorporating a free surface (air-water interface). This type of modelling includes viscous and rotational effects and, by employing sophisticated numerical techniques, surface effects such as wave breaking (with full overturning in plunging breakers) can be simulated. Therefore, when viscous/turbulent effects and/or air entrainment effects cannot be justifiably neglected, the use of CFD models is required for meaningful simulations. The current approach is to employ the CFD model only in the immediate surroundings of the structure due to two primary limiting factors. First, the computational cost of simulating relatively large computational domains is prohibitive. Second, as a consequence of numerical (i.e., non-physical) diffusion, the majority of available CFD models cannot accurately model waves propagating over long distances (e.g., of the order of tens of kilometers).

The goal of this book is to bring together a comprehensive catalogue of state-of-the-art numerical modelling techniques for all key aspects of wave structure interaction within the coastal zone. Each of these approaches has its advantages and drawbacks, and consequently it has advocates and detractors; therefore, we aim to present an unbiased view and let each of the methods speak for themselves. A unitary approach is sought by evaluating different coastal structures subjected to a range of coastal hydrodynamics in each chapter. These vary from wave generation and propagation to the violent wave structure interaction that leads to greenwater overtopping, as well as the key problems of impulsive wave loading on, and scour around, coastal structures.

Chapter 1 addresses the state-of-the-art in numerical wave generation and absorption techniques. The chapter provides a thorough review of wave generation and absorption techniques utilized by modern CFD models. Approaches including internal generation as well as static- and moving-boundary wave generation procedures are studied critically, with the advantages and disadvantages of each technique being highlighted. Passive and active (static-

and moving-boundary) wave absorption techniques, including the most recent advances, are also analyzed. The chapter ends with the discussion on future prospects for numerical wave generation.

Chapter 2 considers the use of the equations of irrotational inviscid flow coupled with CFD to produce far-field boundary conditions. This technique saves on computational time and minimizes the diffusion of wave energy, thus producing highly accurate kinematics that are used as input on the CFD solver, which models the flow close to the structure(s). In the chapter, Buldakov revives, and modernizes, a novel Lagrangian model and provides a convincing argument for the use of one-way coupling, as opposed to full coupling between models.

Chapter 3 describes wave breaking and air entrainment processes from a dual physical-numerical perspective. Lubin introduces the requisite massively parallel simulation tools, which are key to unravel the origin of complex three-dimensional turbulent structures existing under breaking waves. Special attention is paid to vortex filaments, which are only found beneath plunging breaking waves. The chapter finishes with a thorough review of existing work, focussed on discussing the gaps of knowledge and the most important challenges for future works.

In Chapter 4, Ma and coauthors consider the modelling of air compressibility and aeration effects in wave structure interaction, including important phenomena such as cavitation. In particular, the chapter considers the numerical modelling of air entrainment in breaking wave impacts on coastal and offshore structures, during which air pockets are trapped and break up to form bubbles. This study highlights that numerical modelling has the potential to produce valuable insights into these complex processes that are particularly hard to determine from experiments alone.

Chapter 5 presents a detailed description of the numerical treatment for the simulation of violent wave impacts on structures, and the associated wave loading, using the meshless smoothed particle hydrodynamics (SPH) technique. The chapter summarises some of the cutting edge research into SPH undertaken at the Italian Ship Model Basin (INSEAN) in Rome. The emphasis in this chapter is primarily on the advantage of a meshless Lagrangian method for simulating the complex free surface (air-water interface) configurations. Along with the treatment of the free surface, the requirements and effectiveness of the SPH technique in dealing with solid boundaries are also discussed.

Chapter 6 investigates the interaction of waves and porous coastal structures using a set of volume-averaged Navier-Stokes (VARANS) equations implemented in *olaFlow*, an open source model based on OpenFOAM®. The VARANS equations are derived to include gradients of porosity in both space and time to enable the simulation of moving porous media. In the first test case, RANS modelling capabilities for pre-designing structures are demonstrated by testing different alternatives to protect a breakwater against the extreme action of a tsunami-like wave. In the second application, the swash flow of a solitary wave on a beach, comprising thousands of spheres modelled with the discrete element method (DEM), is studied and compared against a smooth slope counterpart.

Chapter 7 describes the state-of-the-art in the CFD based modelling of scour around coastal structures. In the chapter, the authors detail a hybrid Eulerian-Lagrangian approach for three-phase (air-water-sediment) modelling of scour, which is implemented in a heavily modified version of the well-known interFoam solver from the OpenFOAM® package. Consideration is given over to modelling interphase momentum transfer, fluid volume exclusion and sediment packing, as well as suitable turbulence closure models within the Reynolds averaged Navier-Stokes framework. Worthy of note is that the model presented in the book chapter is able to initiate scour on an undisturbed bed for flow around a horizontal cylinder. Promising results are also presented for scour around a complex 3D gravity foundation structure.

In Chapter 8, de Lataillade and coauthors present a viable strategy for coupling fluid and solid dynamic models for modelling Fluid-Structure Interaction applications. The aim of this work is to take some bold steps towards the "holy grail" of modelling fluid structure interaction processes: a robust coupling of CFD and solid dynamic models. The target case study in this occasion concerns the simulation of moored platforms for floating wind turbines. The proposed scheme yields promising results with respect to predicting the motion of, and the forces acting upon, the moored structure.

The book concludes with an outline of selected potential future developments that have not necessarily been covered in the component chapters of the book. Principally, the chapter identifies several numerical techniques that, whilst not currently in the mainstream, appear to be extremely promising for the development of CFD in a coastal engineering framework in the short- to medium-term.

Chapter 1

Wave Generation and Absorption Techniques

Aggelos Dimakopoulos[1],* and *Pablo Higuera*[2],*

1 Introduction

Numerical modelling of wave propagation in the nearshore area has been around for several decades. Research work in this area has resulted in the development of a variety of theoretical and numerical methods and techniques to solve this problem. Initial efforts of modelling coastal wave propagation were performed using depth-integrated approaches, such as the Boussinesq equations (Peregrine (1967); Madsen et al. (1991); Karambas and Koutitas (1992); Nwogu (1993); Wei et al. (1995)) and the shallow water equations (Hibberd and Peregrine (1979), Kobayashi et al. (1987), Titov and Synolakis (1995)), although the latter concerned strictly shallow water wave propagation (wavelength > 20 times the water depth). Additional techniques were also developed to include flow evolution in the vertical (aligned with gravity) direction, initially by using simplified forms of mass and momentum conservation equations, e.g., potential (irrotational) flow coupled with the Boundary Element Method (Grilli et al. (1989, 2001); Belibassakis and Athanassoulis (2004)).

Models that used the Navier-Stokes equations for simulating wave propagation were also developed at about the same time, but were initially limited to 2D-vertical setups due to significant computational requirements (Johns and Jefferson (1980); Lemos (1992); Lin and Liu (1998); and Bradford (2000), among others). With the rapid improvement in computer performance during the 2000s and specially in the early 2010s, the simulation of 3D wave propagation processes using Navier-Stokes approaches became practical. Initial efforts that used three-dimensional flow domains aimed to investigate detailed turbulence structures under wave breaking with normal wave incidence (Christensen (2006); Watanabe et al. (2005)) or to simulate 3D wave transformation and breaking processes under oblique incidence (Huang et al. (2009); Dimakopoulos and Dimas (2011)).

All these early studies provided a template for establishing the concept of the numerical wave tank (NWT), i.e., a particular setup of the computational domain which allows numerical testing and performance evaluation of coastal engineering applications. These studies, nevertheless, did not yet offer a robust methodology for simulating fully developed sea states, which would allow testing realistic natural or man-made features encountered near the coast. Therefore, it was necessary to develop techniques for generating and absorbing regular and random waves at the offshore and landward boundaries in a consistent manner, in particular:

[1] HR Wallingford, Wallingford, Oxon, OX10 8BA, UK.

[2] Department of Civil and Environmental Engineering, Faculty of Engineering, University of Auckland, Auckland, 1010, New Zealand.

* Corresponding authors: A.Dimakopoulos@hrwallingford.com; pablo.higuera@auckland.ac.nz

- at the offshore boundary, simultaneous wave generation and absorption techniques allow the representation of incident wave condition which often come from a wave forecast/transformation model (e.g., SWAN, Booij et al. (1996)) and prevent re-reflections from bathymetric or man-made features; and

- at the landwards boundary, wave absorption techniques allow the complete absorption of the waves to prevent unwanted wave reflection from the outlet boundary.

To some extent, the concept of the NWT is influenced by the layout of physical modelling flumes and basins, which is based on a technology that has been mature for few decades. But, fortunately for the numerical modellers, the absence of real world limitations such as space allocation, material type and strength, and laboratory safety considerations gave rise to a diversity of wave generation and absorption techniques for NWTs that were mostly based on abstract concepts. These techniques essentially enable performing long simulations by maintaining stable and fully developed wave statistics over time and by keeping a constant water level in the NWT.

Wave generation and absorption techniques in depth-integrated models, such as the Bouss2D model (Nwogu and Demirbilek, 2001) or the FUNWAVE model (Shi et al., 2016) were implemented relatively early, as these models were generally least computationally demanding. Similar and in some ways more advanced techniques have also been developed for Navier-Stokes equations, within the context of Computational Fluid Dynamics (CFD) models. NWTs in CFD models are in a maturing process over the last 5–10 years, having as key milestones the works of Jacobsen et al. (2012) and Higuera et al. (2013). These two key publications left a significant legacy within the research community, which is summarised in the following points.

- They both demonstrated that NWTs in CFD models are capable of efficient wave absorption which was at least equal, if not better than in the laboratory.

- They both implemented the techniques in OpenFOAM®, the leading open-source toolkit for CFD applications, thus allowing the research and scientific community to replicate these results and gain confidence that these methods work.

- They presented a set of benchmarks (particularly in Higuera et al. (2013)) that can be used to compare and evaluate similar or substantially different techniques for wave generation and absorption.

The aim of this work is to present an account of the most important concepts and techniques for wave generation and absorption, such as the Dirichlet-type and radiation boundary conditions, the relaxation zone methods, internal mass/momentum source wavemakers and moving paddles. These are presented in more detail in the next section, followed with relevant advantages and disadvantages. In the last sections, an overall assessment of the existing methods is given along with ideas for future development.

2   Review of wave generation and absorption methods

In this section we will review the most important references on wave generation and absorption applied in numerical modelling. Since in most of coastal engineering cases the dynamics of interest are wave-driven, the accuracy of the wave kinematics will have a major impact in the solution, therefore, producing accurate waves starting from wave generation will minimise error propagation and second order effects. Additionally, in most cases we would like to simulate open-sea conditions, in which waves will travel away from the domain of interest.

However, since it is impractical to simulate large domains in CFD modelling, the role of the absorbing methods is to simulate the correct conditions for the wave to propagate outside of the domain, minimising the reflections.

There are several wave generation and absorption approaches that are worth mentioning, thus, this section will be split into four subsections. Taking into consideration practical space limitations, the review presented in this section is not complete, therefore, before reviewing individual techniques it is worth mentioning several recent review papers which can help the reader gather further details on wave generation and absorption methods.

The first reference worth mentioning is Schäffer and Klopman (2000), which is a classical reference in this field. Although this paper is focussed on active wave absorption for laboratory wavemakers, it has been extensively applied in numerical models afterwards, both for static and dynamic wave absorbing conditions.

The second highlighted reference is Miquel et al. (2018). In this paper different combinations of wave generation and absorption methods are tested in terms of performance (reflection coefficient) and the differences between them are reported. The model used in this work is REEF3D, and the results are compared with previous benchmarks published for OpenFOAM®.

Finally, Windt et al. (2019) is the most recent reference. This work includes an extensive assessment of all the wave generation methods available in the OpenFOAM® framework presently, considering multiple implementations of boundary conditions, relaxation zones and moving boundaries. Comparisons are made in terms of accuracy, computational requirements and features available. Differences point out that none of the methods are consistently superior in all three key performance indicators, therefore, Windt et al. (2019) can be used as a guide to select the most suitable wave generation and absorption methods for a particular problem.

2.1 Static boundary wave generation and absorption boundary conditions

Static boundary wave generation is the most straightforward technique that can be implemented in most Eulerian numerical models. In it, the theoretical expressions for free surface elevation and velocities are applied on static boundaries as fixed value (Dirichlet-type) BCs. There are numerous wave theories available with which to calculate the necessary wave kinematics, ranging from the most simple ones (Stokes first order, Stokes (1847)), only applicable to a limited range of conditions, to universal theories of high order (e.g., streamfunction, Dean (1965)) that can be used for nearly-breaking waves. The most suitable theory for a particular wave condition can be obtained with Le Méhauté (1976). Wave absorption can also be performed at the boundaries, either independently or, in some cases, simultaneously with wave generation. The most relevant wave absorption techniques are reviewed in next subsection.

2.1.1 Wave absorption boundary conditions

Probably the simplest wave absorption boundary condition is Sommerfeld radiation condition (Sommerfeld, 1964), also called wave-transmissive or open boundary condition. This technique is based on the analytical solution of radiation problems:

$$\left(\frac{\partial}{\partial t} + c^{\mathrm{out}} \frac{\partial}{\partial x} \right) \Phi^{\mathrm{out}} = 0 \tag{1}$$

in which c^{out} is the phase velocity and Φ^{out} is the potential of the target wave component to absorb. As noted in Dongeren and Svendsen (1997), perfect absorption can be achieved, in principle, for monochromatic waves perpendicular to the boundary and in which the wave celerity is known. Since Equation 1 is a first-order approximation, Engquist and Majda (1977) extended this BC to higher order and derived local approximations, decreasing the reflections for angles of incidence that deviate from the normal. Higdon (1986, 1987) further developed this boundary condition, introducing the incident angle in the formulation. The reader is referred to Givoli (1991); Dongeren and Svendsen (1997) for a comprehensive review of additional developments.

In this chapter we will focus only on one of the most recent advances available in literature. The work by Wellens et al. (2009) and Wellens (2012) aimed to apply open boundary conditions to real offshore problems, in which dispersive irregular wave sea states are required. Since wave components with wave celerities (c^{out}) different from the target in Equation 1 will produce noticeable reflections, Wellens (2012) proposed a rational approximation for them with the form of a digital filter (DF):

$$c^* = \sqrt{g\,h}\,\frac{a_0 + a_1\,(k\,h)^2}{1 + b_1\,(k\,h)^2} \tag{2}$$

in which a_i and b_i are the coefficients of the filter. The digital filter needs to be designed accordingly to the wave celerity range expected, with the goal of minimising the overall reflections. In this sense, Wellens (2012) recommended using the wave celerity associated to the peak spectral frequency in irregular sea states. Extensive work was performed to obtain a set of weights which achieved high performance for real cases, yielding reflections generally below 5%.

Active Wave Absorption (AWA) is another Dirichlet-type BC that can absorb waves at a boundary with minimum reflections. AWA was originally developed for laboratory wavemakers, nevertheless, it is directly applicable to numerical models, both for Dirichlet-type and moving boundaries (later introduced in Section 2.4).

The concept behind AWA is that the boundary responds actively to the incident waves, unlike passive absorption (later introduced in Section 2.2), in which damping does not change with time. In order to do so, the AWA technique requires a hydrodynamic feedback, with which to estimate the waves incident to the boundary, and react accordingly.

To study how this works we will analyse the simplest implementation of AWA, which is based on the assumption of linear wave theory in shallow waters and it has been widely applied in the literature with outstanding results. The starting point is Equation 1 in Schäffer and Klopman (2000),

$$U_c = -\sqrt{\frac{g}{h}}\,\eta \tag{3}$$

which is a very simple digital filter to calculate the velocity correction (U_c) necessary to absorb a disturbance (i.e., incident wave) of magnitude η, only depending on gravity (g), water depth (h) and η itself. The negative sign indicates that positive velocity (i.e., inflow) is required to absorb a negative elevation disturbance (i.e., wave trough), and that outflow will absorb wave crests.

Several more complex commercial implementations of AWA exist. In these, the incident-reflected wave separation is often performed via digital filters (see Antoniou (2006)) from one or several hydrodynamic feedback magnitudes measured inside the wave tank. For example, nonrecursive DF (Finite Impulse Response, FIR) involves a convolution, which introduces a time delay, whereas recursive digital filters (Infinite Impulse Response, IIR), which are composed of decaying exponentials, can be designed to be delay-free. As a result, FIR requires the hydrodynamic feedback to be collected far enough from the wavemaker to allow the system to react in time, e.g., in Christensen and Frigaard (1994) two free surface elevation gauges are placed away from the wavemaker. Alternatively, in Schäffer et al. (1994)

the hydrodynamic feedback is produced by a free surface elevation gauge mounted directly on the wavemaker paddle.

The AWA procedure is suitable to absorb not only normal long-crested waves, but also oblique waves too. This is done by splitting the boundary into smaller pieces, similar to the individual paddles of a laboratory wavemaker, each of which applies the Equation 3 independently. One of the limitations is that this procedure cannot absorb the components of the waves that propagate tangentially to the boundary which are also difficult to absorb in the laboratory.

The simple DF in Schäffer and Klopman (2000) has been applied successfully in literature (Torres-Freyermuth et al., 2007; Higuera et al., 2013; Miquel et al., 2018), and is quite effective, especially for shallow water conditions. AWA has also been applied outside the shallow water regime, with acceptable but decreasing performance as the wave conditions approach deep waters. The reflection coefficients obtained for deep water conditions are generally larger than 20%, thus not acceptable for practical simulations. This is caused by two simplifications that the shallow water conditions assume: a constant velocity profile along the water depth and non-dispersive waves ($c = \sqrt{g\,h}$), whereas in deep waters, the velocity profile decreases to almost zero throughout the water column and wave celerity is calculated by solving the general dispersion relation.

In the recent work of Higuera (2020), AWA absorption has been extended to work efficiently in deep water conditions. Two important shortcomings have been corrected by introducing wave dispersivity and vertically-varying velocity correction profiles. An additional input parameter to the model, the wave period (T), is required to calculate the precise wave celerity according to the dispersion relation. Moreover, the velocity correction profile is adapted to the depth conditions based on the general linear wave theory. Finally, a combination of AWA and passive wave absorption (relaxation zone) has been tested, indicating that each technique can benefit from the other to increase the overall wave absorption performance, as will be discussed in Section 3.

2.1.2 Examples and applications

Sommerfeld (open) boundary conditions have been widely applied in nonlinear shallow water and Boussinesq models (Israeli and Orszag, 1981; Larsen and Dancy, 1983). As Navier-Stokes equations models were introduced, Open BCs were extended to work with free surface flows via the Volume Of Fluid (VOF) technique. Examples of initial applications include the SKYLLA (van Gent et al., 1994), SOLA-VOF (Iwata et al., 1996) and the COBRAS (Lin and Liu, 1999) models. Lin and Liu (1999) report a good performance of the open BC, even for deviations of the celerity around 20%. More recently Wellens et al. (2009); Wellens (2012) have perfected the open BC in ComFLOW. Their approach is only applicable in numerical modelling, because it uses the pressure and velocities and their gradients throughout the water column as input variables, but extends the applicability of open BCs to irregular sea states.

Open boundary conditions have also been recently developed for weakly compressible SPH solvers in Verbrugghe et al. (2019), claiming that it can be extended to the incompressible SPH method too. It can be argued that this approach corresponds to a relaxation zone rather than to a boundary condition, because strictly speaking wave generation and absorption are performed in an area of buffer particles serving as ghost nodes. However, it must also be noted that these areas are much smaller (8 particles wide) than the typical lengths required in classical relaxation zone techniques, on the order of magnitude of a wavelength, therefore being closer to the application characteristics of boundary conditions.

Regarding active wave absorption, this technique was first developed based on the Sommerfeld equations in Van der Meer et al. (1993) and Sabeur et al. (1997). Later, the VOF-break model (Troch and De Rouck, 1999) was implemented for fixed boundaries, based on a commercial AWA system available for laboratory wavemakers. Some differences were introduced in the numerical model version, taking advantage of the flexibility that numerical models offer. First, the feedback was changed from free surface elevation to velocities, which are directly retrievable from the model and do not reduce the absorption performance (Hald and Frigaard, 1996). Second, arbitrarily long filters could be used because the numerical model, unlike the physical system, does not require real-time performance, thus enhancing the overall absorption rates.

A simpler system introduced in Schäffer and Klopman (2000), which does not require designing complex digital filters and only uses the free surface elevation at the wavemaker as input, was implemented and applied for the IH2VOF model in Torres-Freyermuth et al. (2010). The same approach was later extended to 3D simulations in Higuera et al. (2013) for the IHFOAM model based on OpenFOAM®. The 2D version has been implemented in the REEF3D model (Miquel et al., 2018). As mentioned before, the aim of their paper was to explore the performance of different wave generation and absorption methods and their combinations.

The limitations of this method, derived from the initial assumptions of linear waves in shallow waters, have been recently revisited and extended to work at any water depth regime in Higuera (2020). The new method, called extended range active wave absorption (ER-AWA) offers a higher overall performance and has been implemented and released in the open source *olaFlow* model (Higuera, 2017), also based on OpenFOAM®.

2.1.3 Advantages and disadvantages

Dirichlet-type wave generation presents multiple advantages. These include the simplicity of implementing them, simply as a fixed-value boundary condition, and the low computational cost of this procedure, which is often negligible and it is the lowest among the other wave generation techniques reviewed in this chapter. Nevertheless, the main drawback of this method is that it needs to be coupled with wave absorption necessarily, either at the same boundary or elsewhere. This is so because there is an imbalance between the amount of mass (water) introduced in the domain to produce a wave crest and the mass extracted to generate a wave trough. The mass differences accumulate wave by wave, producing a progressive increase of the mean water level in the domain, which needs to be absorbed (Torres-Freyermuth et al., 2007). Many wave absorption methods are suitable to deal with this effect, except for specific passive wave absorption methods which damp momentum and not mass such as momentum damping zones or dissipative beaches. As demonstrated next, AWA is the natural selection to mitigate this effect.

Regarding wave absorption, the main advantage of Sommerfeld BCs is their simplicity, which generally does not produce any significant increase in the computational cost of the model. Moreover, the absorption performance can be extremely high, provided the wave celerity and wave incidence direction is known in advance. The main disadvantages of traditional radiation conditions are that they were formulated to absorb monochromatic waves, therefore, they are not able to absorb effectively irregular (multi-chromatic) sea states. Furthermore, these techniques were conceived to work in purely absorbing boundaries rather than combined with wave generation. Nevertheless, the work by Wellens (2012) corrects these two factors.

The application of AWA in numerical models presents several advantages with respect to the laboratory systems. One convenient feature of the numerical models is the ease of obtaining flow measurements, without disturbing the flow and including variables that cannot be measured directly in the lab. New variables can provide valuable feedback for AWA systems, which by accepting additional features as input can improve their absorption performance or provide additional functionalities. For example, Troch and De Rouck (1999) uses velocities instead of free surface elevation, which are easy to obtain in the numerical model. Moreover, depth-averaged velocity components at the wavemaker boundaries were used in Higuera et al. (2013) to discriminate waves parallel and perpendicular to them, preventing spurious wave generation for non-incident waves.

Another important factor is that, AWA systems in numerical models do not require to operate in real-time, unlike AWA in laboratories. This constraint limits the complexity of the digital filters that can be applied in physical wavemakers, whereas arbitrarily complex digital filters can be used in numerical models, because they would just delay the start of the next time step calculations. The main consequence is that higher absorption performance can be achieved in numerical models, as reported in Troch and De Rouck (1999). Furthermore, eliminating the requirement of having a real-time response permits having a more flexible testing framework in which to develop new ideas.

As mentioned earlier, AWA may be required even if no incident waves to the boundary are expected, in order to prevent the increase in water level produced by the Dirichlet type wave generation. Fortunately, AWA is perfect for inhibiting such mean water level rise, which can be assimilated with a long period wave, because AWA is extremely efficient absorbing long waves. Moreover, AWA can work at the same boundary as wave generation and without significant additional computational cost. Finally, unlike the traditional radiation BCs, AWA can be applied to absorb irregular waves, not only monochromatic waves.

The main disadvantage of AWA is that it usually depends on complex digital filters, which might need to be designed ad-hoc for specific wave conditions. The simplest version of AWA (Schäffer and Klopman, 2000) presents an additional limitation, its application is typically constrained to shallow water conditions. However, the range of applicability has been extended recently in Higuera (2020).

2.2 Relaxation zones

2.2.1 Background and development

The relaxation zone method for generating and absorbing waves is an abstract method that is based on the concept of dedicating a part of the CFD model domain to perform wave generation and/or absorption. The concept can be very broadly summarised in the following equation, which, without loss of generality, is written in 1D form:

$$F(x_r) = F_b\sigma(x_r) + F_d(1 - \sigma(x_r))x_r \in [0, L_r] \qquad (4)$$

where F is a model variable or parameter in broad terms (e.g., velocity, free-surface elevation, pressure, mass/momentum sink scaling parameter, numerical parameters, mesh size, among others), σ is a weighting function that ranges from 0 at the non-relaxed part to 1 at the boundary and can have various forms (e.g., linear, exponential, polynomial), x_r is the coordinate along the wave propagation axis, with $x_r = 0$ at the interface of the numerical domain and the relaxation zone, and L_r the length of the relaxation zone along the x_r-axis. Variables or terms with indices b (boundary) and d (domain), indicate that these are calculated or imposed at the boundary or the internal domain, respectively. The application of this concept in a NWT is shown in Figure 1.

In the context of wave absorption and generation problems, it has been demonstrated that an appropriate form of Equation 4 can be introduced to the flow equations (physics of the model) or the numerical solution procedure (numerics of the model) to relax the solution towards the boundary condition, within the relaxation zone domain. It has been additionally shown that given enough length, this transition becomes smooth and reflections of outgoing waves are minimised. Techniques based on Equation 4 may be referred to as "sponge layer",

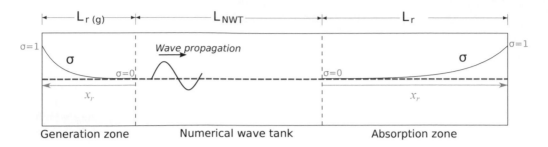

Figure 1: Illustration of relaxation zones in a NWT, according to (Jacobsen et al., 2012).

"numerical beach", "generation/absorption layers" among others. In this paper, all these techniques are considered to fall under the term "relaxation zone methods/techniques". It is worth noting that to classify as such, the relaxation zone must share an interface with the domain boundary, otherwise it falls under the internal wave maker techniques, which will be discussed in the following section.

Early work on passive absorption

Relaxation zone methods for NWTs were first applied to absorb outgoing waves at the inshore/landwards outlet, in combination with a Dirichlet boundary condition or an internal wave maker, rather than used to fully equip a NWT. In this case, the boundary condition at the outlet can be a no-slip wall (as in an actual wave tank) or a free-flow outlet (constant sea level). The purpose of the relaxation zone is to entirely dissipate the flow velocities (passive absorption) before the waves reach the outlet boundary.

The application of the relaxation zone presents a physical analogy with layouts used in the laboratory for absorbing waves such as absorbing foam, screens and beaches. In this case, the absorbing layout usually takes over few or several meters of the experimental flume or tank (as opposed to a paddle, which is the equivalent of a boundary condition in numerical models) and the absorbing layout properties (primarily the resistance it offers to the flow) are gradually increased towards the outlet wall, in a manner consistent with Equation 4. Examples of absorbing layouts in experiments can be found in Tiedeman et al. (2012) and Delafontaine (2016), among others.

An early investigation of water wave absorption using a relaxation zone was conducted by Le Méhauté (1972), by introducing an "internal friction term" in the dynamic (Bernoulli) equation of motion for the pressure, in the context of potential flow equations. The friction term is proportional to the flow potential and is broadly equivalent to a Darcy resistance term in the Navier-Stokes equations. This technique can be considered as "physical forcing" of the relaxation zone. Indeed, the physical analogy presented in Le Méhauté (1972) is parallel plates with varying density aligned with the wave direction. This may be now considered primitive, but it is consistent with the concept of increasing viscous resistance towards the outlet boundary, which is used in modern absorption techniques in the laboratory (e.g., screens, foam or beach). The concept of adding an internal friction term in a relaxation zone to facilitate wave absorption has nevertheless proven timeless, as many modern depth-integrated and CFD models still utilize this idea (Perić and Abdel-Maksoud, 2015; Dimakopoulos et al., 2019).

A generic formulation for using a sponge layer for passive absorption of waves was established by Israeli and Orszag (1981), where it was demonstrated that sponge layers, as opposed to simple radiation conditions, were capable of absorbing multiple wave frequencies.

They also demonstrated that a combination of radiation conditions and relaxation zones can optimise the absorption efficiency. Early applications of this concept can be found in depth integrated or potential flow models which matured earlier than CFD models for ocean and coastal engineering applications. In the Boussinesq model proposed by Larsen and Dancy (1983), a sponge layer method is applied by "dividing, at each timestep, the surface elevation and the flow on a few grid lines next to the boundary by a set of numbers which increase towards the boundary". In essence, they use a weighted denominator to divide the pressure and the free-surface at each timestep, right after these values are calculated by the numerical solution, and it is demonstrated that this approach efficiently absorbs outgoing waves. Their scheme is coupled with an internal source generator and it is argued that active absorption capabilities are not needed, as long as the sponge layer is introduced in all domain boundaries. The concept of introducing intermediate steps in the numerical algorithm to absorb waves, survives in many modern NWT approaches, based on CFD (e.g., in Jacobsen et al. (2012)). This concept will be hereafter mentioned as "numerical forcing", as opposed to "physical forcing" which is based on introducing appropriate source terms in the momentum equations, while preserving the original numerical solution algorithm.

Cao et al. (1993) argued that the formulation of Le Méhauté (1972) is not bounded at positive values and could sometimes result in negative dissipation, which is not desirable. Grilli et al. (1997) proposed an improved implementation of the relaxation zone technique for potential flow equations, within the context of NWTs using the Boundary Element Method (BEM). Their method included a modified dynamic free surface boundary condition for the pressure, resistant to the motion of the free surface. The influence of the modified boundary condition was increased approaching to the boundary by using polynomial weighing function with an order of 2–3. This technique was shown to achieve good absorption characteristics for short waves ($< 3\%$ reflection) achieved with a 2 wavelengths long absorption zone, but relatively poor performance for longer waves (up to 40%). This is because the modified boundary condition affected the wave kinematics close to the free-surface, thus being more appropriate for deep water waves, where the wave energy is concentrated in the upper water layers. To improve performance for longer waves, Grilli et al. (1997) combined the relaxation zone method with an absorbing piston at the outlet based on the Sommerfeld equation, which is known to yield good absorption performance for longer, shallower waves. The relaxation zone approach proposed by Grilli et al. (1997), without radiation conditions, was adopted for a NWT based on a single phase Navier-Stokes model with a moving free-surface boundary in Dimas and Dimakopoulos (2009), where good performance was achieved, for intermediate waves, with a 4 wavelengths long relaxation zone.

The application of the relaxation zone method for passive absorption was also further elaborated in Boussinesq modelling by adding linear resistance terms in the momentum equations (Kirby et al., 1998). A step further was taken in Nwogu and Demirbilek (2001), by adding a linear mass sink term. Whilst this term interferes with mass conservation, Nwogu and Demirbilek (2001) recommend its inclusion, arguing that it further enhances absorption performance. This is the physical equivalent of, e.g., adding appropriate overflow arrangements in a numerical wave tank, with the overflow level reaching the mean water level as the onshore boundary is approached.

A relaxation zone method for wave absorption in CFD models was proposed by various researchers, starting from the late 90's. Mayer et al. (1998) used a "numerical forcing" approach which relaxes the free surface elevation and velocities to prescribed values at each time-step, using a 3rd order polynomial equation for σ. Note that a fractional time-step scheme with pressure correction is employed for solving the Navier-Stokes equations. Whilst Mayer et al. (1998) do not report the absorption efficiency, they demonstrate a good overall comparison with analytical solutions and experimental data, including structures interacting with waves and currents. The COBRAS model (Lin and Liu, 1999) also employed the formulation of the Israeli and Orszag (1981) approach, coupled with an internal source wave maker, and has since been used for numerous applications during the late 90's and 00's (see section on internal wave makers below).

Early work primarily comprised passive absorption schemes, while wave generation was mostly achieved by using internal wave makers. More recently, CFD-based NWTs were developed by fully utilising the relaxation zone method, once this concept was demonstrated to be capable of simultaneously absorbing and generating waves at the boundary Jacobsen et al. (2012). This process has the same result as the concept of the active wave/generation absorption applied to absorbing boundary conditions and numerical wave paddles, but strictly speaking, it cannot be described as active, as the relaxation zone method does not adapt to the local wave properties. In order to distinguish between the two, the term "simultaneous wave absorption and generation" (SWAG) is hereafter used in the context of using relaxation zone method for both wave generation and absorption.

Simultaneous wave absorption and generation

The conversion of the passive absorption zone techniques presented above to SWAG techniques are—in hindsight—relatively straightforward, particularly for the ones that rely on the notion of dissipating waves by forcing the solution to constant or zero target velocity and free surface elevation values. The generation can be achieved by introducing a target velocity or free surface that corresponds to a wave theory. An early demonstration of this idea was performed by Afshar (2010) who, in terms of classification, stands between the relaxation zone and the internal wavemaker concept (see Figure 2). A layout with three relaxation zones is proposed, with the first two located at the offshore/generation boundary (Zone I and Zone II) for "ramping up" and "ramping down" the wave field. A third relaxation zone (Zone III) is introduced at the outlet for wave absorption. Generation and outlet boundaries were considered as solid walls, and in this sense, the overall scheme approaches the idea of an internal wave maker. Pressure, velocities and volume fractions were relaxed with 3rd and 6th order functions, using "numerical forcing". Zone I and II were one wavelength long while Zone III was two wavelengths. The method was tested for linear, standing and nonlinear waves and was demonstrated to be successful at a conceptual level. Several areas of improvements were identified, including the interpolation of forcing terms between cell centres to avoid the appearance of a saw-tooth profile in generation zones and the removal of high air velocities in the relaxation zone.

Jacobsen et al. (2012) devised the waves2Foam library which was written for the OpenFOAM® platform and was fully based on the relaxation zone method, as the relaxation zone concept was combined with a Dirichlet condition for wave generation at the offshore boundary, as in Figure 1. A substantial improvement was achieved over the early concept

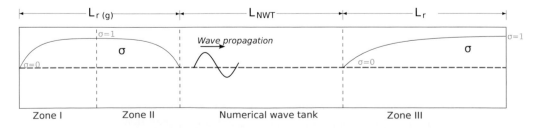

Figure 2: Illustration of relaxation zone from Afshar (2010).

of Afshar (2010). The proposed technique addressed the interpolation issues by projecting the location of the free-surface in a boundary face or a computational cell and calculating the water fraction based on the interpolated volumes. Air velocities were forced to zero through the relaxation zone technique, and pressure forcing was made redundant. The relaxation function had an exponential form, following recommendations from Mayer et al. (1998). The method was shown to be able to efficiently generate and absorb waves of linear and nonlinear regimes with high efficiency, as wave reflections were $< 3\%$ for linear and nonlinear waves and relaxation zone length of two wavelengths.

This idea was largely successful and was picked up and further developed by the research community. Dimakopoulos et al. (2016) verified a similar level of reflection efficiency by testing the waves2Foam library against regular wave conditions proposed by Higuera et al. (2013). In addition, they further demonstrated the capability of the toolkit as a SWAG scheme for generating regular, random, directional waves, with or without currents. The capability of the scheme to simultaneously absorb regular and random waves was demonstrated through numerical tests of wave reflection against solid walls. The evolution of wave height of standing regular and random waves was compared against analytical solutions and agreement was excellent for low-steepness waves. It was also further demonstrated that the method is capable of performing well within the context of a 3D numerical basin, showing a good performance for generating short-crested waves, as well as wave and current interaction in oblique directions.

The relaxation zone technique of Jacobsen et al. (2012), is also adapted to the Reef3D model (Miquel et al., 2018), and includes numerical forcing of the dynamic pressures. Their implementation proved to be highly efficient for random waves absorption ($< 2\%$) and less efficient for regular waves ($< 10\%$), something that seems counter-intuitive and contradicts findings from Jacobsen et al. (2012) and Dimakopoulos et al. (2016). It should nevertheless be noted that the implementation of the method in Reef3D includes relaxation of the pressure terms and the reflection analysis that they use is more likely to be suitable for random rather than regular waves, so this may justify the different findings. Additional models that use the relaxation zone approach are the OpenFOAM®-based toolkit Naval Hydro pack (Jasak et al., 2015) or Star-CCM CFD platform (Perić et al., 2018).

As the relaxation zone technique became increasingly widespread, further analysis was performed regarding the parameters relating to the relaxation zone. Perić and Abdel-Maksoud (2016) used mass and momentum sources by setting $F_b = F_s\sigma(x)(f - f_t)$ and $F_d = 0$ in Equation 4, where f, f_t are the calculated and target flow variables (volume of fluid or velocity) and F_s is a scaling coefficient. Their analysis showed that F_s should be proportional to the wave frequency while the length of the relaxation zone should be proportional to the wave length for the absorption zone to achieve relaxing capabilities. In addition, they demonstrated that the optimal values of these parameters are $F = \pi\omega$ and $x = 2\lambda$ for an exponential type relaxation function. Similar values were consistently used by researchers using trial and error approaches, e.g., see Nwogu and Demirbilek (2001) or Jacobsen et al. (2012). In Perić and Abdel-Maksoud (2018), a theory was developed to predict reflection coefficients for monochromatic waves, given the parameters of the relaxation zone. The theory was found to agree with numerical tests and the researchers or engineers can use it for selecting appropriate parameters for the relaxation zone conditions, given a target reflection coefficient. The authors nevertheless recommend further work on extending the theory to cover irregular or random waves.

In Dimakopoulos et al. (2019), a relaxation zone technique is proposed, which is very similar to the one presented in Perić and Abdel-Maksoud (2016). Dimensional analysis confirms findings from Perić and Abdel-Maksoud (2016) that the scaling factor F_s must be proportional to the wave frequency to achieve good performance and the value of the scaling coefficient is set to 5 times the angular wave frequency, following recommendations from Nwogu and Demirbilek (2001). The method is implemented in the CFD toolkit Proteus (Kees et al., 2011), showing good performance for generating and absorbing random waves, although the approach can be possibly improved by modifying the mass conservation equation, as the method only considers momentum sources.

2.2.2 Examples and applications

The relaxation zone method for CFD models became widespread after the release of the waves2Foam library and it is usually combined with the OpenFOAM® CFD toolkit. The method proposed by Jacobsen et al. (2012) has been used for a wide range of applications, including modelling wave structure interaction processes in seawalls and breakwaters (Richardson et al., 2014; Jensen et al., 2014; Medina-Lopez et al., 2015; Hu et al., 2016; Jensen et al., 2017), marine and offshore structures (Palm et al., 2013; Fuhrman et al., 2014; Eskilsson et al., 2015; Jasak et al., 2015; Moradi et al., 2016; Vyzikas et al., 2017; Chen and Christensen, 2018), naval and ship hydrodynamics (Piro, 2013; Windén et al., 2014; Shen and Korpus, 2015; Vukčević et al., 2017; Kim and Lee, 2017; Liu et al., 2018) and soil, sediment and scour processes (Stahlmann and Schlurmann, 2012; Jacobsen and Fredsoe, 2014; Karagiannis et al., 2016; Cheng et al., 2017; Elsafti and Oumeraci, 2017). The list is not exhaustive as many more researchers and engineers have used the relaxation zone method in the context of the OpenFOAM® CFD toolkit.

The relaxation zone method has been also used for modelling engineering applications in the coastal and marine environments in other models as well, such as the Reef3D model (Bihs et al., 2015; Chella et al., 2015; Kamath et al., 2016; Ong et al., 2017)) and the Proteus CFD toolkit (Cozzuto et al., 2019; de Lataillade et al., 2017; Mattis et al., 2018), covering a broad range of applications, similar to the one listed for OpenFOAM®.

2.2.3 Advantages and disadvantages

The main advantage of the relaxation zone method is primarily the high fidelity of wave generation along with high efficiency in absorption. This is because the method is not susceptible to boundary instabilities due to the sudden change of flow equations, while forcing analytical solutions or linear approximations meaning that evanescent modes and parasitic waves (typical in moving paddles or internal wave makers) are not present or very quickly dissipated.

Another advantage is the ease of implementation. It is relatively straightforward to add source terms to existing equations or to use "numerical forcing", without particular considerations for numerical stability and accuracy, as opposed to, e.g., radiation boundary conditions. For example, the method used in (Dimakopoulos et al., 2019) performed a simple adaptation of the Darcy term in the porous media equation to implement SWAG. This method is therefore suitable to implement in newly and/or rapidly developed models, where for example an adaptation of an existing model component may allow the model to be used as a NWT. The inclusion of the mass source term through the volume of fluid scheme may require some treatment to, e.g., avoid non-physical oscillations at the generation zone (Afshar, 2010; Jacobsen et al., 2012). Whilst this may require some additional efforts, including mass sources is recommended, as it has been reported that it enhances absorption (Nwogu and Demirbilek, 2001) and it is also capable of maintaining the water level in the domain constant, thus not needing corrections for influx due to a mass imbalance between generation of wave troughs and crests.

The relaxation zone also needs a suitable selection of parameters such as relaxation function, length and scaling of the coefficient, thus being subject to uncertainties. Recent research and engineering practice have addressed these issues. Based on the literature consensus, the authors recommend the following:

- Length of generation zone: $L_r \simeq \lambda$ (corresponding to peak frequency for random waves)

- Length of absorption zone: $L_r \simeq 2\lambda$ (corresponding to peak frequency for random waves)

- Scaling parameter: F_s=3-5 ω

- Relaxation function: $\sigma = \frac{\exp^{x_r/L_r} - 1}{\exp^1 - 1}$.

Some of these values are theoretically proven (see Perić and Abdel-Maksoud (2015) for regular waves. For the remaining parameters or wave regimes, these recommendations are demonstrated to have a reasonable performance throughout multiple studies in literature. Note that frequency and wavelength parameters correspond to peak frequency and wavelength for random wave regimes.

Undoubtedly the greatest disadvantage of the method is the additional computational resources required for running engineering applications, compared to other methods presented in this article. Regarding wave absorption, the domain must be extended by about two wavelengths and this can significantly increase the length of the domain. Perić et al. (2018) also demonstrated that a careful selection of the relaxation zone parameters may allow for a reduction of the size of the absorption zone, e.g., using 0.5–1 wavelenghts, rather than two. (Dimakopoulos et al., 2016) have demonstrated that increasing the mesh size with a mesh progression ratio of 10%, the mesh cells in the absorption zone can be reduced by > 95%, whilst maintaining equal or better absorption efficiency. A similar technique can be applied in unstructured meshes (de Lataillade, 2019). The caveat of this approach is that the free surface tracking scheme should be robust enough in order to cope with changes in mesh size along the free surface without introducing spurious oscillations.

Dedicating up to a wavelength for SWAG cannot essentially be considered as an increase of the computational domain, as for all generation methods, it is considered good practice to leave a buffer zone after the generation boundary to allow the waves to develop before entering the main domain (e.g., approaching a slope or a structure). In this case, the role of the buffer zone can be played by the generation zone. It was nevertheless demonstrated by Dimakopoulos et al. (2016) that the computational cost associated with generating non-repeating random wave sequences for, e.g., simulating a storm event may be disproportionate to the overall cost of the simulation. For example, in Dimakopoulos et al. (2016), it was demonstrated that using 500 wave components for generating a random wave series which corresponds to 250 non-repeating waves doubles the simulation time in a NWT of 240,000 mesh cells, when compared to simulations that use 50 components. This is primarily due to the fact that trigonometric and hyperbolic functions are rather expensive to calculate for each cell and each time step in the relaxation zone.

Several approaches have been proposed to address this problem. Dimakopoulos et al. (2016) showed that by replacing native C++ and FORTRAN trigonometric functions with 4th order Taylor approximations results in a reduction of the computational cost by an order of magnitude, whilst the approximation errors remained < 1%. This reduced the overall cost but it did not completely eliminate it, e.g., simulations with 500 components in the NWT mentioned before need 1.4 more times, rather than double with the native functions. In Jacobsen (2017) a stretched distribution of the spectrum is proposed for improving non-repeatability of the series, thus needing less wave components for describing a full storm, but this approach does not offer a consistent discretisation of the spectra in the frequency range away from the peak. Similarly the computational time in Jacobsen (2017) may be further reduced by manipulating the trigonometric and hyperbolic functions to split spatial and temporal terms, storing spatial terms in memory and only calculating temporal terms during time advancement. This has demonstrated to speed up the calculation time needed to compute the signal by 50 times for 1000 wave components. More recently, Dimakopoulos et al. (2019) developed a method based on pre-processing the random wave signal using spectral window decomposition methods. Their technique was demonstrated to be able to generate non-repeating signals of $\mathcal{O}(1000)$ waves, size using $\mathcal{O}(10)$ frequency components, thus achieving similar reductions in computational cost. The technique was tested in the CFD toolkit Proteus.

Overall, the relaxation zone is an accurate and easy technique to implement for generating and absorbing all wave regimes, but the method is indeed computationally expensive. For its merits, thhe relaxation zone is one of the most preferred methods to use in NWT and researchers have been making significant advances in addressing its drawbacks relating to speed. Including these advances in future works will further increase the attractiveness of the technique.

2.3 Internal wave makers

2.3.1 Background and development

The Internal wave maker (IWM) concept is usually implemented by creating a generation line or area inside the domain and adding source terms to the mass or momentum equations to generate waves. This area is expected to be transparent to the wave propagation (i.e., does not reflect waves), therefore, full wave absorption can be achieved as long as passive absorption zones or absorbing boundary conditions are added to inlet/outlet boundaries.

Originally, Larsen and Dancy (1983) proposed an internal generation scheme that is based on adding to or removing some volume of water from the domain by modifying the mass equation of the Boussinesq approach. The source terms were implemented after the equations had been discretised ("numerical forcing"). In their work it was demonstrated that the IWM remained transparent to the propagation of nonlinear regular waves. The proposed implementation was rather rudimentary and required tuning its performance on a case-by-case basis (Wei et al., 1999). It was also demonstrated that it was not suitable for more advanced formulations of the Boussinesq problems (Nwogu, 1993; Wei and Kirby, 1995). The overall concept was nevertheless relatively successful as it was the starting point for further improvements and adaptations in Boussinesq and CFD models.

Wei et al. (1999) proposed a new formulation for IWM, comprising the addition of source functions in the momentum and mass equations before these are discretised ("physical forcing"). The source function is relaxed over a region by using a bell-shaped distribution, rather than being a point source. It seems that the source functions parameters (amplitude, width) require tuning/calibrating against the target wave characteristics. The concept is shown to be transparent to the wave field, as reflected waves can propagate to the other side of the IWM and absorbed at an offshore absorption zone. The physical analogy of this implementation is to use a perforated internal wave maker in a flume/basin, but it is evident that such a concept may be not be fully realisable in a laboratory setting.

In CFD models, Lin and Liu (1999) applied an IWM to their VOF-based model, initiating the concept of a NWT for 2DV numerical flumes. The approach uses a relatively concise source region which is placed internally in the flume and below the free surface. In this region, the continuity equation is locally modified to include inflow and outflow fluxes that cause wave excitation. These terms subsequently drive the momentum equation towards wave generation through the pressure-velocity coupling scheme. The mass source functions

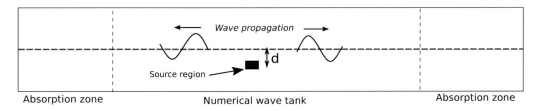

Figure 3: Illustration of internal wave maker in a NWT according to Lin and Liu (1999).

for a monochromatic wave have the following form:

$$s(t) = \frac{cH}{A} \cos\left(\omega t + \phi\right) \tag{5}$$

where ω, H and c is the wave angular frequency, wave height and celerity, respectively, and A is the area of the IWM. The concept can be extended to cover higher order waves, as well as random and directional waves, by linear superimposition of components. A sketch of the concept used by Lin and Liu (1999) is shown in Figure 3. The physical equivalent of this layout would be installing a submerged outlet in an laboratory wave basin and forcing oscillatory discharge to generate waves. Lin and Liu (1999) performed numerical tests with a rectangular source area placed in the middle of the flume, and absorbing (radiation) boundary conditions at the sides. Their approach performed relatively well for the wave regimes tested, including nonlinear and random waves. It was nevertheless found that the position and size of the source area greatly affects the wave generation performance. It was further demonstrated that the optimal location was at about 1/4–1/5 of the depth below the surface and the optimal size was < 5% of the wavelength, meaning that both the size and location of the source wave maker may have to be optimised depending on the wave conditions. The concept was extensively used in the COBRAS model, in tandem with both absorbing (radiation) boundaries, and relaxation zones for passive absorption (see relevant section). This approach was later proposed for establishing a NWT within the PHOENICS finite element model (Hafsia et al., 2009).

Saincher and Banerjee (2017) further assessed the performance of Lin and Liu (1999) IWM for steep waves in deep, intermediate and shallow waters by placing an IWM in the middle of the tank and using absorption zones at the sides to dissipate outgoing waves. They concluded that steep waves generated in deep or shallow waters may experience excessive wave height damping. It is argued that wave height dissipation is caused by vortices generated by the IWM. In deep water waves, the arrival of these vortices to the free surface was associated with wave height dampening (possibly through artificial incipient breaking), and it was noted that the problem was solved by adjusting the location of the IWM towards deeper water and increasing the size of the source area to reduce velocities. Due to the change of location and size of the IWM, additional calibration of the IWM had to be performed. Using a larger IWM to counter vorticity generated by the local flow gradients developed in IWM similar to the ones proposed in Lin and Liu (1999) was originally conceived by Perić and Abdel-Maksoud (2015) to facilitate deep water wave generation. For shallow water waves, relocating the source to lower depths was not particularly effective, so the authors proposed to increase the source size to cover the full water depth. This modification yielded good results, although it can be argued that this method is converging to the one proposed by Wei et al. (1999), as the source area is extending throughout the whole water column.

Choi and Yoon (2009) developed a concept for a momentum source based IWM in ANSYS Fluent. In their model, they adapted the methodology proposed by Wei et al. (1999),

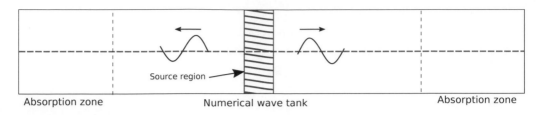

Figure 4: Illustration of internal wave maker in a NWT according to Choi and Yoon (2009).

for application in CFD models and developed a 3D numerical wave tank with capability of generating plane and directional waves. The formulation of the source terms included in the Navier-Stokes equations can be summarised in the following equation:

$$S_x = -2\beta x g \exp{-\beta x^2} \frac{D}{\omega} \sin{(k_y y - \omega t)}$$
$$S_y = g \exp{-\beta x^2} \frac{D}{\omega} \sin{(k_y y - \omega t)}$$

(6)

where x, y are coordinates in the propagation plane, along and perpendicular to the main wave direction, respectively, g is the gravitational acceleration, β is a coefficient depending on the width of the generation area and the wavelength and k_y and ω are the wavenumber on the vertical (y) axis and angular frequency of the waves, respectively. The concept can be extended to cover higher order waves, as well as random and directional waves, by linear superimposition of components. A sketch of the numerical flume set-up can be shown in Figure 4.

The formulation in Choi and Yoon (2009) did not include the variation of the source terms over the vertical axis as in Wei et al. (1999), since it was found that this did not significantly affect performance. During numerical tests of 2D and 3D regular wave propagation, satisfactory agreement was demonstrated against physical modelling results, analytical solutions and bechmark cases previously simulated by Lin and Liu (1999) model. Choi and Yoon (2009) also observed that the influence of evanescent modes due to the vertical profile at the IWM is limited to a distance of 2–3 water depths from the IWM, however this was observed for intermediate to shallow wave conditions, where the assumption of constant velocity over the depth has some validity, as opposed to deep water waves.

Ha et al. (2013) further improved the approach proposed by Choi and Yoon (2009), by implementing it in the NEWTANK model. They argued that the use of both momentum and mass source in the Navier-Stokes equations improved generation performance. They also note that the method has limitations for generating deep water waves, given that it has been originally developed for Boussinesq equations. The IWM was coupled with passive absorption layers and numerical tests of random and multi-directional waves were performed, showing overall good performance for intermediate and shallow waves, but not for deep water waves. It could be nevertheless argued that the latter might have been because they used a relatively coarse mesh, as their results demonstrate that the wave height is correctly generated at the IWM location but quickly dissipates during propagation in the NWT.

Liu et al. (2015) implemented the IWM concept of Choi and Yoon (2009) in the framework of an Incompressible Smoothed Particle Dynamics (ISPH) model (particle-based Langrangian method). They do not use mass source terms, as these would require adding and removing SPH particles from the domain thus causing difficulties and complications for numerical implementation and stability. They observed that the optimal width of the source region is about 20%–50% of the wavelength and that a buffer zone is required to allow the wave signal to stabilise and reach good agreement with theory. Buffer areas were areas from wavelengths away from the generation zone, for intermediate or shallow water waves. The method is proven not to be very efficient for generating deep water waves, despite the rather refined particle resolution.

2.3.2 Example applications

The IWM is widespread in Boussinesq modelling, as modern Boussinesq models employ variations of the internal generation concept proposed by Wei et al. (1999). In Bouss2D (Nwogu and Demirbilek, 2001) internal wave generation is achieved by adding mass and momentum sources in the form of Dirac equations, which are later discretised over two computational cells. Their scheme is a hybrid of "physical" and "numerical" forcing as the source terms are present in the equations, but the smoothing of the Dirac function is dependent on the grid size rather than other properties (e.g., local wave properties). FUNWAVE (Shi et al., 2016) uses the methodology of Wei et al. (1999) and Mike 21 BW (DHI, 2017) employs a similar scheme.

In CFD models, the IWM has been employed in the COBRAS model, especially during the early days for wave structure interaction applications, such as modelling low-mound or submerged porous structures (Garcia et al., 2004; Lara et al., 2006; Losada et al., 2008), floating breakwaters (Koftis et al., 2006), pressure propagation in soils around structures (Zhang et al., 2018b, 2019) and sediment transport processes (Amoudry et al., 2013), to name a few. Although the COBRAS model may be still used by researchers, it is not as widespread as in the mid-00's, and it has been only used for 2DV simulations, which are equivalent to numerical wave flume. The NEWTANK (Ha et al., 2013) model is a fully developed 3D model that uses the IWM concept, frequently combining it with Large Eddy Simulation (LES) approach for turbulence. Representative publications include Ha et al. (2014), where a submerged breakwater was studied under tsunami action, and wave interaction with floating structures Zhang et al. (2018a).

2.3.3 Advantages and disadvantages

The IWM concept has been very successful in its use for depth-integrated and in particular Boussinesq models. In these models, the free surface elevation variable is an integral part of the equations and it is, therefore, straightforward to develop source terms associated with a particular forcing of the free surface elevation. This is not the case in two-phase flow CFD models, where the free surface elevation is implicitly calculated by the air/water interface transport equations.

The IWM is demonstrated to have a good overall accuracy, but often requiring additional treatment for steep waves in deep and shallow waters. As the IWM does not force an analytical solution of a wave theory, buffer areas have to be introduced in the domain to allow smooth development of the waves and to dampen evanescent modes. These areas, depending on the approach and the wave regime, could be from 2–3 times the water depth according to (Choi and Yoon, 2009) up to 2–3 wavelengths (Liu et al., 2015) and can be implicitly associated with increased computational cost.

Due to these discrepancies, it could be argued that in VOF based models, the application of the internal wave maker concept is less advantageous, often requiring adaptation of the size and position of the IWM or calibration of the source terms to achieve the target wave height. These adaptation and calibration procedures on a case-by-case basis would require multiple simulations to achieve a target wave height, thus increasing computational cost. In Schmitt et al. (2019), a calibration procedure is designed for a momentum source wave maker, similar but perhaps simpler to the one proposed by Choi and Yoon (2009). Schmitt et al. (2019) argue that their procedure can be performed in 2DV tanks with a relatively low computational cost. However, their method is assessed for regular waves or wave groups, rather than for random wave series that correspond to full storms, for which much longer simulations would be expected. Taking into account the need for buffer areas and the calibration, it could be argued that on some occasions, the IWM method could be as costly as the relaxation zone, which is generally considered an expensive method. It should be nevertheless stated that on some occasions, wave calibration could be desired regardless of the wave generation method, for, e.g., quality assurance purposes or for direct comparison with experimental data. With respect to computational cost for calculations of trigonometric functions (see discussion in the relaxation zone section), these should not be as important because the extent of the IWM remains relatively small compared to the wavelength and the overall size of the flume. Optimisation methods such as the ones discussed in (Dimakopoulos et al., 2016, 2019) could nevertheless benefit the methods and help maintain efficiency in relatively refined meshes.

Regarding absorption efficiency, this will depend on the wave absorption scheme. Researchers have employed both radiation boundary conditions and absorption zones, with a preference to the latter, as it is more aligned with the theoretical background of the IWM (adding source/sink terms to momentum equations). Using an absorption zone in tandem with an IWM could be both efficient and relatively cost-effective, as long as the mesh is coarser in the absorption zone, following recommendations from Dimakopoulos et al. (2016); Saincher and Banerjee (2017).

To summarise, whilst the IWM method can be conveniently implemented through the mass or momentum equations, its overall use is not as straightforward as, e.g., the relaxation zone method or radiation conditions, which require little or no wave calibration. For highly nonlinear and deep water waves in particular, the IWM may require more intensive calibration procedures and treatment (e.g., resizing the source area, or scaling source terms). Absorption efficiency can be in par with the relaxation zone method. In terms of computational cost, although the IWM will increase the size of the computational domain per se, addition of an absoprtion zone at the upstream boundary, and the establishment of buffer areas between the generation zone and the area of interest, may reach up to 2–3 wavelengths.

As a conclusion, it could be stipulated that whilst the IWM method is the norm for Boussinesq models, its use is not as well established in CFD models, whereas the relaxation zone and the radiation boundary conditions are increasingly becoming widespread. Notwithstanding the disadvantages of the method, this could be because the IWG method needs to be complemented either with an absorption technique based on "competing" techniques, such as the radiation condition or the relaxation zone method. In terms of implementation, it is sometimes more convenient and efficient for CFD developers, who often deal with highly complex software codes, to convert these "competing" techniques to accommodate simultaneous wave generation, rather than design a different technique for generation from scratch.

2.4 Moving wavemakers with active absorption

Using moving wavemakers to generate and absorb waves in numerical models presents the advantage that they can replicate wave modelling operations in a laboratory environment, i.e., performing the exact movement of the physical wavemakers. This behaviour may pose significant benefits in seiche or harbour resonance studies, as long waves are especially sensitive to changes in the length of the domain, which varies due to the displacement of the wavemaker (Higuera et al., 2015).

2.4.1 Moving wavemaker types

Normally, the most popular devices simulated in the literature are of piston or flap (including variable draft) type. However, devices that are not that common, as double-, triple-flap (Halfiani et al., 2015), mixed type (piston + flap) or plunging wavemakers (typically wedge-shaped) (Gadelho et al., 2015) can also be replicated. Even arbitrarily-deforming boundaries (Koshizuka et al., 1998) have been tested.

Piston-type wavemakers (e.g., Huang et al. (1998)) are probably the most extended in coastal engineering studies, both in experimental and numerical modelling references. The velocity profile generated with this category of wavemakers is constant throughout the water depth, thus it matches well the kinematics of waves in shallow waters. Alternatively, flap wavemakers (e.g., Antuono et al. (2011)) are better suited to replicate the hydrodynamic conditions of waves in deep waters, thus, they are widely used for offshore engineering applications. In both cases, piston and flap wavemakers support second order wave generation for irregular waves, a breakthrough feature that suppresses the generation of spurious free waves. Second order generation was developed in Schäffer (1996) in 2D and later extended to 3D in Schäffer and Steenberg (2003).

A number of procedures for performing active wave absorption (AWA) with moving wavemakers are available in literature. In all cases a displacement correction to the original wavemaker signal is calculated using some hydrodynamic feedback as input. The new calculated movement absorbs the incoming waves to the boundary, preventing a high degree of reflections. Furthermore, AWA can be connected to perform independently in purely absorbing boundaries or simultaneously to the target wave generation.

As mentioned before, wave absorption in CFD models was adopted from systems initially developed for laboratory wavemakers. The two main AWA programmes AWACS and AwaSys, already introduced in Section 2.1, are straightforward to apply on numerical moving wavemakers. The procedure is equivalent to AWA on static boundaries, but the correction velocity is used to correct the movement of the wavemaker. For example, for a piston-type wavemaker, considering that velocity is the time derivative of displacement, the displacement correction is calculated as:

$$U_c = \frac{dX_c}{dt} \rightarrow X_c = \int U_c dt = U_c \Delta t \tag{7}$$

The simplest version of AWACS (Schäffer and Klopman, 2000), introduced in Equation 3, has been applied successfully to absorb waves in Lara et al. (2011); Higuera et al. (2015). Reasonable results are reported outside the range of application (shallow water wave regime) of this simplified approach.

The way to generate directional waves with moving wavemakers is based on what is usually called "snake movement". This means that each individual paddle of the wavemaker moves with a lag with respect to its neighbours. This method is also applicable to absorb oblique waves, but it cannot absorb the wave components propagating tangentially to the wavemaker, though.

AWA has also been developed for wedge-type wavemakers. Bullock and Murton (1989) presented a system for a physical wavemaker in which the solid wedge moved obliquely up and down. The feedback came from a gauge located at the front wall of the wavemaker, mounted on rails so that it was always at the same level, and was converted via an analog recursive filter. The authors are not aware of an implementation of this method in CFD models, however, plunging wedge-shaped wavemakers (without AWA) have been modelled in Gadelho et al. (2015).

Another completely different approach for AWA is presented in Spinneken and Swan (2009a,b). In these works the feedback is not obtained from free surface elevation gauges, but from forces measured at the wavemaker. The absorption correction is then calculated via an infinite impulse response (IIR) filter. This method presents the advantage that the feedback signal is not obtained from point measurements, but from an integrated magnitude instead (force is calculated by integrating fluid pressures at the wavemaker), therefore, the output is less sensitive to local disturbances. The force-feedback AWA is applied numerically in Spinneken et al. (2014). Even though the numerical model is a nonlinear Boundary Element Method (BEM) and the method is applied for first order wavemaker theory, there are no limitations for porting it to any CFD package and with the second order formulation. However, some shortcomings are pointed out in Spinneken et al. (2014). The most important is the limitation to absorb high-frequency or high-nonlinearity waves because of the extremely high accelerations required (i.e., "excessive amount of added mass"), although this should not pose a major challenge in numerical models.

2.4.2 Mesh movement algorithms

Replicating moving wavemakers in CFD models requires mesh-movement algorithms, as waves are generated by prescribing a deterministic displacement on a solid, given by the methods presented in the previous subsection. However, we would like to stress that the techniques that are going to be reviewed next can also be applied for a wide variety of other cases, as discussed later.

Generally speaking, wavemakers can be emulated in numerical models by two procedures: Moving Boundaries (MB) and Immersed Boundary Methods (IBM). A complete diagram showing the most relevant algorithms and examples of applications is presented in Figure 5.

The MB approach is able to represent the prescribed movement of a solid in which one or more of its faces correspond to the mesh boundaries. This can be the case of a piston or flap wavemaker (a single boundary moving to replicate the device) or an embedded body which has been meshed around (e.g., a plunger wavemaker with several faces), similar to the one portrayed in Figure 5. Usually, the MB procedure involves a set of boundary conditions (BCs) to replicate the action of the body movement, which will deform the mesh. The procedure can be outlined as follows. First, for each time step the BCs set the displacement of the individual nodes at the moving boundaries and keep the rest of the nodes at the static boundaries in the same position. Then, the mesh is deformed accordingly. This step can be performed in a number of ways, some of which are introduced in Jasak and Tuković (2006). For example, the displacement of the internal nodes can be calculated by solving a Laplacian equation:

$$\nabla \cdot \left(k \nabla \frac{\Delta \mathbf{d}}{\Delta t} \right) = 0 \tag{8}$$

where $\nabla = \frac{\partial}{\partial x_i}$; k is the diffusivity, i.e., a scalar field, either constant or variable in space, which helps redistribute the node displacements across the mesh; $\Delta \mathbf{d}$ is the displacement vector; and Δt is the time step. Further details on this mesh movement technique can be found in Higuera et al. (2015). Next, the system needs to re-distribute the quantities (e.g., VOF, velocities...) and fluxes to adjust to the new cell volumes and face areas from the deformed mesh. Finally, the BCs are imposed to the fluid to solve the Navier-Stokes equations as usual. The moving boundaries are represented as moving walls in this step, with a no-slip boundary condition and no-flux across them, therefore, the fluid in contact with a wall will have the same velocity as the wall.

The MB procedure discussed is only valid for small displacements. The larger the displacements experienced, the further the mesh quality will decrease, until the case will fail for violating the mesh quality requirements for the finite volume discretization. If displacements are large, the mesh evolution stage might require topological changes (TC), such as

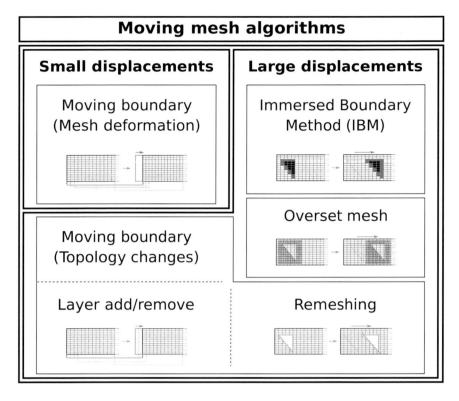

Figure 5: Moving mesh algorithms available for wave generation with moving boundaries.

remeshing, to maintain the mesh quality. This approach includes an interesting technique which might be useful for simulating long-stroke wavemakers: to insert or remove layers of cells in front of the piston as it moves back and forth. This procedure is applied already when simulating pistons in turbomachinery (Montorfano et al., 2014), however, the authors are not aware of any applications to piston wavemakers published yet.

Methodologies for mitigating issues with mesh quality over large mesh deformations are also discussed in de Lataillade (2019) in the context of floating structures. These methodologies may be used to maintain an acceptable mesh quality without the need of introducing TC in the mesh. In this work, the application of Arbitrary Eulerian Langragian (ALE) techniques (Hirt et al., 1974) are discussed for mesh motion, namely the linear elastostatics model (LEM, e.g., Dwight (2009)) and using mesh adaptivity (without TCs) with monitoring functions for mesh quality (e.g., Grajewski et al. (2005)). Both methods perform well in the context of floating structures, but it is further commented that for long simulations involving nonlinear oscillations, the LEM could allow mesh quality to deteriorate after several dozens of cycles. The authors are not aware of any applications to piston wave makers published yet, but it could be argued that such an application could be straightforward, due to the proximity of the moving/floating structures with the numerical paddle problem.

Another very important technique that enables the simulation of large displacements of solids, this time on a fixed grid, is the Immersed Boundary Method (IBM) (Mittal and Iaccarino, 2005). Although there are countless IBM implementations in literature, the underlying idea is to apply a momentum source within the mesh cells to simulate an interface between the fluid and the solid. This enables representing a complex solid moving inside a mesh that is not body-fitted (Kim and Choi, 2019). This is the main reason why the IBM is widely applied to solve Fluid-Structure Interaction (FSI) problems, as for example ship hydrodynamics (Yang and Stern, 2009), rigid floating objects (Calderer et al., 2014; Bihs and Kamath, 2017) or with flexible elements mimicking vegetation (Chen and Zou, 2019).

There are two main types of IBM depending on how the forcing is applied. Continuous forcing implies that the IB force term is included in the Navier-Stokes equations before the discretization, whereas discrete forcing involves applying the IB force to the discretized equations (Mittal and Iaccarino, 2005). Most of the modern implementations if IBM, such as the Fractional Area-Volume Obstacle Representation (FAVOR) (Wei, 2005), the Cartesian cut-cell method with openness coefficients (Lara et al., 2011) or the ghost-fluid (Bihs and Kamath, 2017) methods are discrete forcing methods. One of the common points in most methods is the way to represent the solids, by means of the area and volume fractions that each solid occupies for each face and cell, respectively.

The IBM technique presents advantages for large displacements, thus it has been applied to simulate long-stroke piston-type wavemakers (Usui et al., 2016), which are capable of simulating long tsunami-like waves. The main disadvantage is that it is not straightforward to acquire flow variables at the boundary, necessary to feed the active absorption algorithm because these are likely to be retrieved by interpolation, which will lead to increasing numerical errors.

The last mesh movement algorithm reviewed in this chapter is overset (or Chimera) grids (Panahi and Shafieefar, 2009; English et al., 2013). In overset grids (OG), the flow is solved in independent but overlapping meshes simultaneously. There are no restrictions on any of the coexisting meshes, i.e., they can be moving or static, structured or unstructured, deforming or fixed. Generally, the background mesh is static, significantly larger than others and usually simple (i.e., structured). Additional meshes can have different applications, such as to provide a finer resolution than the base mesh locally, or to simulate fluid-solid interactions. Such meshes are usually body-fitted to internal solids/obstacles and may employ curvilinear grids, allowing having a proper boundary layer discretization if required. Similar to IBM grids, obstacles can be given a prescribed movement, or response to the action of the flow (6 degrees of freedom).

The peculiarity of the OG method is that the moving mesh does not necessarily undergo any deformations, but it is generally subjected to rigid movements and rotations only, which is ideal for simulating relative motion of different solid components. In this sense OG has been employed to simulate very complex problems such as Wave Energy Converters (WECs) (Windt et al., 2018) or ship self-propulsion and manoeuvring (Shen et al., 2015).

Due to the multi-mesh approach, all the individual meshes need to be coupled. Coupling takes place over the regions of overlap, generally near the border of the smaller meshes. Therefore, the CFD code requires a specific library or routines and additional steps to calculate the connectivity between meshes, perform 3D interpolation to map and exchange the flow fields data and establish locations that can be excluded from the solution, to avoid duplicating calculations.

Although OG have been extensively applied in the marine industry, they are still a very novel approach and the authors are not aware of any works in which OG are applied to simulate moving wavemakers. Possibly this is because OG methods require a more significant effort in development and implementation than the other MB methods discussed here, and the moving paddle problem is relatively simple, so such an effort may not be justified.

In short, all the techniques reviewed in this section present advantages and disadvantages, therefore, the user needs to decide which method to apply in case several are applicable. Moving boundaries present a clear advantage, since no topological changes are required for small displacements, the mesh deformation takes relatively less time and mass conservation is easily achievable. The possible complexity to create the mesh and the limitation to small displacements are the main drawbacks. If large displacements are required, additional algorithms need to be applied to monitor the mesh quality and adapt the mesh (with or without topological changes). In case of topological changes, the remeshing steps would take longer time, errors can be introduced due the grid-to-grid interpolation and mass conservation might be violated because remeshing changes the topology (e.g., introducing or deleting cells). When the algorithms involve monitoring quality and adapting the mesh, these particular errors may not be present, but other errors or computational overheads may be present, depending on the complexity and the efficiency of the monitoring algorithm. As the main advantage, moving mesh methods with a scheme for monitoring quality offer more flexibility and can preserve a better mesh quality.

Immersed boundary methods offer the same capabilities to simulate large displacements without the need to perform mesh deformation or deal with topology changes. The main advantage of IBM is that the base mesh can often be very simple and, at the same time, the method can deal easily with very complex geometries. However, the simple treatment of the solids uses a castellated (saw-tooth) approach, which is not suitable to simulate near wall dynamics correctly. This issue can be overcome to some extent by increasing the resolution locally. Nevertheless, as outlined in Kalitzin and Iaccarino (2003) and Kim and Choi (2019), the application of IBM to high Reynolds number cases still poses challenges, because gradients are large in turbulent boundary layer flows and body-fitted meshes are desirable as compared to castellated meshes.

Overset grids are probably the most advanced technique to simulate moving objects. The advantages of this method are very similar to those of the IBM: extreme flexibility to simulate long displacements, plus enabling the simulation of relative motion between solids. Furthermore, "overhead grids" do not pose problems to resolving the boundary layer flow if required, as IBM does, since proper body-fitted meshes can be used around solids. The main disadvantage of OG method is the additional complexity in setting up the simulations, and the additional computational costs associated with the steps required to couple the meshes. Nevertheless, as remarked by Windt et al. (2018), "overhead grids" are a suitable option when moving boundaries would present numerical instabilities and the results are not needed as fast as with other methods.

2.4.3   Examples and applications

Moving wavemaker procedures have been widely applied in literature for different CFD modelling techniques, for commercial and research models alike.

Piston wavemakers can be traced back to (Monaghan, 1994), the first paper to present the SPH method applied to model inviscid and incompressible free surface flows. Some years later a special moving wavemaker with arbitrary deformation was presented in Koshizuka et al. (1998). In this paper the numerical model is implemented with the Moving Particle Semi-Implicit (MPS) method, and the wavemaker displacement amplitude varies along the water depth to mimic the velocity profile of cnoidal waves.

The increase in adoption of moving wavemakers is linked to the general adoption of Eulerian Navier-Stokes models in the last 10 years. Lal and Elangovan (2008) simulated a numerical wave tank with a flap-type wavemaker in ANSYS CFX as a moving boundary. They studied different combinations of water depths, oscillation periods and amplitudes, and found good agreement with theory. In Lara et al. (2011), a piston wavemaker was replicated in the model IH2VOF using the IBM and openness coefficients. The goal of this paper was to simulate the propagation of infragravity waves generated by a short focused wave group breaking over a sloping beach. Waves were generated with a second order theory and active wave absorption was implemented following the method in Schäffer and Klopman (2000) and comparisons with experimental results were excellent.

Piston-type wavemakers have also been implemented in Flow3D (Vanneste and Troch, 2015). In this solver the paddle is represented by a moving solid IBM using the FAVOR technique (Wei, 2005). Vanneste and Troch (2015) simulated regular waves interacting with a porous breakwater. Results showed good accuracy when compared to experimental measurements. Long time series were simulated in this work, therefore, active wave absorption has been reported to play an important role in achieving stable simulations. The AWA methodology is the same applied in Troch and De Rouck (1999), in which the feedback is obtained from the horizontal and vertical velocities at a fixed location in front of the wavemaker, but applied to the moving wavemaker.

All the references presented so far in this section were 2D. However, Wu et al. (2016) performed a 3D simulation of a piston wavemaker to investigate the effects of the gap between the moving paddle and the walls of the flume. Solitary waves were generated with the classical method developed by Goring (1978) and a previous formulation (Wu et al., 2014), which yields a cleaner solitary wave. The flow leaks around the edges of the wavemaker was found to impact the wave height of the wave generated, decreasing significantly the wave amplitude, as reported in Madsen (1970) and later by Chwang (1983).

Higuera et al. (2015) presented an implementation in OpenFOAM® to replicate piston wavemakers with the moving boundary approach. The first case presented in this paper replicates an unpublished experiment from the set referenced in Lara et al. (2011) as a basic validation test. Similarly to the work in Lara et al. (2011), the AWA procedure implemented in Higuera et al. (2015) is based on Schäffer and Klopman (2000). The major novelty introduced in this work is an implementation of a multi-paddle piston wavemaker. This solver used the time series extracted from a laboratory wavemaker with 10 paddles as input and was able to successfully replicate a 3D wave focussing. Several enhancements to the original developments in Higuera et al. (2015) have been added to *olaFlow* model (Higuera, 2017), which now includes the capability to reproduce flap-type wavemakers too. Most recently, *olaFlow* has been applied to simulate the evolution of breaking and near-breaking wave groups (Stagonas et al., 2018; Buldakov et al., 2019), demonstrating excellent capabilities to simulate accurate wave kinematics.

In Usui et al. (2016), a piston wavemaker is simulated in the model CADMAS-SURF 3D. Since the goal of this work was to replicate tsunamis and storm surge, the wavemaker presents a relatively long stroke, thus it required being simulated with the immersed boundary method. Zhang et al. (2017) presented an application to replicate a piston wavemaker with the moving boundary technique for ANSYS Fluent. His implementation includes second order wave generation, which is found advantageous with respect to first order in terms of preventing the development of spurious free waves, as expected. Chella et al. (2017) presented an application of flap wavemakers for the model REEF3D to simulate focused waves breaking over a slope. Comparisons of free surface elevations and the overall breaking wave shape were compared with experimental data successfully.

Finally, a piston-type wavemaker has been implemented recently in a Lattice-Boltzmann method model (Davarpanah et al., 2018), combined with nested grids to obtain higher mesh resolutions locally to the free surface.

2.4.4 Advantages and disadvantages

The application of moving wavemakers in numerical models presents several advantages with respect to the laboratory systems. As previously discussed in Section 2.1, the ease of access to all the variables and not having the requirement to operate in real-time offer a higher degree of flexibility for AWA. Such conditions enable testing formulations, as complex as desired, which can be more efficient than their laboratory counterparts.

Within the methods, the moving boundary technique conserves the mass of the system. This has important implications, as in Dirichlet-type wave generation or other flux-based wave generation methods, the mass imbalance between crests and troughs can accumulate and make the mean water level increase in time, requiring absorption to be rectified. In a BM NWT, the only variations in the mean water level are induced by the shortening/enlarging of the tank itself due to the wavemaker movements, as happens in physical wave tanks.

Another advantage of numerical models is that they can mimic even the smallest and detailed features of the experimental facilities. For example, models can simulate any degree of leakage of water between the edges of the wavemaker and the flume walls, including the watertight condition too. Such fine details can be proven important to ensure fidelity between the modelled and simulated results, as shown in Wu et al. (2016).

Finally, numerical wavemakers present no mechanical or inertial effects, hence, they respond instantly to the prescribed displacement and/or velocity. Furthermore, numerical wavemakers always move in the same way, irrespectively of the environmental conditions, unlike physical systems in which factors such as the temperature of the system can have a significant impact. This means that the numerical models produce perfectly repeatable simulations, and they can replicate in an exact way a time series of displacements provided.

There are several disadvantages of applying moving wavemakers. Two of them are caused by mechanical limitations, therefore they can be circumvented in the numerical model. First is the limitation to generate or to absorb long waves, as such will require long strokes, i.e., on the order of magnitude of the target wavelength, and generally, the total stroke of physical wavemakers is quite limited. Another issue is the limitation to generate and to absorb large steepness waves, which usually requires large accelerations that the wave generation machines often cannot provide.

Another issue is inherent to the method itself, therefore, it appears for physical and numerical experiments alike. This problem arises because waves are seldom perfect sinusoids, and small but systematic deviations from a zero-mean wavemaker movement will produce a constant drift of the wavemaker, as noted in Schäffer and Klopman (2000). Once the movement of the wavemaker has reached either of the limits, waves cannot continue to be generated or absorbed normally, since the wavemaker will stop at the minimum or maximum displacement positions. This behaviour is called saturation and can be prevented by forcing the wavemaker to return automatically to its initial position. The time frame for this additional movement needs to be short enough to prevent saturation, but long enough at the same time to be able to absorb longer waves to some extent and prevent the development of seiches in the flume (Schäffer and Klopman, 2000).

Finally, another of the disadvantages associated with moving wavemakers in numerical models is the larger computational costs associated with the moving-mesh techniques. Reports available in literature state that longer simulation times, between 20% to 40%, can be expected when comparing the moving wavemaker generation mode versus Dirichlet-type wave generation (Higuera et al., 2015).

3   Discussion

Moving boundary is the wave generation procedure that best matches the well-understood laboratory wavemakers, both for wave generation and absorption. At the same time, thanks to the capabilities of numerical models, some of the shortcomings of laboratory wave generation can be solved. For example, numerical models do not experience mechanical or inertial effects and allow perfect repeatability. Furthermore, the flexibility of numerical frameworks permits testing new and more complex forms of wave generation and absorption, as arbitrarily-deforming boundaries, which are not feasible to develop physically. Other restrictions, such as real time operation, can also be circumvented numerically, offering the possibility to obtaining more efficient wave absorption rates. However, moving wavemaker generation and absorption usually require longer computational times than other techniques, including relaxation zones, because the numerical model performs complex moving mesh steps, which significantly increase the computational cost. In short, moving wavemakers in numerical models offer higher fidelity for reproducing laboratory tests, at the cost of increased model complexity which may result in longer simulation times.

It is not always convenient to generate waves with moving boundaries, as this approach can be complex and quite limited to represent real wave conditions. In this sense, Dirichlet-type wave generation is more flexible, as it offers a way to generate waves using theoretical formulations for the hydrodynamic variables (e.g., free surface elevation, velocities, pressures...). Wave absorption, open BC or AWA, can also be implemented in a similar way and in some cases be applied simultaneously with wave generation. This approach is the most computationally efficient, as it takes place at a fixed boundary, hence, no mesh movement is required, and the domain boundaries can be located close to the area of interest, further minimising the computational costs.

There are several disadvantages for Dirichlet-type wave generation. The most noticeable one is that this method produces a progressive increase in the mean water level, as explained before. Fortunately, linking wave generation with AWA solves this problem without significant increase in computational costs. Another limitation is generating very steep waves, as experience tells that they are more prone to start breaking at the boundary below the expected breaking limit. Regarding wave absorption, both open BCs and AWA work more efficiently for monochromatic waves. Moreover, absorption may be sub-optimal if all the incident wave information is not available beforehand, or in cases of time-varying wave conditions (e.g., irregular wave sea states) and in most cases the performance largely depends on the digital filter that has been designed. AWA, which appears to be the most used fixed-value method presently, is very sensitive to relative water depth conditions. This technique is especially effective for absorbing long waves and it has recently been extended to absorb waves in deep water conditions as efficiently.

Relaxation zones are other effective and widely used methods to generate and absorb waves within a domain, rather than at a boundary, since they are essentially a mean of "diffusing" the boundary condition over a model subdomain, rather than applying the boundary condition locally. The procedure of "diffusing" the boundary conditions can be achieved either by introducing appropriately scaled source or sink terms in the flow equations ("physical forcing") or by relaxing the flow variables toward the desired boundary condition within the numerical solution algorithm ("numerical forcing"). The inherent property of "diffusing" the boundary condition is responsible for both the most significant advantages and disadvantages of the method. The method is generally considered superior in fidelity of generation for high steepness waves and more efficient in absorbing waves, but it is also more expensive, especially when compared to open BC or AWA. Ongoing research is looking to address computational cost issues and substantial steps have been made towards this direction.

Internal wave makers (IWM) are based on the concept of dedicating internal areas in the mesh for wave generation, while relying on absorbing boundaries or relaxation zones for absorption. This method was proven particularly successful for depth integrated models. The method has been used also in CFD models showing good overall performance. For CFD models, the method requires a substantial calibration procedure and its implementation does not seem as straighforward as other methods and relies on the implementation of Dirichlet-type conditions or the relaxation zone for passive absorption for use in the context of NWT. In addition, IWM can be costly in terms of computational cost and it is very often comparable to the relaxation zone method. Whilst IWMs remain widespread for depth integrated models, it is the authors impression that recent CFD modelling applications increasingly Dirichlet type conditions or the relaxation zone method are more effective IWM for NWTs, due to the aforementioned reasons.

As a conclusion, significant advances have been made in wave generation and absorption techniques for numerical models over the last decades. Developments linked to physical wave tanks have been the main driver pushing the state of the art forward for a long time. This situation has reversed in the last decade, as now the scientific community has wider access to powerful computational clusters. Consequently, future new developments will likely be computational-based, powered by the flexibility that numerical models and other computational tools offer.

In this sense, hybrid modelling is an extremely promising approach. Hybrid models rely on coupling to blend different modelling approaches. For example, wave propagation can be performed with a simplified (e.g., potential flow or Boussinesq-type) model to study the wave transformations before arriving to the area of interest, in which CFD modelling is performed. This approach reduces the overall computational cost but requires a proper link between the models to allow a seamless transition not compromising the wave kinematics. Recent applications include a two-way coupling between the fully nonlinear and dispersive potential flow model OceanWave3D (Engsig-Karup et al., 2009) and the CFD NWT waves2Foam in OpenFOAM® (Jacobsen et al., 2012), developed with relaxation zones. Another example is the one-way coupling of a Lagrangian model solving the mass and vorticity conservation equations (Buldakov, 2013, 2014) and the CFD NWT *olaFlow* (Higuera, 2017), also developed in OpenFOAM®, via a boundary condition.

Another area to explore is the combination of existing wave absorption methods to increase the performance of wave absorption while minimising the increase of computational costs. Although this topic has already been explored in the past quite successfully, e.g., Israeli and Orszag (1981); Clément (1996); Grilli and Horrillo (1997) combining open BCs and relaxation zones for absorption, this research line has been discontinued for some time, and new formulations have been introduced and perfected since. For example, some initial tests combining AWA and relaxation zones have been performed in Higuera (2020), showing that this approach can reduce the overall reflections in most cases.

Although to the authors knowledge there has not been an attempt towards this direction, AWA and SWAG could be combined to create a hybrid scheme and improve wave generation at the offshore boundary. For example, the AWA condition could be relaxed over a layer of cells at the generation area, which ideally would be much shorter than the typical length of the generation zone used in the conventional relaxation zone technique. This approach could be combining the better of two worlds, e.g., low computational cost, high absorption efficiency. It could be nevertheless argued that this technique would be solving two equations in the same domain, so challenges for developing this approach may be similar to the ones encountered when developing multi-scale models.

References

Afshar, M. A. (2010). Numerical Wave Generation in OpenFOAM. Master's thesis, Chalmers University of Technology.

Amoudry, L. O., Bell, P. S., Thorne, P. D. and Souza, A. J. (2013). Toward representing wave-induced sediment suspension over sand ripples in RANS models. Journal of Geophysical Research: Oceans, 118(5): 2378–2392.

Antoniou, A. (2006). Digital Signal Processing: Signals, Systems and Filters. McGraw-Hill.

Antuono, M., Colagrossi, A., Marrone, S. and Lugni, C. (2011). Propagation of gravity waves through an SPH scheme with numerical diffusive terms. Computer Physics Communications 182(4): 866–877.

Belibassakis, K. A. and Athanassoulis, G. A. (2004). Three-dimensional Green's function for harmonic water waves over a bottom topography with different depths at infinity. Journal of Fluid Mechanics 510(4): 267–302.

Bihs, H. and Kamath, A. (2017). A combined level set/ghost cell immersed boundary representation for floating body simulations. International Journal for Numerical Methods in Fluids 83(12): 905–916.

Bihs, H., Kamath, A., Chella, M. A. and Arntsen, Ø. A. (2015). CFD simulations of roll motion of a floating ice block in wave using REEF3D. In Proceedings of the International Conference on Port and Ocean Engineering Under Arctic Conditions.

Booij, N., Holthuijsen, L. H. and Ris, R. C. (1996). The "SWAN" wave model for shallow water. In Proceedings of the 25th International Conference on Coastal Engineering (ICCE), Orlando, USA, Volume 1.

Bradford, S. F. (2000). Numerical simulation of surf zone dynamics. Journal of Waterway, Port, Coastal and Ocean Engineering 126(1): 1–13.

Buldakov, E. (2013). Tsunami generation by paddle motion and its interaction with a beach: Lagrangian modelling and experiment. Coastal Engineering 80: 83–94.

Buldakov, E. (2014). Lagrangian modelling of fluid sloshing in moving tanks. Journal of Fluids and Structures 45: 1–14.

Buldakov, E., Higuera, P. and Stagonas, D. (2019). Numerical models for evolution of extreme wave groups. Applied Ocean Research 89: 128–140.

Bullock, G. N. and Murton, G. J. (1989). Performance of a wedge-type absorbing wavemaker. Journal of Waterway, Port, Coastal and Ocean Engineering 115(1): 1–17.

Calderer, A., Kang, S. and Sotiropoulos, F. (2014). Level set immersed boundary method for coupled simulation of air/water interaction with complex floating structures. Journal of Computational Physics 277: 201–227.

Cao, Y., Beck, R. F. and Schultz, W. W. (1993). An absorbing beach for numerical simulations of nonlinear waves in a wave tank. In Proc. 8th Intl. Workshop Water Waves and Floating Bodies, pp. 17–20.

Chella, M. A., Bihs, H., Myrhaug, D. and Muskulus, M. (2015). Breaking characteristics and geometric properties of spilling breakers over slopes. Coastal Engineering 95: 4–19.

Chella, M. A., Bihs, H., Pâkozdi, C., Myrhaug, D. and Arntsen, Ø. A. (2017). Simulation of breaking focused waves over a slope with a CFD based numerical wave tank. In Proceedings of VII International Conference on Computational Methods in Marine Engineering (MARINE 2017), Nantes, France.

Chen, H. and Christensen, E. D. (2018). Simulating the hydrodynamic response of a floater-net system in current and waves. Journal of Fluids and Structures 79: 50–75.

Chen, H. and Zou, Q. -P. (2019). Eulerian-Lagrangian flow-vegetation interaction model using immersed boundary method and OpenFOAM. Advances in Water Resources 126: 176–192.

Cheng, Z., Hsu, T. -J. and Calantoni, J. (2017). SedFoam: A multi-dimensional Eulerian two-phase model for sediment transport and its application to momentary bed failure. Coastal Engineering 119: 32–50.

Choi, J. and Yoon, S. B. (2009). Numerical simulations using momentum source wave-maker applied to RANS equation model. Coastal Engineering 56: 1043–1060.

Christensen, E. D. (2006). Large eddy simulation of spilling and plunging breakers. Coastal Engineering 53(5-6): 463–485.

Christensen, M. and Frigaard, P. (1994). Design of absorbing wave-maker based on digital filters. In Proceedings of the International Symposium: Waves—Physical and Numerical Modelling: University of British Columbia, Vancouver, Canada, pp. 100–109.

Chwang, A. T. (1983). A porous-wavemaker theory. Journal of Fluid Mechanics 132: 395–406.

Clément, A. (1996). Coupling of two absorbing boundary conditions for 2D time-domain simulations of free surface gravity waves. Journal of Computational Physics 126(1): 139–151.

Cozzuto, G., Dimakopoulos, A., De Lataillade, T., Morillas, P. O. and Kees, C. E. (2019). Simulating oscillatory and sliding displacements of caisson breakwaters using a coupled approach. Journal of Waterway, Port, Coastal, and Ocean Engineering 145(2): 04019005.

Davarpanah, E., Anbarsooz, M. and Rajabiani, E. (2018). Numerical simulation of a pistontype wavemaker using Lattice-Boltzmann method with moving nested grids. Amirkabir Journal of Mechanical Engineering 50(4): 799–812.

de Lataillade, T., Dimakopoulos, A., Kees, C., Johanning, L., Ingram, D. and Tezdogan, T. (2017). CFD modelling coupled with floating structures and mooring dynamics for offshore renewable energy devices using the Proteus simulation toolkit.

de Lataillade, T. d. (2019). High-Fidelity Computational Modelling of Fluid-Structure Interaction for Moored Floating Bodies. PhD thesis, IDCORE Doctoral Training Centre, Edinburg (submitted).

Dean, R. G. (1965). Stream function representation of nonlinear ocean waves. Journal of Geophysical Research 70(18): 4561–4572.

Delafontaine, M. (2016). Experimental study on the performance of different wave absorbers in a wave flume.

DHI (2017). FUNWAVE-TVD fully nonlinear Boussinesq wave model with TVD solver-Documentation and User's Manual (Version 3.0). Mike 21 BW-Boussinesq model user guide.

Dimakopoulos, A. S., Cuomo, G. and Chandler, I. (2016). Optimized generation and absorption for three-dimensional numerical wave and current facilities. Journal of Waterway, Port, Coastal, and Ocean Engineering 142(4): 06016001.

Dimakopoulos, A. S., de Lataillade, T. and Kees, C. E. (2019). Fast random wave generation in numerical tanks. Proceedings of the Institution of Civil Engineers-Engineering and Computational Mechanics 172(1): 1–11.

Dimakopoulos, A. S. and Dimas, A. A. (2011). Large-wave simulation of three-dimensional, cross-shore and oblique, spilling breaking on constant slope beach. Coastal Engineering 58(8): 790–801.

Dimas, A. A. and Dimakopoulos, A. S. (2009). Surface roller model for the numerical simulation of spilling wave breaking over constant slope beach. Journal of Waterway, Port, Coastal, and Ocean Engineering 135(5): 235–244.

Dongeren, A. R. V. and Svendsen, I. A. (1997). Absorbing-generating boundary condition for shallow water models. Journal of Waterway, Port, Coastal, and Ocean Engineering 123(6): 303–313.

Dwight, R. P. (2009). Robust mesh deformation using the linear elasticity equations. In Computational Fluid Dynamics, pp. 401–406.

Elsafti, H. and Oumeraci, H. (2017). Analysis and classification of stepwise failure of monolithic breakwaters. Coastal Engineering 121: 221–239.

English, R. E., Qiu, L., Yu, Y. and Fedkiw, R. (2013). Chimera grids for water simulation. In Proceedings of the 12th ACM SIGGRAPH/Eurographics Symposium on Computer Animation (SCA '13), Anaheim, California, USA, pp. 85–94.

Engquist, B. and Majda, A. (1977). Absorbing boundary conditions for the numerical simulation of waves. Mathematics of Computation 31(139): 629–651.

Engsig-Karup, A. P., Bingham, H. B. and Lindberg, O. (2009). An efficient flexible-order model for 3D nonlinear water waves. Journal of Computational Physics 228: 2100–2118.

Eskilsson, C., Palm, J., Kofoed, J. P. and Friis-Madsen, E. (2015). CFD study of the overtopping discharge of the Wave Dragon wave energy converter. Renewable Energies Offshore, 287–294.

Fuhrman, D. R., Baykal, C., Sumer, B. M., Jacobsen, N. G. and Fredsøe, J. (2014). Numerical simulation of wave-induced scour and backfilling processes beneath submarine pipelines. Coastal Engineering 94: 10–22.

Gadelho, J. F. M., Lavrov, A., Guedes Soares, C., Urbina, R., Cameron, M. P. and Thiagarajan, K. P. (2015). CFD modelling of the waves generated by a wedge-shaped wave maker. In Maritime Technology and Engineering, Taylor & Francis Group, London, pp. 993–1000.

Garcia, N., Lara, J. L. and Losada, I. J. (2004). 2-D numerical analysis of near-field flow at low-crested permeable breakwaters. Coastal Engineering 51(10): 991–1020.

Givoli, D. (1991). Non-reflecting boundary conditions. Journal of Computational Physics 94(1): 1–29.

Goring, D. G. (1978). Tsunami: the propagation of long waves onto a shelf. Technical report, Report No. KH-R-38, W. M. Keck Laboratory of Hydraulics and Water Resources, California Institute of Technology, Pasadena, California, USA.

Grajewski, M., Köster, M., Kilian, S. and Turek, S. (2005). Numerical analysis and practical aspects of a robust and efficient grid deformation method in the finite element context. Submitted to SISC.

Grilli, S. T., Guyenne, P. and Dias, F. (2001). A fully non-linear model for three-dimensional overturning waves over an arbitrary bottom. International Journal for Numerical Methods in Fluids 35(7): 829–867.

Grilli, S. T. and Horrillo, J. (1997). Numerical generation and absorption of fully nonlinear periodic waves. Journal of Engineering Mechanics 123(10): 1060–1069.

Grilli, S. T., Skourup, J. and Svendsen, I. A. (1989). An efficient boundary element method for nonlinear water waves. Engineering Analysis with Boundary Elements 6(2): 97–107.

Grilli, S. T., Svendsen, I. A. and Subramanya, R. (1997). Breaking criterion and characterisitics for solitary waves on slopes. Journal of Waterway, Port, Coastal and Ocean Engineering 123(3): 102–112.

Ha, T., Lin, P. and Cho, Y.-S. (2013). Generation of 3D regular and irregular waves using Navier-Stokes equations model with an internal wave maker. Coastal Engineering 76: 55–67.

Ha, T., Yoo, J., Han, S. and Cho, Y.-S. (2014). Numerical study on tsunami hazard mitigation using a submerged breakwater. The Scientific World Journal, 2014.

Hafsia, Z., Hadj, M. B., Lamloumi, H. and Maalel, K. (2009). Internal inlet for wave generation and absorption treatment. Coastal Engineering 56(9): 951–959.

Hald, T. and Frigaard, P. (1996). Performance of active wave absorption systems: Comparison of wave gauge and velocity meter based systems. In Proceedings of the 7th International Offshore and Polar Engineering Conference, Honolulu, Hawaii, USA, volume ISOPE-I-97-289.

Halfiani, V., Muarif and Ramli, M. (2015). Triple-flap wavemaker based on linear wave theory. Bulletin of Mathematics 7(1): 51–63.

Hibberd, S. and Peregrine, D. H. (1979). Surf and run-up on a beach: a uniform bore. Journal of Fluid Mechanics 95(2): 323–345.

Higdon, R. L. (1986). Absorbing boundary conditions for difference approximations to the multi-dimensional wave equation. Mathematics of Computation 47(176): 437–459.

Higdon, R. L. (1987). Numerical absorbing boundary conditions for the wave equation. Mathematics of Computation 49(179): 65–90.

Higuera, P. (2017). olaFlow: CFD software for wave modelling. https://doi.org/10.5281/zenodo.1297013.

Higuera, P. (2020). Enhancing active wave absorption in RANS models. Applied Coastal Research 94.

Higuera, P., Lara, J. L. and Losada, I. J. (2013). Realistic wave generation and active wave absorption for Navier-Stokes models: Application to OpenFOAM. Coastal Engineering 71: 102–118.

Higuera, P., Losada, I. J. and Lara, J. L. (2015). Three-dimensional numerical wave generation with moving boundaries. Coastal Engineering 101: 35–47.

Hirt, C. W., Amsden, A. A. and Cook, J. L. (1974). An arbitrary Lagrangian-Eulerian computing method for all flow speeds. Journal of Computational Physics 14(3): 227–253.

Hu, Z. Z., Greaves, D. and Raby, A. (2016). Numerical wave tank study of extreme waves and wave-structure interaction using OpenFoam R. Ocean Engineering 126: 329–342.

Huang, C. -J., Zhang, E. -C. and Lee, J. -F. (1998). Numerical simulation of nonlinear viscous wavefields generated by piston-type wavemaker. Journal of Engineering Mechanics 124: 1110–1120.

Huang, Z. C., Hsiao, S. C., Hwung, H. H. and Chang, K. A. (2009). Turbulence and energy dissipations of surf-zone spilling breakers. Coastal Engineering 56(7): 733–746.

Israeli, M. and Orszag, S. A. (1981). Approximation of radiation boundary conditions. Journal of Computational Physics 41: 115–135.

Iwata, K., Kawasaki, K. and Kim, D. (1996). Breaking limit, breaking and post-breaking wave deformation due to submerged structures. 3: 2338–2351.

Jacobsen, N. G. (2017). waves2Foam Manual. Deltares, The Netherlands.

Jacobsen, N. G. and Fredsoe, J. (2014). Formation and development of a breaker bar under regular waves. Part 2: Sediment transport and morphology. Coastal Engineering 88: 55–68.

Jacobsen, N. G., Fuhrman, D. R. and Fredsøe, J. (2012). A wave generation toolbox for the open-source CFD library: OpenFOAM. International Journal for Numerical Methods in Fluids 70(9): 1073–1088.

Jasak, H. and Tuković, Ž. (2006). Automatic mesh motion for the unstructured finite volume method. Transactions of FAMENA 30(2): 1–20.

Jasak, H., Vukčević, V. and Gatin, I. (2015). Numerical simulation of wave loading on static offshore structures. In CFD for Wind and Tidal Offshore Turbines, pp. 95–105.

Jensen, B., Jacobsen, N. G. and Christensen, E. D. (2014). Investigations on the porous media equations and resistance coefficients for coastal structures. Coastal Engineering 84: 56–72.

Jensen, B., Liu, X., Christensen, E. D. and Rønby, J. (2017). Porous media and immersed boundary hybrid-modelling for simulating flow in stone cover-layers. In Coastal Dynamics, 2017.

Johns, B. and Jefferson, R. J. (1980). The numerical modeling of surface wave propagation in the surf zone. Journal of Physical Oceanography 10(7): 1061–1069.

Kalitzin, G. and Iaccarino, G. (2003). Toward immersed boundary simulation of high Reynolds number flows. Center for Turbulence Research Annual Research Briefs, pp. 369–378.

Kamath, A., Bihs, H., Chella, M. A. and Arntsen, Ø. A. (2016). Upstream-cylinder and downstream-cylinder influence on the hydrodynamics of a four-cylinder group. Journal of Waterway, Port, Coastal, and Ocean Engineering 142(4): 04016002.

Karagiannis, N., Karambas, T. and Koutitas, C. (2016). Numerical simulation of scour in front of a breakwater using OpenFoam. In Sustainable Hydraulics in the Era of Global Change: Proceedings of the 4th IAHR Europe Congress (Liege, Belgium, 27–29 July 2016), volume 1, page 115.

Karambas, T. V. and Koutitas, C. (1992). A breaking wave-propagation model based on the Boussinesq equations. Coastal Engineering 18(1-2): 1–19.

Kees, C. E., Farthing, M. W. and Bazilevs, Y. (2011). Parallel computational methods and simulation for coastal and hydraulic applications using the Proteus toolkit. In Supercomputing11: Proceedings of the PyHPC11 Workshop.

Kim, W. and Choi, H. (2019). Immersed boundary methods for fluid-structure interaction: A review. International Journal of Heat and Fluid Flow 75: 301–309.

Kim, Y. J. and Lee, S. B. (2017). Effects of trim conditions on ship resistance of KCS in short waves. Journal of the Society of Naval Architects of Korea 54(3): 258–266.

Kirby, J. T., Wei, G., Chen, Q., Kennedy, A. B. and Dalrymple, R. A. (1998). FUNWAVE 1.0: fully nonlinear Boussinesq wave model-Documentation and user's manual. Research report NO. CACR-98-06.

Kobayashi, N., Otta, A. K. and Roy, I. (1987). Wave reflection and run-up on rough slopes. Journal of Waterway, Port, Coastal and Ocean Engineering 113(3): 282–298.

Koftis, T. H., Prinos, P. and Koutandos, E. (2006). 2D-V hydrodynamics of wave-floating breakwater interaction. Journal of Hydraulic Research 44(4): 451–469.

Koshizuka, S., Nobe, A. and Oka, Y. (1998). Numerical analysis of breaking waves using the moving particle semi-implicit method. International Journal for Numerical Methods in Fluids 26(7): 751–769.

Lal, A. and Elangovan, M. (2008). CFD simulation and validation of flap type wave-maker. World Academy of Science, Engineering and Technology 46(1): 76–82.

Lara, J. L., Garcia, N. and Losada, I. J. (2006). RANS modelling applied to random wave interaction with submerged permeable structures. Coastal Engineering 53: 395–417.

Lara, J. L., Ruju, A. and Losada, I. J. (2011). Reynolds averaged Navier–Stokes modelling of long waves induced by a transient wave group on a beach. Proceedings of the Royal Society A 467: 1215–1242.

Larsen, J. and Dancy, H. (1983). Open boundaries in short wave simulations—A new approach. Coastal Engineering 7(3): 285–297.

Le Méhauté, B. (1972). Progressive wave absorber. Journal of Hydraulic Research 10(2): 153–169.

Le Méhauté, B. (1976). Introduction to Hydrodynamics and Water Waves. Springer-Verlag, New York.

Lemos, C. M. (1992). A simple numerical technique for turbulent flows with free surfaces. International Journal for Numerical Methods in Fluids 15(2): 127–146.

Lin, P. and Liu, P.-F. (1998). A numerical study of breaking waves in the surf zone. Journal of Fluid Mechanics 359: 239–264.

Lin, P. and Liu, P. L. -F. (1999). Internal wave-maker for Navier–Stokes equations models. Journal of Waterway, Port, Coastal and Ocean Engineering 125: 207–215.

Liu, C., Wang, J., Wan, D. et al. (2018). CFD computation of wave forces and motions of DTC ship in oblique waves. International Journal of Offshore and Polar Engineering 28(02): 154–163.

Liu, X., Lin, P. and Shao, S. (2015). ISPH wave simulation by using an internal wave maker. Coastal Engineering 95: 160–170.

Losada, I. J., Lara, J. L., Guanche, R. and Gonzalez-Ondina, J. M. (2008). Numerical analysis of wave overtopping of rubble mound breakwaters. Coastal Engineering 55(1): 47–62.

Madsen, O. S. (1970). Waves generated by a piston-type wavemaker. In Proceedings of the 12th Conference on Coastal Engineering, Washington D.C., USA, pp. 589–607.

Madsen, P. A., Murray, R. and Sørensen, O. R. (1991). A new form of the Boussinesq equations with improved linear dispersion characteristics. Coastal Engineering 15(4): 371–388.

Mattis, S. A., Kees, C. E., Wei, M. V., Dimakopoulos, A. and Dawson, C. N. (2018). Computational model for wave attenuation by flexible vegetation. Journal of Waterway, Port, Coastal, and Ocean Engineering 145(1): 04018033.

Mayer, S., Garapon, A. and Søresen, L. S. (1998). A fractional step method for unsteady free-surface flow with applications to non-linear wave dynamics. International Journal for Numerical Methods in Fluids 28(2): 293–315.

Medina-Lopez, E., Allsop, W., Dimakopoulos, A. and Bruce, T. (2015). Conjectures on the failure of the OWC breakwater at Mutriku. In Coastal Structures and Solutions to Coastal Disasters Joint Conference, Boston, Massachusetts.

Miquel, A. M., Kamath, A., Chella, M. A., Archetti, R. and Bihs, H. (2018). Analysis of different methods for wave generation and absorption in a CFD-based numerical wave tank. Journal of Marine Science and Engineering 6(2).

Mittal, R. and Iaccarino, G. (2005). Immersed boundary methods. Annual Review of Fluid Mechanics 37: 239–261.

Monaghan, J. J. (1994). Simulating free surface flows with SPH. Journal of Computational Physics 110(2): 399–406.

Montorfano, A., Piscaglia, F. and Onorati, A. (2014). A moving mesh strategy to perform adaptive large eddy simulation of IC engines in OpenFOAM. In International Multidimensional Engine Modeling Users Group Meeting.

Moradi, N., Zhou, T. and Cheng, L. (2016). Two-dimensional numerical study on the effect of water depth on resonance behaviour of the fluid trapped between two side-by-side bodies. Applied Ocean Research 58: 218–231.

Nwogu, O. (1993). Alternative form of Boussinesq equations for nearshore wave propagation. Journal of Waterway, Port, Coastal and Ocean Engineering 119(6): 618–638.

Nwogu, O. and Demirbilek, Z. (2001). BOUSS-2D: A Boussinesq wave model for coastal regions and harbors. ERDC. In Technical Report. US Army Engineer Research and Development Center Vicksburg, MS.

Ong, M. C., Kamath, A., Bihs, H. and Afzal, M. S. (2017). Numerical simulation of free-surface waves past two semi-submerged horizontal circular cylinders in tandem. Marine Structures 52: 1–14.

Palm, J., Eskilsson, C., Paredes, G. M. and Bergdahl, L. (2013). CFD simulation of a moored floating wave energy converter. In Proceedings of the 10th European Wave and Tidal Energy Conference, Aalborg, Denmark, volume 25.

Panahi, R. and Shafieefar, M. (2009). Application of overlapping mesh in numerical hydrodynamics. Polish Maritime Research 16: 24–33.

Peregrine, D. H. (1967). Long waves on a beach. Journal of Fluid Mechanics 27(4): 815–827.

Perić, R. and Abdel-Maksoud, M. (2015). Generation of free-surface waves by localized source terms in the continuity equation. Ocean Engineering 109: 567–579.

Perić, R. and Abdel-Maksoud, M. (2016). Reliable damping of free-surface waves in numerical simulations. Ship Technology Research 63(1): 1–13.

Perić, R. and Abdel-Maksoud, M. (2018). Analytical prediction of reflection coefficients for wave absorbing layers in flow simulations of regular free-surface waves. Ocean Engineering 147: 132–147.

Perić, R., Vukčević, V., Abdel-Maksoud, M. and Jasak, H. (2018). Tuning the case-dependent parameters of relaxation zones for flow simulations with strongly reflecting bodies in free-surface waves. arXiv preprint arXiv:1806.10995.

Piro, D. J. (2013). A Hydroelastic Method for the Analysis of Global Ship Response Due to Slamming Events. PhD thesis, University of Michigan.

Richardson, S., Cuomo, G., Dimakopoulos, A. and Longo, D. (2014). Coastal structure optimisation using advanced numerical methods. In From Sea to Shore–Meeting the Challenges of the Sea: (Coasts, Marine Structures and Breakwaters 2013), pp. 1184–1194.

Sabeur, Z. A., Allsop, N. W., Beale, R. G. and Dennis, J. M. (1997). Wave dynamics at coastal structures: Development of a numerical model for free surface flow. In Proceedings of the 25th International Conference on Coastal Engineering (ICCE), Orlando, Florida, USA, pp. 389–402.

Saincher, S. and Banerjee, J. (2017). On wave damping occurring during source-based generation of steep waves in deep and near-shallow water. Ocean Engineering 135: 98–116.

Schäffer, H. A. (1996). Second order wavemaker theory for irregular waves. Ocean Engineering 23(1): 47–88.

Schäffer, H. A. and Klopman, G. (2000). Review of multidirectional active wave absorption methods. Journal of Waterway, Port, Coastal and Ocean Engineering 126(2): 88–97.

Schäffer, H. A. and Steenberg, C. M. (2003). Second-order wavemaker theory for multidirectional waves. Ocean Engineering 30(10): 1203–1231.

Schäffer, H. A., Stolborg, T. and Hyllested, P. (1994). Simultaneous generation and active absorption of waves in flumes. In Proceedings of the International Symposium: Waves-Physical and Numerical Modelling: University of British Columbia, Vancouver, Canada, pp. 100–109.

Schmitt, P., Windt, C., Davidson, J., Ringwood, J. V. and Whittaker, T. (2019). The efficient application of an impulse source wavemaker to CFD simulations. Journal of Marine Science and Engineering 7(3): 71.

Shen, Z. and Korpus, R. (2015). Numerical simulations of ship self-propulsion and maneuvering using dynamic overset grids in OpenFOAM. In Tokyo 2015 A Workshop on CFD in Ship Hydrodynamics. Presented at the Tokyo 2015 A Workshop on CFD in Ship Hydrodynamics, Tokyo, Japan.

Shen, Z., Wan, D. and Carrica, P. M. (2015). Dynamic overset grids in OpenFOAM with application to KCS self-propulsion and maneuvering. Ocean Engineering 108: 287–306.

Shi, F., Kirby, J., Tehranirad, B., Harris, J., Choi, Y.-K. and Malej, M. (2016). FUNWAVETVD fully nonlinear Boussinesq wave model with TVD solver-Documentation and User's Manual (Version 3.0). Research report NO. CACR-11-03.

Sommerfeld, A. (1964). Lectures on Theoretical Physics. Academic Press, San Diego.

Spinneken, J., Christou, M. and Swan, C. (2014). Force-controlled absorption in a fully nonlinear numerical wave tank. Journal of Computational Physics 272: 127–148.

Spinneken, J. and Swan, C. (2009a). Second-order wave maker theory using force-feedback control. Part I: A new theory for regular wave generation. Ocean Engineering 36(8): 539–548.

Spinneken, J. and Swan, C. (2009b). Second-order wave maker theory using force-feedback control. Part II: An experimental verification of regular wave generation. Ocean Engineering 36(8): 549–555.

Stagonas, D., Higuera, P. and Buldakov, E. (2018). Simulating breaking focused waves in CFD: a methodology for controlled generation to 1st and 2nd order. Journal of Waterway, Port, Coastal and Ocean Engineering 144(2).

Stahlmann, A. and Schlurmann, T. (2012). Investigations on scour development at tripod foundations for offshore wind turbines: Modeling and application. Proceedings 33rd International Conference on Coastal Engineering, Santander, Spain.

Stokes, G. G. (1847). On the theory of oscillatory waves. Transactions of the Cambridge Philosophical Society 8: 441–455.

Tiedeman, S. A., Allsop, W., Russo, V. and Brown, A. (2012). A demountable wave absorber for wave flumes and basins. Proceedings 33rd International Conference on Coastal Engineering, Santander, Spain.

Titov, V. V. and Synolakis, C. E. (1995). Modeling of breaking and nonbreaking longwave evolution and runup using VTCS-2. Journal of Waterway, Port, Coastal and Ocean Engineering 121: 308–316.

Torres-Freyermuth, A., Lara, J. L. and Losada, I. J. (2010). Numerical modelling of shortand long-wave transformation on a barred beach. Coastal Engineering 57: 317–330.

Torres-Freyermuth, A., Losada, I. J. and Lara, J. L. (2007). Modeling of surf zone processes on a natural beach using Reynolds-Averaged Navier-Stokes equations. Journal of Geophysical Research 112(C9).

Troch, P. and De Rouck, J. (1999). An active wave generating-absorbing boundary condition for VOF type numerical model. Coastal Engineering 38(4): 223–247.

Usui, A., Hiratsuka, Y., Aoki, S. and Kawasaki, K. (2016). Proposal of a self-propelled wavemaker for tsunami and storm surge experiments. Journal of Japan Society of Civil Engineers, Ser. B2 (Coastal Engineering) 72(2): 31–36. (In Japanese).

Van der Meer, J. W., Petit, H. A. H., Van den Bosch, P., Klopman, G. and Broekens, R. D. (1993). Numerical simulation of wave motion on and in coastal structures. In Proceedings of the 23rd International Conference on Coastal Engineering (ICCE), Venice, Italy, pp. 1772–1784.

van Gent, M. R. A., Tonjes, P., Petit, H. A. H. and van den Bosch, P. (1994). Wave action on and in permeable structures. In Proceedings of the 24th International Conference on Coastal Engineering (ICCE), Kobe, Japan 24: 1739–1753.

Vanneste, D. and Troch, P. (2015). 2D numerical simulation of large-scale physical model tests of wave interaction with a rubble-mound breakwater. Coastal Engineering 103: 22–41.

Verbrugghe, T., Domínguez, J. M., Altomare, C., Tafuni, A., Vacondio, R., Troch, P. and Kortenhaus, A. (2019). Non-linear wave generation and absorption using open boundaries within DualSPHysics. Computer Physics Communications, In Press.

Vukčević, V., Jasak, H., Gatin, I. and Uroić, T. (2017). Ship scale self propulsion CFD simulation results compared to sea trial measurements. In VII International Conference on Computational Methods in Marine Engineering, MARINE 2017.

Vyzikas, T., Deshoulières, S., Giroux, O., Barton, M. and Greaves, D. (2017). Numerical study of fixed oscillating water column with RANS-type two-phase CFD model. Renewable Energy 102: 294–305.

Watanabe, Y., Saeki, H. and Hosking, R. J. (2005). Three-dimensional vortex structures under breaking waves. Journal of Fluid Mechanics 545: 291–328.

Wei, G. (2005). A fixed-mesh method for general moving objects in fluid flow. Modern Physics Letters B 19(28-29): 1719–1722.

Wei, G. and Kirby, J. T. (1995). Time-dependent numerical code for extended Boussinesq equations. Journal of Waterway, Port, Coastal and Ocean Engineering 121(5): 251–261.

Wei, G., Kirby, J. T., Grilli, S. T. and Subramanya, R. (1995). A fully nonlinear Boussinesq model for surface waves. Part 1. Highly nonlinear unsteady waves. Journal of Fluid Mechanics 294: 71–92.

Wei, G., Kirby, J. T. and Sinha, A. (1999). Generation of waves in Boussinesq models using a source function method. Coastal Engineering 36(4): 271–299.

Wellens, P. (2012). Wave Simulation in Truncated Domains for Offshore Applications. PhD thesis, Delft University of Technology.

Wellens, P., Luppes, R., Veldman, A. E. P. and Borsboom, M. J. A. (2009). CFD simulations of a semi-submersible with absorbing boundary conditions. In Proceedings of the 28th International Conference on Offshore Mechanics and Arctic Engineering (OMAE), Honolulu, Hawaii, USA, volume 5, pages 385–394.

Windén, B., Turnock, S. and Hudson, D. (2014). A RANS modelling approach for predicting powering performance of ships in waves. International Journal of Naval Architecture and Ocean Engineering 6(2): 418–430.

Windt, C., Davidson, J., Akram, B. and Ringwood, J. V. (2018). Performance assessment of the overset grid method for numerical wave tank experiments in the OpenFOAM environment. In Proceedings of the 37th International Conference on Offshore Mechanics and Arctic Engineering (OMAE), Madrid, Spain, volume 10 (Ocean Renewable Energy).

Windt, C., Davidson, J., Schmitt, P. and Ringwood, J. V. (2019). On the assessment of numerical wave makers in CFD simulations. Journal of Marine Science and Engineering 7(2).

Wu, N. -J., Hsiao, S. -C., Chen, H. -H. and Yang, R. -Y. (2016). The study on solitary waves generated by a piston-type wave maker. Ocean Engineering 117: 114–129.

Wu, N. -J., Tsay, T. -K. and Chen, Y. -Y. (2014). Generation of stable solitary waves by a piston-type wave maker. Wave Motion 51(2): 240–255.

Yang, J. and Stern, F. (2009). Sharp interface immersed-boundary/level-set method for wave-body interactions. Journal of Computational Physics 228(17): 6590–6616.

Zhang, H., Liu, S., Li, J. and Tian, X. (2017). Numerical wave flume based on second order wavemaker theory. In Proceedings of the 27th International Ocean and Polar Engineering Conference, San Francisco, USA, volume ISOPE-I-17-171.

Zhang, J., He, G., Chen, L., Wang, J. et al. (2018a). Effect of vertical degree of freedom on hydrodynamic resonance from narrow gap between twin floating bodies. In The 28th International Ocean and Polar Engineering Conference. International Society of Offshore and Polar Engineers.

Zhang, J., Song, S., Zhai, Y., Tong, L. and Guo, Y. (2019). Numerical study on the wave-induced seabed response around a trenched pipeline. Journal of Coastal Research 35(4): 896–906.

Zhang, J., Tong, L., Zheng, J., He, R. and Guo, Y. (2018b). Effects of soil-resistance damping on wave-induced pore pressure accumulation around a composite breakwater. Journal of Coastal Research 34(3): 573–585.

Chapter 2

Wave Propagation Models for Numerical Wave Tanks

Eugeny Buldakov

1   Introduction

Most wave-structure interaction models are based on a numerical solution of boundary value problems for partial differential equations. For such models the solution accuracy depends not only on the quality of the numerical approximation of the equations, but also on the accuracy of the boundary conditions. Apart from defining the physical boundaries of a fluid domain, they are used to specify the incoming wave conditions. A common numerical tool for modelling wave-structure interaction is a numerical wave tank (NWT), where the fluid domain is bounded and waves are generated by a numerical wavemaker, e.g., by specifying wave kinematics and elevation at the incoming boundary. Difficulties experienced by NWT users are very similar to difficulties of wave generation in experimental wave facilities. These include the accuracy of incoming wave generation and reflections from the boundaries of the fluid domain.

Effective absorption of reflected waves normally requires large absorbing zones and thus larger computational domains. At the same time, a sufficient distance from the wavemaker to the test section is recommended to allow the natural development of the waves. Depending on the absorption method and type of a wavemaker, this can lead to a considerable increase in the size of the NWT. On the other hand, to achieve higher computational efficiency, it is necessary to minimise the size of the computational domain around a structure. This is particularly important because of the high demands of modern computational fluid dynamics (CFD) models for computing resources. Furthermore, for the direct comparison between experiments and calculations and for the execution of computer-assisted experiments, it is useful to model an entire experimental wave tank with the exact replication of the wavemaker shape and the position of the model. This again leads to a much larger domain size than the region of interest around the structure. Often, the optimal NWT size for accurate wave input is impractical for CFD models in terms of computational efficiency. As can be seen, a numerical wave-structure interaction model should satisfy conflicting requirements and it would be natural to apply different models in different regions of a computational domain or to simulate different aspects of the process. For example, a simpler and faster model can be used to simulate wave evolution in the far field, and the region close to the structure can be modelled by a more sophisticated, but slower, CFD model.

UCL, Department of Civil Engineering, Gower Street, LONDON, WC1E 6BT, UK.

The idea of hybrid models has received considerable attention recently and numerous hybrid models have been developed. The most popular couple for creating a hybrid model are a boundary element model (BEM) as a computationally efficient component and volume of fluid (VoF) as an advanced component (e.g. Lachaume et al., 2003; Kim et al., 2010; Guo et al., 2012). However, coupling of other models has also been attempted, for example BEM with SPH (Landrini et al., 2012) and a finite element method (FEM) with a meshless Navier-Stokes solver (Sriram et al., 2014). More examples can be found in the introduction to Sriram et al. (2014). The hybrid methods revive simple but computationally efficient wave propagation models as important elements of wave-structure interaction modelling tools. Over the years, many numerical models of nonlinear water waves have been developed. Descriptions of numerical methods for water wave modelling and reviews of numerical simulation of water waves can be found in Tsai and Yue (1996); Fenton (1999); Kim et al. (1999); Dias and Bridges (2006); Lin (2008); Ma (2010).

In this chapter we introduce inviscid models, starting with a brief historical review of models based on the fully nonlinear potential flow theory (FNPT) given in Section 2. In comparison with CFD models, based on solving Reynolds Averaged Navier-Stokes equations, these methods use much simpler governing equations with smaller number of variables. As a result, FNPT models are more efficient computationally. At the same time, they are able to reproduce the principal physical phenomena important for wave propagation, namely non-linearity and dispersion. As a result, such models describe propagation of highly nonlinear waves up to breaking with good accuracy, as demonstrated by multiple comparisons with experiments (e.g. Dommermuth et al., 1988; Skyner, 1996; Seiffert et al., 2017). We will not consider models based on further simplifications, such as depth-averaged models (shallow water and Boussinesq equations) or models of spectral evolution (nonlinear Schrödinger equation, Zakharov equation). Since we consider the wave propagation problem, we will not review literature related to application of FNPT models to interaction with structures, floating bodies, etc.

FNPT models differ by particular methods of solving Laplace equation for velocity potential in the fluid domain and by methods of specifying a fully nonlinear boundary condition on a moving free surface. There are three main classes of numerical methods used to solve wave problems in potential formulation: boundary element methods (BEM), finite element methods (FEM) and high-order spectral methods (HOS). They are discussed in separate parts of Section 2. We do not consider various finite difference methods, which use a wide range of approaches to discretise a moving domain, including boundary-fitted coordinates and σ-transform. They can not be considered the mainstream for inviscid models and an interested reader is referred to the reviews of Tsai and Yue (1996); Fenton (1999) and Kim et al. (1999). Section 2 concludes with a brief review of Lagrangian wave models, which offer certain advantages and can be considered as an alternative to conventional FNPT models.

The rest of the chapter discusses a wave propagation model based on Lagrangian description of fluid motion and is organised as follows. Section 3 gives a detailed description of the Lagrangian wave model, presenting the mathematical and numerical formulation of the model. Then, it introduces a method of numerical treatment for breaking waves, discusses the computational efficiency of the model and validates the model with experimental results. In Section 4 the model is applied to simulate the evolution of steep breaking wave groups in a wave flume. In Section 5 the model formulation for waves over sheared currents is introduced and a numerical wave-current flume is constructed and applied to simulate wave groups over following and opposing currents. Both Section 4 and 5 include comparisons between experimental and numerical results. Finally, brief concluding remarks are given in Section 6, where model coupling for wave-structure interaction problems is discussed.

2   Historical development

2.1   BEM models

Boundary element methods can be traced back to the work of Longuet-Higgins and Cokelet (1976) who considered the evolution of two-dimensional space-periodic waves in deep water. They work in complex variables and use conformal mapping to transform a semi-infinite periodic domain into internals of a closed contour representing the mapped free surface. The material time derivatives of surface coordinates and of surface potential are then expressed via the normal derivative of the potential. To find the normal derivative, the Dirichlet problem of finding the normal gradient of a harmonic function from its values on a closed contour is formulated and solved by using Green's theorem. This leads to an integral equation relating the normal gradient of the potential with its values at the mapped free surface. The integral equation is solved numerically and the normal derivative of the potential is used to calculate the time derivatives. Then material coordinates of the free surface and the surface potential at the incremented time are calculated by using a fourth-order finite difference technique. The numerical scheme demonstrated a weak saw-toothed instability which was suppressed by applying polynomial smoothing. The method was applied to simulate the evolution of high periodic waves and development of overturning profiles at initial stages of wave breaking. Since the method uses the Lagrangian approach to track the evolution of the free surface, it is referred to as a mixed Eulerian-Lagrangian method (MEL). This becomes a common feature of boundary-element methods. Later, the method was extended to the case of constant finite depth (New et al., 1985). Vinje and Brevig (1981) suggested an alternative method where Cauchy's integral theorem is applied to a complex flow potential in physical space. Working in physical variables allowed to construct solutions with vertical solid boundaries (fixed or moving) and to simulate waves in a wave flume with a piston wavemaker (Dommermuth et al., 1988).

Using complex variables restricts the application of the boundary integral formulation to 2D problems. A more flexible formulation was therefore developed with Green's theorem applied in physical space. Apart from being applicable to both 2D and 3D problems, such formulation also allows flexible treatment of fixed and moving solid boundaries of arbitrary shape and is suitable for development of efficient numerical wave tanks. Examples of BEM-based 2D numerical wave tanks, which differ by details of numerical realisation and methods of wave generation and absorption can be found in Grilli et al. (1989); Ohyama and Nadaoka (1991); Wang et al. (1995). Though the first works on application of BEM to 3D waves appeared relatively early (e.g., Isaacson, 1982), it took long time to develop robust and flexible models suitable for wide range of applications. Apart from the more difficult formulation for 3D geometry, this was caused by the drastic increase of computational cost for such models. Simplified formulations were often suggested to deal with these problems, which restricted models applicability. For example, Isaacson (1982) used a Green function that assumes the symmetry of the solution about the flat sea bed (method of images). This restricts the application of the method to constant depth. Xue et al. (2001) considered deep water waves periodic in both horizontal directions. This simplification allowed to perform high-resolution simulations and to obtain valuable results on dynamics and evolution of 3D breaking waves. Considerable efforts have finally resulted in the development of 3D numerical wave tanks capable of simulating general highly nonlinear waves on arbitrary bathymetry Grilli et al. (2001).

The solution method used by the models mentioned above and in fact by most contemporary BEM models can be briefly summarised as follows. By applying Green's theorem with an appropriate selection of the Green function, the Laplace equation can be reduced to an integral equation defined on a boundary of the computational domain. The equation relates the values of the potential and its normal derivative at the boundary. The free surface boundary conditions are written in the mixed Eulerian-Lagrangian form and connect the material time derivatives of the surface potential and of the surface position with the gradient of the potential (velocity) at the free surface. If both the potential and the normal derivative are known, the time derivatives can be calculated. The boundary conditions on solid surfaces specify the normal derivatives of the potential at these boundaries (Neumann boundary condition). For the free surface, the surface potential is known either from the initial conditions or from a previous calculation step (Cauchy boundary condition). Then, the boundary integral equation can be used to calculate the normal derivatives at the free surface and the potential at the solid boundary. After discretisation of the integral equation, a system of linear algebraic equations is solved to find the unknown values of the potential and the normal derivative at the surface nodes. The free surface boundary conditions can then be used to update the free surface position and the surface potential by applying an appropriate time stepping technique.

Major improvements had been made in terms of the method accuracy and stability (e.g., Grilli and Svendsen, 1990). More recently, considerable efforts are concentrated on improving computational efficiency of BEM models (Fochesato and Dias, 2006; Yan and Liu, 2011; Jiang et al., 2012) and to the adaptation of the method for parallel computing (Nimmala et al., 2013). Over the years BEM models have been applied to a wide range of water wave problems including problems of propagation of extreme waves directly relevant to this chapter (e.g., Fochesato et al., 2007; Ning et al., 2009). For further reference, the up-to-date formulation for a 3D BEM numerical wave tank with a review of earlier work and examples of applications can be found in Grilli et al. (2010).

2.2 FEM models

Finite Element Methods (FEM) are extensively used for a wide variety of fluid mechanics problems (e.g., Zienkiewicz et al., 2005). They are more universal than BEM and can be applied to solving nonlinear equations in domains of complex shapes using unstructured meshes. Realisation of a numerical wave tank by FEM within the FNPT approach seems relatively straightforward by the discretisation of a bounded fluids domain with Neumann boundary conditions on solid boundaries. The first known FEM model for nonlinear waves represents a fully developed 2D numerical wave tank (Wu and Eatock Taylor, 1994).

Most FEM implementations for water waves treat the moving surface using the MEL approach described above for BEM models. A typical FEM solution for a water wave problem includes the following steps. The free surface boundary conditions are represented in the mixed Eulerian-Lagrangian form, as previously described. The velocity potential in the computational domain is known from the initial conditions or from the previous step of the solution. The velocity at the free surface can then be recovered and the free surface conditions can be used to update the position of the free surface and the value of the surface potential using a suitable time stepping technique. This leads to the formulation of a mixed boundary value problem for the Laplace equation at the next time step with the Cauchy boundary conditions on the free surface and the Neumann boundary conditions on fixed or moving solid surfaces. The boundary value problem is then solved using a finite element approach, e.g., the Galerkin method. This approach was successfully applied for constructing both 2D (Wu and Eatock Taylor, 1994) and 3D (Ma et al., 2001; Wu and Hu, 2004) numerical wave tanks.

The FEM formulation requires the discretisation of the entire fluid domain. This leads to a much larger number of unknowns compared to BEM, where only the domain boundary is discretised. However, the discretisation for BEM is based on surface integrals, which leads to linear systems with dense non-symmetric matrices On the other hand, FEM discretisation procedure leads to sparse linear systems because of the local nature of discrete differential operators, which include only a few spatially close nodes. As a result, the number of non-zero matrix elements for FEM is much smaller. In addition, for regular meshes, the resulting matrix has a diagonal structure. If a typical number of discrete elements in one of the spatial dimensions is N, then for a 3D problem the number of nodes for FEM is $\sim N^3$ and for BEM it is $\sim N^2$. The corresponding numbers of matrix elements are $\sim N^6$ and $\sim N^4$ and the number of non-zero elements being receptively $\sim N^3$ and $\sim N^4$. This implies higher computational efficiency of FEM compared to BEM. Wu and Eatock Taylor (1995) compared BEM and FEM models for a 2D wave problem and reported that in many cases the finite element method may be more efficient but admitted that this conclusion is based on limited experience. A discussion of numerical efficiency of FEM in comparison to BEM can also be found in Cai et al. (1998). Complementary advantages of BEM discretisation of moving domains near bodies and higher efficiency of FEM far from the body were used in coupled BEM-FEM models by Wu and Eatock Taylor (2003) in 2D and by Eatock Taylor et al. (2008) in 3D.

Although FEM models rely on the MEL formulation for updating the free surface, the difficulties in generating a computational mesh in highly deformed domains did not allow the application of early FEM models to overturning waves. Another serious disadvantage of these models is that the generation of an unstructured mesh is required at each time step. The repeated generation of such a mesh increases the computational time considerably. To overcome this difficulty, a moving mesh method was developed by Ma and Yan (2006). In this method the mesh is generated only once. After that, the mesh is deformed at each step with a simple algorithm. The original idea of mobile mesh came from the arbitrary Lagrangian-Eulerian (ALE) formulation for Navier-Stokes equations. However, a different mesh deformation algorithm is used, which does not use fluid velocities to move mesh nodes. The method is therefore called the quasi-arbitrary Lagrangian-Eulerian finite element method (QALE-FEM). The new positions of the free surface nodes are found by following the surface particles and then relocating the nodes. Relocation distributes the nodes more evenly over the free surface and does not allow them to move too close or too far apart. Then, the positions of the internal nodes are calculated using the spring analogy method. Further development of the QALE-FEM model allowed simulating overturning waves (Yan and Ma, 2010).

There is ongoing work on improving numerical implementations of FEM for water wave models to achieve better accuracy, stability, and computational efficiency. This includes using different types of a finite element method, for example the method of spectral elements (Robertson and Sherwin, 1999; Engsig-Karup et al., 2016). The reader can find details of FEM water wave formulation and application examples in Ma and Yan (2010) and Wang and Wu (2011).

2.3 Spectral models

High-order spectral methods are undeniably the most computationally efficient methods for modelling nonlinear waves, being capable of simulating 3D random sea states at linear scales of tens of wavelengths during tens of wave periods (Ducrozet et al., 2007).

In an early application of a spectral method, Fenton and Rienecker (1982) represented the potential and surface elevation in a 2D periodic domain of constant depth via Fourier expansion by basic functions satisfying the Laplace equation and boundary conditions. If the initial values of the Fourier coefficients are known, the kinematic free-surface condition can be used to advance the surface elevation using a finite difference approximation of the time derivative. All spatial derivatives are computed in the Fourier space. Inverse Fourier transforms are then used to perform a time step in the physical space. This simple approach, however, can not be applied to advance the potential. Instead, the dynamic free surface condition is used to calculate the time derivatives of each Fourier coefficient, which are then used to find the values of the coefficients at the next time step. Calculating derivatives requires solving a large system of simultaneous equations, which is responsible for the low computational efficiency of the method.

This problem was solved in high-order spectral methods (Dommermuth and Yue, 1987a; West et al., 1987). In this method, the potential is expressed as an asymptotic expansion by a small steepness parameter. In addition, the free surface potential is expanded in a Taylor series around the mean water level, and a double expansion is used to represent the surface potential. The known initial values of surface potential and surface elevation define a Dirichlet boundary value problem for each term of the expansion in the domain below the mean water level. The solution of these problems is sought in the form of a Fourier expansion by modal functions satisfying the Laplace equation and the boundary conditions at side boundaries and the bottom. This makes it possible to express the components of the vertical velocity at the free surface via modal coefficients, which themselves are defined by the surface elevation and the surface potential. This closes the evolution equations provided by the free surface conditions and allows to update the surface values. Fast Fourier Transforms are used to switch between spectral and physical spaces. The shape of the domain should be selected to define a simple spectral basis to expand the velocity potential. Therefore, either periodic domains in both horizontal dimensions or rectangular tanks are usually used.

An alternative approach uses the Dirichlet-Neumann (DN) operator, which expresses the normal surface velocity in terms of velocity potential at the surface. If such an operator is defined, the water wave problem is reduced to the integration over time of free-surface boundary conditions with unknown functions evaluated only at the free surface. The non-linear DN operator is expanded in terms of a convergent Taylor expansion about the mean water level. This method was introduced by Craig and Sulem (1993) for 2D waves and extended by Bateman et al. (2001) to 3D cases. Schäffer (2008) demonstrated that different variants of HOS methods and methods that used DN operator are either identical or have only minor differences. The use of the additional potential allowed the modelling of a wavemaker (Ducrozet et al., 2012b) and a variable bathymetry (Gouin et al., 2016). This makes the HOS approach acceptable for numerical wave tanks. High efficiency and accuracy of spectral methods compare to other methods for wave propagation was demonstrated by Ölmez and Milgram (1995) and Ducrozet et al. (2012a).

One of the drawbacks of the spectral methods is that they can not model the overturning waves. However, this can not be considered as a serious disadvantage compared to the BEM and FEM models, if we consider their application to wave propagation. Neither of the models considered here is able to continue calculations after wave breaking. However, to model severe sea states and extreme waves, a model should continue calculations after waves break and provide a reasonable prediction of energy dissipation due to breaking. Seiffert and Ducrozet (2018) solved this problem by introducing eddy viscosity as a diffusive term to the free surface boundary conditions to simulate breaking waves in a HOS model. Breaking onset is determined by a breaking criterion. The model demonstrated an impressive comparison with experiments on the propagation of surging wave groups.

More details on formulation and application of HOS models can be found in Bonnefoy et al. (2010).

2.4 Fully Lagrangian models

Another method to describe water waves is to use equations of fluid motion in the Lagrangian formulation. These equations are written in coordinates moving with the fluid. Each material point of the fluid continuum is labelled with a specific label, and the labels in the fluid-occupied domain create a continuous set of coordinates. These are Lagrangian coordinates or Lagrangian labels. Equations of fluid motion are solved in a fixed Lagrangian domain with the free surface represented by a fixed domain boundary. Some numerical methods use elements of the Lagrangian description. For example MEL free surface treatment by BEM and FEM described above. The smooth particle hydrodynamics (SPH) approach can be considered as fully Lagrangian. In this method, the fluid domain is represented by a set of material particles which serve as physical carriers of fluid properties. An integral operator with a compact smoothing kernel is used to represent the average properties of the fluid at a certain location, which are used to satisfy the equations of fluid motion. Each particle interacts with nearby particles from a domain specified by the smoothing kernel (e.g., Gomez-Gesteira et al., 2010; Violeau and Rogers, 2016). However, SPH does not directly refer to the equations of fluid motion in Lagrangian coordinates and should be distinguished from the methods where Lagrangian equations are directly applied to solve water wave problems.

The initial works on discrete approximation of equations of fluid motion in Lagrangian formulation with applications to water wave problems appeared in the early 70s. Brennen and Whitney (1970) used kinematic equations of mass and vorticity conservation for internal points of a domain occupied by an ideal fluid. Flow dynamics were determined by a free-surface dynamic condition. According to Fenton (1999) this approach apparently had not been followed and there are just a few works in which it was used (e.g., Nishimura and Takewaka, 1988). An alternative approach was developed by Hirt et al. (1970) who applied the equations of motion of viscous fluid in material coordinates moving together with the fluid. The next step was the development of an Arbitrary Lagrangian-Eulerian (ALE) formulation (Chan, 1975). ALE formulation uses a computational mesh moving arbitrarily within a computational domain to optimise the shape of computational elements and the problem is formulated in moving coordinates connected to the mesh. At certain regions of a computational domain the formulation can be reduced either to Eulerian (fixed mesh) or to fully Lagrangian (mesh moving with fluid) depending on the problem requirements. The Lagrangian models mentioned so far use quadrangular numerical cells. These models are subject to "alternating errors" and "even-odd" instability (Hirt et al., 1970; Chan, 1975), which is similar to the saw-tooth instability of the ALE approach. Moreover, application of fully-Lagrangian models to viscous problems has serious limitations. Boundary layers, wakes, vortices and other viscous effects lead to complicated deformations of fluid elements and large variations of physical coordinates over cells of a Lagrangian computational mesh. To address these problems the method was generalised for irregular triangular meshes (Fritts and Boris, 1979) and used for development of finite element models (e.g., Ramaswamy and Kawahara, 1987). This method however remains out of the mainstream and only occasionally appears in the literature (e.g., Kawahara and Anjyu, 1988; Radovitzky and Ortiz, 1998; Staroszczyk, 2009). Implementation of a finite element approach with irregular triangular meshes for ALE formulation (Braess and Wriggers, 2000) led to the development of a sophisticated method capable of solving complicated problems with interfaces including surface waves and fluid-structure interaction. A detailed description of the ALE method, examples of application and comprehensive bibliography can be found in Souli and Benson (2013). Finite element Lagrangian models and especially ALE models are complicated in both formulation and numerical realisation and are missing the main advantage expected from a Lagrangian method: simplicity of representing computational domains with moving boundaries. For many problems solved within the framework of ideal fluid, the deformation of the fluid domain remains comparatively simple. These problems can be efficiently approached by much simpler Lagrangian models similar to the original model of Brennen and Whitney

(1970). Recent examples of application of such a model include tsunami waves in a wave flume (Buldakov, 2013), violent sloshing in a moving tank (Buldakov, 2014) and evolution of breaking wave groups (Buldakov et al., 2019).

A particular advantage of the Lagrangian models compared to the FNPT models is the ability to model vortical flows and, therefore, waves over sheared currents. Potential formulation assumes an irrotational flow and can, therefore, only be applied to a uniform current (Ryu et al., 2003; Chen et al., 2017). Potential flow methods can also be generalised to flows with constant vorticity, which preserves the linearity of the problem. This allows the modelling of currents with linear profiles (e.g., Da Silva and Peregrine, 1988). On the other hand, in the inviscid Lagrangian formulation, vorticity does not change over time and can be generally defined as a function of Lagrangian labels. This allows a simple application to waves on arbitrary sheared currents. An example of such application can be found in Buldakov et al. (2015) and Chen et al. (2019). This feature of the Lagrangian formulation can also be useful for simulation of wave behaviour after breaking, which generates intensive vortical motion beneath the surface.

3   Lagrangian numerical wave model

Later in this chapter we consider a two-dimensional fully Lagrangian finite-difference wave model. The model follows the approach of the early Lagrangian models originally introduced by Brennen and Whitney (1970) and was further developed in Buldakov (2013, 2014) and Buldakov et al. (2019). Before continuing, let us first examine some aspects of the Lagrangian description that affect the application of discrete numerical methods, such as finite differences.

One of the main advantages of the Lagrangian approach is that the domain occupied by the fluid in Lagrangian coordinates remains the same during the fluid motion. The form of the Lagrangian domain is arbitrary. The only restriction is that mapping from Lagrangian to physical coordinates should not be singular. Therefore, the computational domain can be chosen by considering the convenience of numerical analysis. For example, a rectangular Lagrangian domain with sides parallel to the axes of the Lagrangian coordinate system can be selected. This greatly simplifies a numerical formulation since the finite difference approximation does not include cross terms. On the other hand, other aspects of the Lagrangian approach make its implementation more difficult, for example, there are situations where the boundary conditions of a part of the boundary change. This is the case for the self-contact of the different parts of the free surface during wave breaking, for the problems of entry and exit of solid bodies, the impacts of wave peaks with high structures, etc. However, for a large class of flows, a particle originally on a specific type of boundary (e.g., a free surface or a solid surface) remains on that boundary and the type of boundary condition does not change. This assumption provides a significant simplification in the formulation of the problem and is used in the Lagrangian formulation considered in this work. In physical coordinates, the fluid domain can significantly change its original form. While strong local deformations, e.g., at the peaks of high waves, do not present a problem for a Lagrangian model, the continuous deformation of the entire volume of the fluid can present significant difficulties in the practical realisation of a Lagrangian formulation. This can be the case for domains with open boundaries. Examples are the Stokes' drift of a regular wave train or waves propagating over sheared currents. To overcome this problem one can apply relabelling, when a physical domain of a suitable shape is mapped into a new space of Lagrangian labels. A practical realisation of this approach is demonstrated later in this chapter, when we consider the application of the Lagrangian model to waves over sheared currents. For a compact travelling wave group the total deformation of the initial fluid volume is finite and such problems are ideal for application of Lagrangian wave formulation.

3.1 Mathematical formulation

Fluid motion in Lagrangian method is described by tracing marked fluid particles. For two-dimensional motion we have

$$x = x(a, c, t); \quad z = z(a, c, t),$$

where (x, z) are Cartesian coordinates of a particle marked by Lagrangian labels (a, c) at time t. Due to volume conservation for an incompressible fluid, the Jacobian J of a mapping $(x, z) \to (a, c)$ is a motion invariant: $\partial J / \partial t = 0$. This leads to the following Lagrangian form of the continuity equation:

$$J = \frac{\partial(x, z)}{\partial(a, c)} = J(a, c), \tag{1}$$

where $J(a, c)$ is a given function of Lagrangian coordinates.

Equations of motion of an inviscid, incompressible fluid in Lagrangian coordinates (a, c) can be obtained using Hamilton's variational principle (e.g., Herivel, 1955). Let us represent the density of the Lagrangian in the following form

$$\mathcal{L} = \mathcal{T} - \mathcal{U} + \rho P(a, c, t) (J - J(a, c)),$$

where the kinematic continuity condition (1) is enforced by the Lagrange multiplier P, and ρ is the fluid density. The densities of the kinetic and potential energies of the fluid are

$$\mathcal{T} = \rho (x_t^2 + z_t^2)/2; \quad \mathcal{U} = \rho g z.$$

According to Hamilton's principle, the variation of the action integral

$$\delta I = \delta \int_{t_1}^{t_2} dt \iint_D \mathcal{L} \, da \, dc = 0$$

is zero, where the integration takes place in the Lagrangian space over a domain D occupied by the fluid. Taking the variation leads to the following equations describing dynamics of the fluid inside D

$$x_{tt} + \frac{\partial(P, z)}{\partial(a, c)} = 0; \quad z_{tt} + \frac{\partial(x, P)}{\partial(a, c)} + g = 0.$$

The Lagrange multiplier P can be recognised as the ratio of pressure to density and the boundary condition on the free surface $c = 0$ is $P = 0$. These equations can be resolved with respect to the spatial pressure derivatives and rewritten in the following form (Lamb, 1932)

$$\frac{\partial P}{\partial a} + g \, z_a = -x_{tt} x_a - z_{tt} z_a; \quad \frac{\partial P}{\partial c} + g \, z_c = -x_{tt} x_c - z_{tt} z_c. \tag{2}$$

The terms on the left hand sides of (2) are gradient components of a certain scalar function in the label space. Taking the curl of both sides of (2) we find that the value

$$\Omega = \nabla_a \times (x_t x_a + z_t z_a, x_t x_c + z_t z_c)$$

is a motion invariant: $\partial \Omega / \partial t = 0$, where $\nabla_a \times$ is the curl operator in (a, c)-space. This gives the second kinematic condition in addition to (1)

$$\Omega = \frac{\partial(x_t, x)}{\partial(a, c)} + \frac{\partial(z_t, z)}{\partial(a, c)} = \Omega(a, c), \tag{3}$$

where $\Omega(a, c)$ is a given function. This is the Lagrangian form of vorticity conservation and for irrotational flows $\Omega = 0$. Functions $J(a, c)$ and $\Omega(a, c)$ from (1) and (3) are defined by the initial conditions. $J(a, c)$ is defined by the initial positions of fluid particles associated with labels (a, c) and $\Omega(a, c)$ by the velocity field at $t = 0$.

The Lagrangian formulation does not require a kinematic free-surface condition, which is satisfied by specifying a fixed boundary of Lagrangian fluid domain corresponding to a free surface, e.g., $c = 0$. The dynamics of the flow are described by a dynamic free-surface condition which can be obtained from the first Equation in (2). For a case of constant pressure on the free surface $c = 0$ we have

$$x_{tt} x_a + z_{tt} z_a + g\, z_a \big|_{c=0} = 0 \,. \tag{4}$$

This condition has a simple physical meaning. The left-hand side of (4) can be written as a dot product of two vectors $\mathbf{a} = (x_{tt}, z_{tt} + g)$ and $\mathbf{t} = (x_a, z_a)$. The first vector is the acceleration of a fluid particle with subtracted gravity acceleration, and the second vector is tangential to the free surface. Therefore, the condition $\mathbf{a} \cdot \mathbf{t} = 0$ means that part of the acceleration of a fluid particle on the free surface produced by other fluid particles is normal to the free surface. The general formulation of the problem consists, therefore, of the continuity Equation (1), the vorticity conservation Equation (3), the free-surface condition (4) and suitable conditions on the bottom and side boundaries. Positions and velocities of fluid particles must be supplied as initial conditions.

One of the advantages of Lagrangian formulation is that the Lagrangian domain and the original correspondence between the physical and Lagrangian coordinates is arbitrary and can be chosen from convenience of numerical or analytical analysis. The only restriction is that the Jacobian J of the original mapping from Lagrangian to physical coordinates $(a, c) \rightarrow (x, z)|_{t=0}$ is not singular. It is often convenient to use a rectangular Lagrangian domain

$$a_{\min} \le a \le a_{\max}; \quad -h \le c \le 0,$$

where $c = -h$ corresponds to the bed, $c = 0$ to the free surface, $a_{\min}$ and $a_{\max}$ to the side boundaries (finite or infinite) of a physical domain and h is a characteristic depth, for example the mean still water depth.

A specific problem within the general formulation (1, 3, 4) is defined by the boundary and initial conditions specified for the Lagrangian domain. For a wave propagation problem, the boundary and initial conditions are used for wave generation. It is, for example, possible to specify the initial shape of a physical fluid domain and initial velocities corresponding to the kinematics of a spatially periodic wave and the corresponding boundary conditions on side boundaries. This, however, requires knowledge of wave kinematics, which in case of nonlinear waves is not normally known and generating such kinematics is one of the primary aims of solving the Lagrangian wave propagation problem. It is more convenient to solve a problem of wave evolution in a wave tank. Though this approach may have problems with reflections from boundaries of a finite domain similar to those of physical wave tanks, its numerical realisation is relatively simple and direct modelling of physical wave flumes makes it possible to have direct comparison between numerical and experimental results. The initial conditions in such an approach can be still water conditions and waves can be generated by moving boundaries.

The boundary conditions for a wave tank problem can be formulated as follows. The known shape of the bottom provides the condition on the lower boundary $c = -h$ of the Lagrangian domain

$$F(\, x(a, -h, t), z(a, -h, t)\,) = 0 \,, \tag{5}$$

where F is a given function. If we consider waves in a wave tank, conditions on the left and right boundaries of the Lagrangian domain $a = a_{\max}$ and $a = a_{\min}$ specify the shape of the basin walls

$$
\begin{aligned}
x(a_{\min}, c, t) &= X_L(\, z(a_{\min}, c, t),\, t\,); \\
x(a_{\max}, c, t) &= X_R(\, z(a_{\max}, c,\, t), t\,),
\end{aligned}
\tag{6}
$$

where X_L and X_R are given functions of z and t. The dependence from x can be used to define the shape and the dependence from t the motion of a wavemaker.

3.2   Numerical scheme

For convenience of numerical realisation, we modify the original problem (1, 3 ,4) and write it in the following form

$$
\Delta_t \left(\frac{\partial(x, z)}{\partial(a, c)} \right) = 0; \qquad
\Delta_t \left(\frac{\partial(x_t, x)}{\partial(a, c)} + \frac{\partial(z_t, z)}{\partial(a, c)} \right) = 0,
\tag{7}
$$

where the operator Δ_t denotes the change between time steps. Equations (7) mean that values in brackets at two time steps are equal to one another. This formulation does not require explicit specification of functions $J(a, c)$ and $\Omega(a, c)$, which are specified implicitly by initial conditions. We also modify the dynamic free surface condition by adding various supplementary artificial and physical terms on the right-hand side

$$
x_{tt} x_a + z_{tt} z_a + g\, z_a = -RHS(a, t) \Big|_{c=0}.
\tag{8}
$$

We use the following set of additional terms

$$
RHS(a, t) = k\,(x_t\, x_a + z_t z_a) + \sigma\, \frac{\partial}{\partial a}\, \frac{\partial \kappa}{\partial t} + \gamma\, \frac{\partial \kappa}{\partial a} + P_a(a, t),
\tag{9}
$$

where κ is the surface curvature. The first term in (9) introduces damping of displacement of surface particles with the damping coefficient $k(a)$. This term is used for absorbing waves approaching a boundary of a numerical wave tank opposite to a wavemaker and minimising reflections. The second term represents damping of surface curvature and $\sigma(a)$ is the corresponding damping coefficient. As described later in this chapter, it is used to simulate the dissipative effects of wave breaking. The third term represents surface tension. The coefficient γ is the ratio of the surface tension over density. In calculations presented in this work we use the value of $\gamma = 0.00073\,\mathrm{m}^3/\mathrm{s}^2$ corresponding to fresh water at $20°C$. The last term is the prescribed surface pressure gradient, which can be used to create a pneumatic wave generator. The set of Equations (7) with boundary conditions (8, 5, 6) is solved numerically using a finite-difference technique described below.

Since Equations (7) for internal points of a computational domain include only first order spatial derivatives, a compact four-point Keller box scheme (Keller, 1971) can be used for finite-difference approximation of these equations. For our selection of the Lagrangian computational domain the stencil box can be chosen with sides parallel to the axes of the Lagrangian coordinate system, which significantly simplifies the final numerical scheme. The values of the unknown functions x and z on the sides of the stencil box are calculated as averages of the values at adjacent points and then used to approximate the derivatives across the box by first-order differences. The scheme provides the second-order approximation for the box central point. Time derivatives in the second Equation in (7) are approximated by second-order backward differences. It should be noted that the same scheme can be constructed by applying conservation of volume and circulation to elementary rectangular contours with linear approximation of unknown functions on boundaries of elementary volumes. Spatial derivatives in the free-surface boundary condition (8) are approximated by second-order central differences and second-order backward differences are used to approximate time derivatives. As demonstrated below, this leads to a stable numerical scheme with weak dissipation. The overall numerical scheme is of second order accuracy in both time and space.

For finite-difference approximation of a boundary-value problem the number of algebraic mesh equations must be equal to the number of unknown values of functions at grid nodes. Let us consider a finite-difference approximation of the problem (7, 8, 5, 6) on a rectangular $N \times M$ mesh, where N and M are the numbers of mesh points in a and c directions. We are required to calculate values of unknown functions x and z at each mesh point. This gives $2 \times N \times M$ unknowns which should be determined by solving the same number of finite difference equations. Keller-box approximation of (7) gives two equations for every mesh cell or $2 \times (N - 1) \times (M - 1)$ equations. Approximation of boundary conditions (8, 5, 6) at internal points of corresponding boundaries results in $2 \times (N - 2) + 2 \times (M - 2)$ equations. Numerical tests demonstrated that both bottom condition (5) and vertical wall conditions (6) should be satisfied at lower corner points to provide their stability (4 equations). The last two equations are provided by vertical boundary conditions (6) at upper corner points. Altogether, this adds up to the required $2 \times N \times M$ equations.

A fully-implicit time marching is applied and the Newton-Raphson method is used on each time step to solve the nonlinear algebraic difference equations. It is important to note that the scheme uses only 4 mesh points in the corners of the box for internal points of the fluid domain. Therefore, the resulting Jacobi matrix used by nonlinear Newton iterations has a sparse 4-diagonal structure and can be effectively inverted using algorithms which are faster and less demanding for computational memory than general algorithms of matrix inversion. The current version of the solver uses a standard routine for inversion of general sparse matrices (NAG, 2016). To reduce calculation time, the inversion of a Jacobi matrix is performed at a first step of Newton iterations and if iterations start to diverge. Otherwise, a previously calculated inverse Jacobi matrix is used. Usually only one matrix inversion per time step is required. To start time marching, positions of fluid particles at three initial time steps should be provided, which specifies initial conditions for both particle positions and velocities. An adaptive mesh is used in the horizontal direction with an algorithm based on the shape of the free surface in Lagrangian coordinates $z(a, 0, t)$ to refine the mesh at each time step in regions of high surface gradients and curvatures. Constant mesh refinement near the free surface is used in the vertical direction.

Since the finite-difference approximation presented above uses quadrangular mesh cells, it may be subjected to the so called alternating errors caused by non-physical deformations of cells (Hirt et al., 1970; Chan, 1975). This deformation preserves the overall volume (area) of a cell, as prescribed by finite difference approximation of the Lagrangian continuity equation. However, volumes of triangles built on cell vertexes can change. The area of one of the triangles increases while the area of the second triangle decreases by the same value. For a cell which is originally rectangular, this corresponds to trapezoidal deformation. This effect finally leads to alternating trapezoidal distortions of neighbouring cells occupying the whole domain, resulting in instability of short-wave disturbances with a wavelength equal to two cell spacing. In earlier models it was usual to implement artificial smoothing to suppress this instability (e.g., Chan, 1975). For the current model, efficient suppression of the instability caused by alternating errors is provided by an adaptive mesh. After several time steps the solution is transferred to a new mesh using quadratic interpolation. The effect of this procedure is similar to regridding used by Dommermuth and Yue (1987b). It reduces short-wave disturbances and does not produce unwanted damping. Alternating errors still remain the main reason for a breakdown in calculations. However, this happens for a large deformation of the computational domain associated with break in continuity of a physical domain caused, for example, by wave breaking. Otherwise, the numerical scheme proves to be stable with respect to this type of instability.

3.3   Numerical dispersion relation and dispersion correction

Special attention must be paid to approximation of second time derivatives in the free surface condition (4) since it defines the form of the numerical dispersion relation and is crucial for the overall stability of the scheme. For simplicity let us first consider a case of continuous spatial field in (1, 3, 4) combined with implicit discrete time approximation in (4). Let us approximate second derivatives by 3-point backward differences and expand this approximation to Taylor series with respect to a small time step τ. We get

$$\frac{f(t - 2\,\tau) - 2\,f(t - \tau) + f(t)}{\tau^2} = f''(t) - \tau\,f'''(t) + O(\tau^2)\,. \tag{10}$$

As can be seen, the approximation is of the first order with the leading term of the error proportional to the third derivative of a function, which gives the main contribution to the error of the dispersion relation. Under an assumption of small perturbations of original particle positions we represent unknown functions in the form

$$x = a + \varepsilon\,\xi(a, c, t); \quad z = c + \varepsilon\,\zeta(a, c, t)$$

and keep only linear terms of expansions with respect to the small displacement amplitude $\varepsilon \to 0$. Introducing a displacement potential ϕ

$$\xi = \partial\phi/\partial a; \quad \zeta = \partial\phi/\partial c$$

we satisfy the vorticity conservation (3) to the first order as $\varepsilon \to 0$ and the corresponding approximation of the continuity Equation (1) is the Laplace equation for ϕ. The dynamic surface condition (4) becomes

$$\phi_a'' + g\,\phi_{ac} - \tau\,\phi_a''' = O(\tau^2)\,, \tag{11}$$

where primes denote time derivatives and only the leading term of the approximation error from (10) is taken into account. To derive the numerical dispersion relation we are looking for a solution in the form of a regular wave in deep water:

$$\phi = e^{i\,k\,a}\,e^{k\,c}\,e^{i\,\omega\,t}\,,$$

which satisfies the Laplace equation. The dynamic condition (11) is satisfied when ω and k are related by a dispersion relation. Similar analysis can be performed for higher orders of approximation of the derivatives. Below is the summary of dispersion relations obtained for orders $n = 1$ to 4:

$$\omega/\sqrt{gk} = \pm 1 + \frac{1}{2}\,i\,\hat{\tau} + O(\hat{\tau}^2)\,; \tag{12a}$$

$$\omega/\sqrt{gk} = \pm 1 \mp \frac{11}{24}\,\hat{\tau}^2 + \frac{1}{2}\,i\,\hat{\tau}^3 + O(\hat{\tau}^4)\,; \tag{12b}$$

$$\omega/\sqrt{gk} = \pm 1 - \frac{5}{12}\,i\,\hat{\tau}^3 + O(\hat{\tau}^4)\,; \tag{12c}$$

$$\omega/\sqrt{gk} = \pm 1 \pm \frac{137}{360}\,\hat{\tau}^4 - \frac{19}{24}\,i\,\hat{\tau}^5 + O(\hat{\tau}^6)\,. \tag{12d}$$

We use a nondimensional expansion parameter $\hat{\tau} = \sqrt{gk}\,\tau$, which is the measure of the problem discretisation representing the ratio of the time step to a typical problem period. As can be seen, the first-order scheme (12a) introduces numerical viscosity proportional to $\hat{\tau}$ which leads to fast non-physical decay of perturbations. The higher-order schemes (12c,12d) include terms proportional to $-i$, leading to growth of perturbations, making the numerical scheme unstable. We therefore use the second-order scheme (12b), which incorporates a numerical error to dispersion at the second order $\hat{\tau}^2$ and a weak third order $(\hat{\tau}^3)$ dissipation term.

Let us now include spatial discretisation according to the numerical scheme described above with discretisation steps $(\delta a; \delta c)$. As before, we consider a linearised approximation for a regular travelling wave in deep water. However, differential approximation of the discretised field Equations (1, 3) includes higher spatial derivatives and the solution can not be represented in the form of the displacement potential. A suitable form of the solution is

$$\xi = i\,e^{ika}\,e^{\varkappa c}\,e^{i\omega t}; \quad \zeta = e^{ika}\,e^{\varkappa c}\,e^{i\omega t},$$

where the constant for the exponential decay of displacement with depth $\varkappa$ is not equal to the wavenumber. The expression for $\varkappa$ to satisfy the discrete versions of (1) and (3) can be found as an expansion by discretisation parameters. As before, the expansion for ω defines the numerical dispersion relation and is used to satisfy the free-surface condition (4). The corresponding expansions are found to be

$$\varkappa/k = 1 - \frac{1}{24}\left(\delta\hat{a}^2 + \delta\hat{c}^2\right) + O(\delta\hat{a}^4; \delta\hat{c}^4) \tag{13}$$

and

$$\omega/\sqrt{gk} = \pm 1 \mp \frac{1}{24}\left(11\,\hat{\tau}^2 + 2\,\delta\hat{a}^2\right) + \frac{1}{2}\,i\,\hat{\tau}^3 + O(\hat{\tau}^4; \delta\hat{a}^4),$$

where the nondimensional discretisation steps $\hat{\tau} = \sqrt{gk}\,\tau$, $\delta\hat{a} = k\,\delta a$ and $\delta\hat{c} = k\,\delta c$ are used. It is interesting to note that the dispersive error is affected only by the horizontal discretisation step, while the vertical discretisation affects wave kinematics. Therefore, if we are interested in the evolution of the waveform alone, we can use relatively few vertical mesh points.

Validation tests presented later in this section show that the dispersion error is crucial for travelling waves, and the achieved convergence rate for the second-order dispersion approximation is not sufficient. To increase the approximation order for the dispersion relation we introduce dispersion correction terms to the free surface boundary condition. These terms should satisfy the following conditions: (i) to have the order of $O(\tau^2; \delta a^2)$; (ii) to be linear; (iii) to not include high derivatives; (iv) to use the same stencil as the original scheme and (v) to reduce the order of the dispersion error. It has been found that the free surface boundary condition (4) with the dispersion correction term satisfying these conditions can be written as follows

$$x_{tt}x_a + z_{tt}z_a + g\,z_a + \left(\frac{1}{6}\delta a^2\,x_{aa,tt} - \frac{11}{12}g\,\tau^2 z_{a,tt}\right) = 0 \Big|_{c=0}, \tag{14}$$

where the dispersion correction term is given in parentheses. The term x_{aa} with the second-order spatial derivative leads to high-wavenumber nonlinear instability for large wave amplitudes. To suppress this instability, we apply 5-point quadratic smoothing to the function $x(a)$ before applying the finite-difference operator. The smoothing is applied to this term in (14) only, and at the future time layer to ensure that the scheme remains fully implicit. It can be shown that the numerical dispersion relation becomes

$$\omega/\sqrt{gk} = \pm 1 + \frac{1}{2}\,i\,\hat{\tau}^3 \pm \left(\frac{361}{480}\hat{\tau}^4 - \frac{1}{360}\delta\hat{a}^4\right) - \frac{13}{12}\,i\,\hat{\tau}^5 + O(\hat{\tau}^6; \delta\hat{a}^6). \tag{15}$$

We now have weak numerical dissipation at 3-rd order and the dispersion error at 4-th order. A term with weak negative dissipation at 5-th order should also be noted. This term can potentially lead to solution instability for large time steps.

3.4   Numerical treatment of breaking

A disadvantage of numerical models for wave propagation considered in this chapter is their inability to model spilling breakers. The discontinuity of the free surface that develops at the spilling crest leads to a singularity in the numerical solution, resulting in a breakdown of the calculations. Models based on Lagrangian representation of the free surface (MEL and fully Lagrangian) can simulate overturning waves and, therefore, with sufficient spatial and temporal resolution they can resolve micro-plungers originating at wave crests during the initial stages of spilling breaking. However, they are unable to continue the calculations after self-contact of the free surface occurs and the resulting solution becomes non-physical. This makes impossible applying such models to steep travelling waves and severe sea states.

Removing a singularity at the breaking crest can help in continuing calculations with only a minor effect on the overall wave behaviour. This can be achieved by implementing artificial local dissipation in the vicinity of a wave crest prior to breaking. With this approach all small-scale local features would disappear from the solution, but the overall behaviour of the wave would still be represented with good accuracy. Practical implementation of the method includes using of a breaking criterion to initiate dissipation right before breaking occurs. The dissipation is usually enforced by including damping terms in the free-surface boundary conditions (Haussling and Coleman, 1979; Subramani et al., 1998; Guignard and Grilli, 2001). Recently, a breaking model based on an advanced breaking criterion and an eddy viscosity dissipation model was developed by Tian et al. (2012) and implemented in a spectral model of wave evolution (Tian et al., 2012; Seiffert and Ducrozet, 2018). The method demonstrates a good comparison with the experiments and allows to apply spectral models to simulate the evolution of severe sea states with breaking waves.

In this chapter we use a method of treatment of spilling breaking which uses the same basic concept but differs in the details of realisation. The method includes dissipative suppression of the breaker and correction of crest shape to provide accurate post-breaking behaviour of the wave. There are several conditions such a method should satisfy: (i) to act locally in the close vicinity of a developing singularity without affecting the rest of the flow; (ii) to simulate energy dissipation caused by breaking; (iii) to be mesh-independent, that is, the change of the effect with changing mesh resolution should be within the accuracy of the overall numerical approximation and (iv) to be naturally included into a problem formulation representing an actual or artificial physical phenomenon. The development of a spilling breaker is associated with a rapid growth of surface curvature. Therefore, the local dissipation effect satisfying these conditions can be created by adding a term $-\sigma \, \partial/\partial a \, (\partial \kappa / \partial t)$ to the right hand side of the free surface dynamic condition (8, 9). This term with a small coefficient σ introduces artificial dissipation due to the change of surface curvature κ which acts locally at the region of fast curvature changes and suppresses breaker development without affecting the rest of the wave. To minimise the undesirable effect of dissipation, the action of the damping term is limited both in time and in space. Breaking dissipation is triggered when the maximal acceleration of fluid particles at the crest exceeds a specified threshold a_{on} and is turned off when the maximum acceleration falls below a second lower value a_{off}. Spatially, the action of the breaking model is limited by the half-wavelength between the ascending and descending zero-crossing points delimiting a breaking wave crest.

The activation of the damping term makes it possible to continue the calculation beyond the breaking event. However, the resulting shape of the wave crest is different from the actual crest after the breaking. Since local dissipation suppresses breaking, the local behaviour of the wave crest is different from the real one. Overturning of the crest does not occur, and for a sufficiently intense breaking, it can significantly affect the shape of the entire wave around the crest. To account for this difference, we increase the surface tension around the crest. Numerical tests show that large surface tension produces an effect similar to that of the peak overturning, changing the shape of the crest and reducing the error in the profile of the post-breaking wave. This effect is achieved by the surface tension term in the right side of the free surface boundary condition (8,9) with an appropriately selected coefficient γ_{br} added to the natural value of γ. It should be emphasised that the desired effect is only possible for values of the γ_{br} much larger than the actual ones and it is used only in the regions and during the periods when the breaking model operates.

To summarise, the intensity of the dissipation (σ), the acceleration thresholds to activate and deactivate the dissipation (a_{on} and a_{off}) and the strength of the surface tension for the correction of the shape of the crest (γ_{br}) constitute the four parameters of the breaking model. The functions of the parameters of the model are as follows. Parameter a_{on} defines the beginning of the breaking process, σ regulates the rate of energy dissipation, a_{off} controls the duration of the breaking and the total amount of energy dissipated and γ_{br} corrects the shape of the breaking crest. It should be noted that being a heuristic model, the breaking model requires calibration of parameters to achieve optimal performance for each particular case. For this chapter, the parameters were selected by running a small number of numerical cases for different parameters values. The following values of the parameters are selected for calculations presented below: $a_{on} = g$, $a_{off} = g/2$, $\sigma = 0.0002\,\mathrm{m^3/s}$ and $\gamma_{br} = 0.01\,\mathrm{m^3/s^2}$. Additional work is required for systematic study of the effects of the parameters on model performance and to establish a rational procedure to select the parameters for different breaking conditions.

3.5   Numerical efficiency

An important question is the computational efficiency of the model. In the implicit time marching scheme implemented in the Lagrangian solver, the most time consuming element is the inversion of a Jacobi matrix used by the Newton iterations for solving nonlinear grid equations. The required calculation time grows fast with increased matrix dimension, which equals to the number of mesh equations and is proportional to the product of the dimensions of a numerical mesh. For a matrix inversion algorithm used in this work, the inversion time is approximately proportional to the square of the dimension of the matrix. For a 2D problem with a constant value of N_z much smaller than N_x, the matrix dimension is proportional to N_x. To provide a uniform discretisation error in space and time, the value of time step should be proportional to δx which implies that the number of steps is proportional to N_x. Thus, the overall calculation time grows with a rate proportional to N_x^3, as illustrated in Figure 1 for the cases presented in Table 1.

High demand for computational resources for large scale problems is a well recognised disadvantage of implicit schemes, which often outweighs their advantages in numerical stability. The radical method of increasing computational efficiency is using parallel computing. Implicit solvers have a single standard time-consuming operation and are, therefore, suitable for efficient parallelisation. A parallel version of an implicit solver can be created with minimal changes to the original code by replacing a matrix inversion subroutine with a parallel analogue. This feature is particularly useful in light of recent advances in GPU-based matrix inversion algorithms, which are much faster than conventional parallelisation using multiple CPUs (e.g., Sharma et al., 2013).

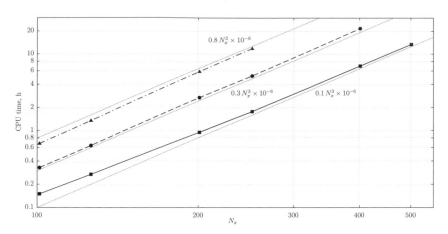

Figure 1: Computation time for modelling of 20 sec of wave propagation by the Lagrangian solver on a single 2.4 GHz CPU for different mesh sizes: $N_z = 11$ (solid); $N_z = 16$ (dashed); $N_z = 21$ (dash-dotted).

Table 1: Numerical cases for convergence and validation tests of the Lagrangian model.

N_x	N_z			$\delta t, sec$	N_x	N_z			$\delta t, sec$
101	11	16	21	0.010	251	11	16	21	0.004
126	11	16	21	0.008	401	11	16		0.0025
201	11	16	21	0.005	501	11			0.002

3.6 Model validation

We validate numerical results against a set of experimental data on propagation of focussed wave groups obtained in a wave flume of the Civil Engineering department at UCL. The flume has a width of 45 cm and the length of the working section between two piston wavemakers is 12.5 m (see Figure 2). A paddle on the right end of the flume is used as a wave generator and the opposite paddle as an absorber. Water depth over the horizontal bed of the flume was set to $h = 40$ cm. We use the centre of the flume as the origin of the coordinate system with the x-axis directed towards the wave generator positioned at $x = 6.25$ m. The vertical z-axis with the origin at the mean water surface is directed upwards. The wavemaker uses a control system with force feedback operating in a frequency domain, which allows precise control and partial absorption of incident waves to reduce reflections. The input of the control system is the desired linear amplitude spectrum of the wave at the centre of the flume. The control system uses discrete spectrum and generates periodic paddle motions. For our experiments we used an overall return period of 128 sec, which is the time between repeating identical events produced by the paddle. This means that the wavemaker generates the discrete spectrum with frequencies $n/128$ Hz, where n is an integer. We use $n = 1 \ldots 512$ to generate 512 frequency components in the range from 1/128 to 4 Hz. The waves generated were monitored by a series of resistance wave probes measuring surface elevation, and an ultrasonic sensor was used to record paddle motion.

We apply an iterative procedure (Buldakov et al., 2017) to generate a Gaussian wave group with peak frequency of $f_p = 1$ Hz focussed at the centre of the flume with the linear

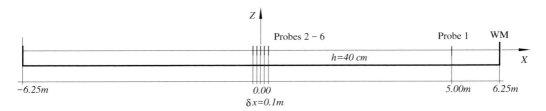

Figure 2: Wave flume layout and positions of wave probes.

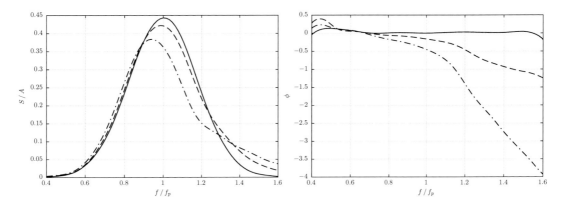

Figure 3: Linearised spectra of experimental wave groups at the focus point ($x = 0$). Amplitude (left) and phase (right). Solid– $A = 2.5$ cm; dashed– $A = 5$ cm; dash-dotted– $A = 7.5$ cm.

focus amplitude of 2.5 cm. Then, we use the resulting input spectrum to generate higher amplitude waves by multiplying the input by factors 2 and 3 without further focussing or spectrum corrections. We, therefore, obtain additional waves with linear focus amplitudes $A = 5$ cm and $A = 7.5$ cm. The linearised spectra of generated waves at $x = 0$ are shown in Figure 3. As can be expected, non-linear defocussing and transformation of the spectrum can be observed for higher amplitude waves. The linearisation is done using the spectral decomposition method described in Buldakov et al. (2017), which requires generation of waves with 4 constant phase shifts $\Delta\phi = 0$, $\pi/2$, π, $3\pi/2$. In this section we are using results with $\Delta\phi = 0$, which corresponds to a peak-focussed linear wave. The resulting waves are of three distinct qualitative types. The small amplitude wave ($A = 2.5$ cm) has weakly non-linear features. The wave with $A = 5$ cm can be described as a strongly nonlinear non-breaking wave. And the high amplitude wave ($A = 7.5$ cm) exhibits two intensive breaking events, while it propagates along the flume.

We reconstruct the experimental setup using a numerical wave tank based on the previously described Lagrangian numerical model. The dimensions of the NWT are the same as the dimensions of the experimental tank. The wave is generated by implementing the boundary condition (6) on the right boundary of the computational domain with $X_R(z, t)$ specified by the paddle displacement recorded in the experiment. To account for gaps between the paddle and walls and bottom of the experimental wave tank and for frictional losses at the wavemaker the amplitude of the numerical wave generator is reduced by 18.5%. The surface displacement damping term in (9) is activated near the tank wall opposite the wavemaker to absorb the reflections. Calculations are performed for all experimental cases, including 3 amplitude values and 4 phase shifts, and repeated with different time steps and horizontal and vertical numbers of mesh points. The time step has been reduced with the increasing number of horizontal mesh points to maintain the dispersion error levels due to temporal and spatial discretisation as given by (15), approximately equal to each other. This provides uniform convergence by both parameters. The summary of parameters for the numerical cases is given in Table 1.

The convergence is tested using an L_2 norm of the difference between the experimental and calculated spectral components of the surface elevation at the linear focus point $x = 0$. The norms for spectral amplitudes and phases are calculated as

$$\|E(a)\| = \sqrt{\sum_i (a_i - a_{i_{\mathrm{EXP}}})^2}\,; \quad \|E(\phi)\| = \sqrt{\sum_i (\phi_i - \phi_{i_{\mathrm{EXP}}})^2}\,,$$

where the sum is taken over discrete spectral components in the range $0.5\,\mathrm{Hz} < f < 1.5\,\mathrm{Hz}$ from the set generated by the wavemaker ($n = 64\dots192$), where the amplitude components are large enough not to be affected by experimental errors. It should be noted that full convergence of the calculated results to the experimental measurements can not be expected. The numerical model is based on a set of assumptions that are satisfied with limited precision. In addition, the measurements include some experimental errors. Therefore, the difference between the experimental and numerical results converges to a certain small value and does not change with an additional increase in the resolution of the numerical model.

The selected results of the convergence tests are presented in Figures 4-6. As can be seen in the top row of Figure 4, the results with and without dispersion correction converge to the

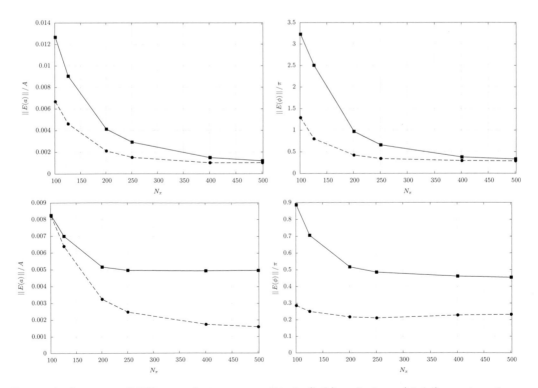

Figure 4: L_2-norm of difference between amplitude (left) and phase (right) spectra of a wave group at $x = 0$ measured in experiment and calculated by the Lagrangian solver. Top – $A = 5\,\mathrm{cm}$; without (solid) and with (dashed) dispersion correction. Bottom – $A = 7.5\,\mathrm{cm}$; breaking control without (solid) and with (dashed) crest correction. $N_z = 11$.

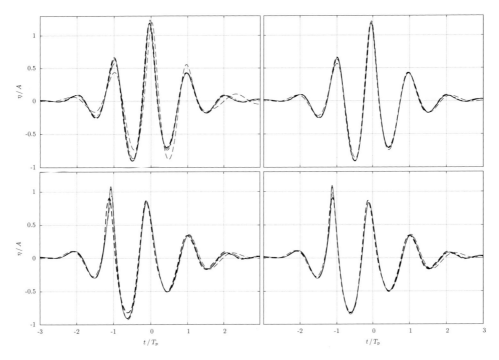

Figure 5: Convergence of time history of surface elevation for a wave group at $x = 0$. Experiment (thick dashed) and Lagrangian solver: $N_x = 101$ (dashed); $N_x = 201$ (dash-dotted); $N_x = 401$ (solid); $N_z = 11$. Top – $A = 5cm$ without (left) and with (right) dispersion correction. Bottom – $A = 7.5cm$, breaking control without (left) and with (right) crest correction.

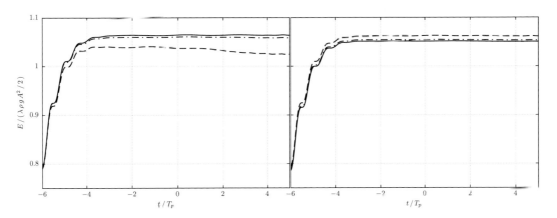

Figure 6: Convergence of the normalised total energy for the Lagrangian solver with dispersion correction. $A = 5$ cm. Left – increasing horizontal mesh resolution: $N_x = 101$ (dashed); $N_x = 201$ (dash-dotted); $N_x = 401$ (solid); $N_z = 11$. Right – increasing vertical mesh resolution: $N_z = 11$ (dashed); $N_z = 16$ (dash-dotted); $N_z = 21$ (solid); $N_x = 251$.

same solution. However, the introduction of a dispersion correction considerably increases the speed of convergence for both amplitudes and phases. The experimental results can be reproduced with sufficient accuracy for a relatively small number of horizontal mesh points and a relatively large time step. For a breaking wave (Figure 4, bottom row), the shape correction of the crest introduces a new physical process. For this reason, the numerical results with and without crest correction converge to different solutions, and the numerical result with the correction shows a much better comparison with the experiment. A general impression of convergence and accuracy of the different versions of the numerical model can be obtained from the graphs of the time history of surface elevation presented in Figure 5.

Figure 6 shows the behaviour of the total wave energy in the wave tank calculated for different temporal and horizontal resolutions and different resolutions of the vertical mesh. Wave absorption is disabled for energy tests. The energy is normalised by the energy of one wave length λ of a linear regular wave with the frequency and the amplitude equal to the peak spectral frequency the linear focus amplitude of the wave group. The kinetic energy is calculated by numerical integration over the entire Lagrangian fluid domain using a bi-linear interpolation of the velocities of the fluid particles inside mesh cells. This provides the second order approximation of the integral with respect to the mesh resolution. According to (13), the error of the velocity profiles within the fluid domain, and thus the kinetic energy, is determined by the horizontal and vertical resolution of the mesh. The potential energy is calculated as an integral of the potential energy density at the free surface and is not directly affected by the discretisation of the vertical mesh. As can be seen in Figure 6, the total energy in the tank increases up to $t \approx 4\,\text{sec}$ due to the energy generated by the wave paddle. For low horizontal mesh resolution, the dissipative term in the numerical dispersion relation leads to the reduction of the energy of the propagating wave. For higher resolutions, energy conservation is satisfied with high accuracy. The right graph of Figure 6 shows the rapid convergence of energy with increased vertical resolution. This includes both the convergence of the numerical integral used to calculate kinetic energy and the convergence of the numerical solution for wave kinematics, as indicated by expansion (13). Overall, the convergence tests show that the numerical results converge towards a solution, which approximates the experiment with a good accuracy. The implementation of the dispersion correction term greatly increases the convergence rate and provides accurate results with a smaller number of mesh points and a larger time step.

4   Model application to the evolution of extreme wave groups

After validation of the model in the previous section, this section considers the performance of the model on the simulation of extreme wave groups with a focus on modelling the wave breaking process. All results presented in this section are calculated using a 401×16 mesh and the time step $\delta t = 0.0025\,\text{sec}$. Figure 7 shows the evolution of the total wave energy for different wave amplitudes and phase shifts $\Delta\phi = 0$ and $\Delta\phi = \pi$, which correspond to opposite wavemaker input signals. The wave energy increases from the initial zero level starting at $t \approx -14\,\text{sec}$, when the wavemaker begins to operate. After the wavemaker stops at $t \approx -3\,\text{sec}$, the total energy remains constant until wave breaking occurs for high amplitude waves. As expected, the energy of non-breaking waves is the same for both phase shifts. It can be observed that for $\Delta\phi = 0$ there are two breaking events of similar intensity which are symmetrical with respect to $t = 0$ and with respect to the centre of the tank $x = 0$. For $\Delta\phi = \pi$, an intensive breaking event occurs near the centre of the tank close to $t = 0$ and a much smaller event later. This is confirmed by wave observations during the experiments. Later in this section, we will limit our attention to the large breaking event at the tank centre and consider the wave with $A = 7.5\,\text{cm}$ and $\Delta\phi = \pi$.

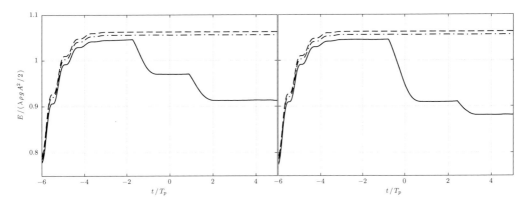

Figure 7: Normalised total wave energy for different wave amplitudes: $A = 2.5\,\mathrm{cm}$ (dashed); $A = 5\,\mathrm{cm}$ (dash-dotted) and $A = 7.5\,\mathrm{cm}$ (solid). Left – phase shift $\Delta\phi = 0$. Right – phase shift $\Delta\phi = \pi$.

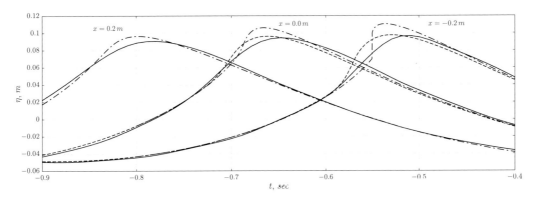

Figure 8: Time histories of wave crest evolution at different positions along the flume, $A = 7.5\,\mathrm{cm}$, $\Delta\phi = \pi$. Experiment (solid); Lagrangian model (dash-dotted); Lagrangian model with breaking model (dashed).

Figure 8 shows the time history of wave crest elevation at three locations along the tank. The corresponding wave profiles can be seen in Figure 9. The difference between measured and calculated wave crests can be observed in Figure 8 near the top of the crest. It should be noted that for high wave peaks experimental measurements by wave probes are unreliable. High speed flow at the crest creates a cavity around the wave probe wires, which generates an error that reduces the recorded crest elevation. However, for the main part of the crest, the calculated and experimental results are compared with good accuracy. The rounded end of the overturning wave observed in Figure 9 is explained by the surface tension which produces a considerable effect on the relatively small experimental scale. At $x = 0.2\,\mathrm{m}$ the breaking model is not yet operational and the Lagrangian solutions with and without the breaking model are identical. At $x = 0$, the top of the wave crest begins to deform. It moves faster than the main body of the crest and eventually overturns. At this stage, the breaking model turns on the damping, and the shape of the crest with and without the breaking model becomes different. Farther along the tank, at $x = -0.2\,\mathrm{m}$, the top of the crest begins to overturn and forms a vertical front. The shape of the crest with the operating breaking model differs considerably from the shape of the crest that evolves

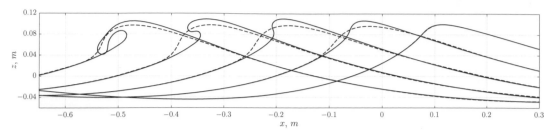

Figure 9: Breaking crest evolution between $t = -0.76\,\mathrm{sec}$ and $t = -0.36\,\mathrm{sec}$. Lagrangian model (solid); Lagrangian model with breaking model (dashed).

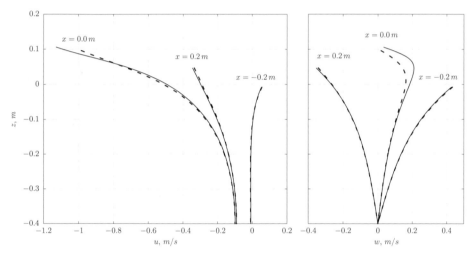

Figure 10: Horizontal (left) and vertical (right) velocity component profiles at $t = -0.665\,\mathrm{sec}$ and different positions along the flume. Lagrangian model (solid); Lagrangian model with breaking model (dashed). The time corresponds to the maximal surface elevation at $x = 0$.

freely. However, the differences are located near the top of the crest and the rest of the wave is unaffected by the breaking model. Examples of horizontal and vertical velocity profiles at the wave crest and on the front and rear slopes are given in Figure 10. The profiles are presented at the moment when surface elevation reaches its maximum at $x = 0$ and when the overturning crest begins to develop. The difference between calculated velocities with and without breaking model can be observed near the top of the wave crest but quickly disappears everywhere else. Calculations without the breaking model are continued until the self contact of the overturning wave occurs at $t = -0.36\,\mathrm{sec}$ (Figure 9). After that, the results produced by the Lagrangian model without the breaking model cease to be physically meaningful. The breaking model allows continuing calculations beyond this point. As can be seen in Figure 9, the solution with the breaking model reproduces relatively well the shape of the self-contacting wave crest, which provides a good starting point for further simulation of the post-breaking wave.

5   Model application to waves on sheared currents

One of the advantages of the Lagrangian formulation is that it offers a simple treatment of flows with vorticity and is, therefore, suitable for modelling waves on sheared currents. A sheared current can be defined by specifying the vorticity that depends solely on the vertical Lagrangian coordinate c. For our choice of the Lagrangian labels the parallel current can be specified as $x = a + V(c)\,t$; $z = c$, where $V(c) = V(z_0)$ is the current profile. Substitution to (3) gives

$$\Omega(a,c) = \Omega(c) = V'(c)\,. \tag{16}$$

Therefore, the waves on a sheared current with an undisturbed profile $V(z_0)$ are described by Equations (1, 3) with the free surface boundary condition (4), the bottom condition (5) and the vorticity distribution given by (16). As before, the numerical implementation is based on a form of the governing equations given by (7) and the free surface boundary condition with the dispersive correction (14) with additional terms on the right-hand side. For wave-current calculations we use the following additional terms

$$RHS = k(a)\,(\,(\,x_t - V(c)\,)\,x_a + z_t z_a) + P_x(x,t)\,x_a\,. \tag{17}$$

The first term of (17) is the modified dissipation term and the second term is the time varying surface pressure gradient that is used for wave generation. The breaking model is not implemented in the wave-current version of the solver.

The numerical wave-current flume is created by specifying inlet and outlet boundary conditions, distribution of the surface dissipation $k(a)$ and the surface pressure gradient $P_x(x,t)$. The NWT design should provide free current inflow and outflow to and from the computational domain, wave generation on the current, and absorption of waves incident to the domain boundaries to eliminate reflections. The dissipation coefficient is set to zero in the working section of the flume. It gradually increases to a large value near the input and output boundaries to ensure a stable horizontal free surface that remains at the initial position $z = 0$. This provides parallel input and output flows and serves a double purpose. Firstly, the reflections from the boundaries are significantly reduced. Secondly, the inlet and outlet boundary conditions can be specified as the undisturbed velocity profile at the inlet and as a parallel flow at the outlet

$$x_t(a_{in},c,t) = V(c); \quad z_a(a_{out},c,t) = 0\,.$$

The waves are generated by creating an area in front of one of the wave absorption zones where the pressure distribution of a prescribed shape is defined. The time varying amplitude of the pressure disturbance is used as the control input for this pneumatic wave generator. It should be noted that the generated waves propagate in both directions, but the waves propagating backwards are damped by the first absorption zone. Figure 11 illustrates the setup of the Lagrangian numerical wave-current flume.

An additional difficulty with the numerical realisation of the Lagrangian formulation on sheared currents is the continuous deformation of the original physical domain. The accuracy of the calculations for highly deformed meshes decreases considerably. If the deformation is too strong, this can lead to calculations breakdown. To avoid these difficulties, we carry out shear deformation of the Lagrangian domain to compensate for the deformation of the physical domain. The deformation takes place after several time steps and brings the boundaries of the physical domain back to the vertical lines. After that, we re-label the fluid particles with new values of Lagrangian coordinates in order to preserve the rectangular shape of the Lagrangian computational domain with the vertical and horizontal lines of the computational grid. The procedure is illustrated in Figure 12.

We use the numerical wave-current flume to reproduce the results of an experimental study of focused wave groups over sheared currents. The experimental flume is 1.2 m wide and the distance between two piston wavemakers is about 16 m. The depth for all tests is $h = 0.5$ m. A recirculation system with three parallel pumps and vertical inlets 13 m apart is

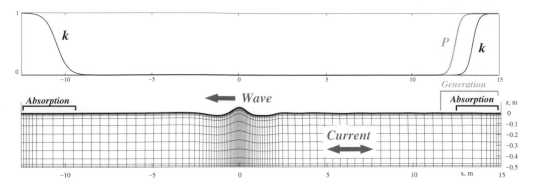

Figure 11: Schematic representation of the Lagrangian numerical wave-current flume. The upper graph demonstrates the shapes of distributions for the surface pressure P and the dissipation coefficient k from Equation (17). Wave and current directions and wave generation and absorption zones are indicated.

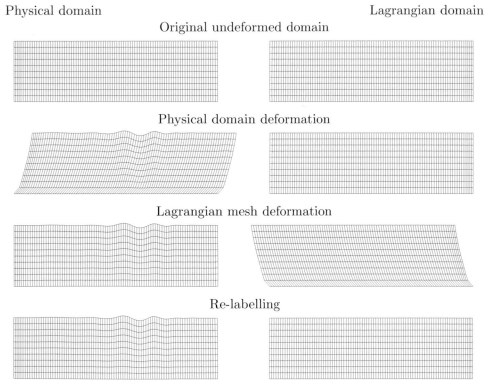

Figure 12: Diagram of the procedure of deformation and re-labelling of the Lagrangian mesh.

used to create a current. A paddle on the right end of the flume is used as a wave generator and the opposite paddle as an absorber. Trapezoidal wire mesh blocks are installed above the inlet and outlet to condition the flow and create a desired current profile. The surface elevation at selected points along the flume is measured by resistance wave probes and a PIV system is used to measure flow kinematics. An iterative procedure (Buldakov et al., 2017) is used to focus the wave group at a prescribed time and place. We use the same coordinate system as previously with the origin on the water surface at the centre of the flume, the x-axes directed towards the wave generator and the z-axis directed upwards. The wave probe at position $x = 4.7\,\mathrm{m}$ is used to match the linearised amplitude spectrum with the target spectrum, and the wave probe at $x = 0$ for focussing the phase of the generated wave group. A broadband Gaussian spectrum with peak frequency $f_p = 0.6\,\mathrm{Hz}$ is used as the target spectrum. Wave groups having different linearised focus amplitudes A on opposing and following sheared currents with different surface velocities V_0 are generated in the experimental study. We use a moderately steep wave with $A = 7\,\mathrm{cm}$ propagating on currents with $V_0 \approx 0.2\,\mathrm{m/s}$ as a test case for comparison with numerical results. More details of the experimental setup and methodology can be found in Stagonas et al. (2018a).

Since the experimental and numerical wave flumes have different wave generators and the flow conditioner can not be modelled adequately, direct replication of the experiment in the numerical flume is not possible. We, therefore, apply in the numerical flume the same iterative wave generation procedure as in the experimental flume using the linearised experimental spectrum as a target. This makes it possible to generate the wave which reproduces the linearised experimental wave with the accuracy of the iterative procedure. This also ensures that the higher order bound wave components are also modelled with the corresponding accuracy. However, the higher order spurious components generated by the experimental and numerical wavemakers are different. This is one of the main sources of difference between experimental and numerical results. The current profiles applied in the Lagrangian model are obtained from PIV measurements of the current velocity, as shown in Figure 13. PIV data only cover the upper part of the water depth ($z > -0.3\,\mathrm{m}$). The shape of the lower part of the profiles is reconstructed from the ADV measurements available for currents with a slightly higher discharge.

The comparison of the numerical results with the experiment is shown in Figures 14 and 15. Both the surface elevation (Figure 14) and the combined wave and current velocity profiles (Figure 15) demonstrate good agreement. The contribution of spurious free components to the difference between the measured and calculated surface elevation is clearly visible in Figure 14. For the opposing current, one can also observe the effect of the dispersive error on the phase difference of the results at $x = 4.7\,\mathrm{m}$. Because the wave is focused at the centre of the flume both in the experiment and in the calculations, the dispersive error increases with the increasing distance from the focus position $x = 0$. This error is higher for the opposing current due to the longer effective path of the wave travelling against the current.

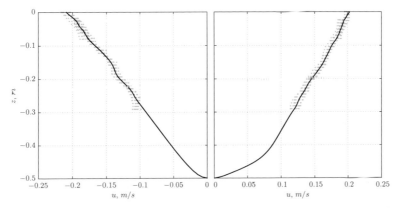

Figure 13: Horizontal velocity profiles for following (left) and opposing (right) currents. PIV measurements (grey dots) and profiles used as the input to the Lagrangian model (solid).

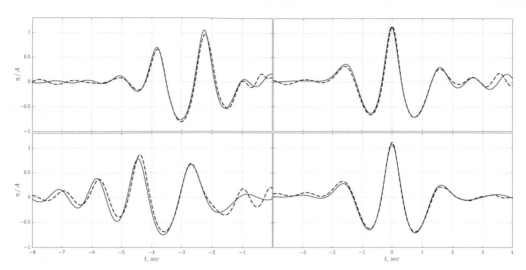

Figure 14: Time history of surface elevation for a wave group at $x = 4.7$ m (left) and $x = 0$ (right) over a following (top) and opposing (bottom) currents. Experiment (dashed) and Lagrangian solver (solid).

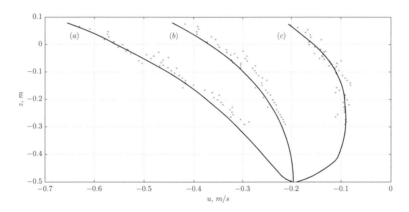

Figure 15: Horizontal velocity profiles under the crest of a focussed wave group over a sheared current. PIV measurements (grey dots) and Lagrangian calculations (solid). (a) – following current; (b) – no current; (c) – opposing current.

6   Concluding remarks

As brief conclusions, we would like to offer reflections on the practical application of wave propagation models to generate incoming waves in wave-structure interaction calculations. It is obvious that the application of a fast but accurate numerical model in a large domain to generate a wave input for a CFD wave-structure interaction solver operating in a much smaller domain offers considerable savings in computational resources. There are two ways to implement this approach. The first involves modelling a random sea state over a large area and a long period of time. Surface elevation and kinematics of selected events, e.g., extreme waves, are then used as input for a CFD solver. HOS models have a clear advantage for such an approach. The benefits of this method are recognised by the industry. See, for example, the recent feature article from DNV-GL where an application of a HOS model is reported to provide a realistic nonlinear wave input for the CFD wave-structure interaction code (Bitner-Gregersen, 2017).

In the second approach, a NWT is used in a manner similar to an experimental wave tank for generating a preselected wave event or for replicating a physical wave tank experiment. Any computationally efficient NWT based on an appropriate wave model can be used for this purpose, including the models described in this chapter. None of them seems to have obvious advantages except for the Lagrangian model in the case of waves on sheared currents. For this method, accurate generation of a desired wave event or reconstruction of experimental conditions are important because the errors due to incorrect wave input may be larger than the errors of a numerical scheme. The application of an iterative wave generation technique may be recommended as an effective solution to this problem. Such techniques are common in wave tank experiments and can be similarly applied in numerical wave tanks (e.g., Fernandez et al., 2014; Buldakov et al., 2017; Stagonas et al., 2018b). An example of an accurate reconstruction of the experimental wave conditions by an advanced iterative technique in a Lagrangian NWT with an application as input to a wave-structure interaction CFD model can be found in Higuera et al. (2018).

The methods described above use the so called one-way coupling between the wave-propagation and CFD models. This means that the wave propagation model is used independently and is not influenced by the CFD model. We believe that for wave-structure interaction problems, this method of communication between models is preferable to real-time two-way coupling, especially when iterative wave generation is used. The computationally demanding CFD component is not executed during the iterative wave generation phase, and the wave propagation component is not executed when the wave interaction with a structure is simulated. Moreover, once generated, wave input can be used in different CFD models using different numerical methods and applied to different structures. One of the technical problems to be solved when applying the one-way coupling approach is to not allow the waves reflected by the structure to be reflected back to the domain by external domain boundaries. At the same time, the precise transition of the wave kinematics generated by the propagation model to the CFD domain must be ensured. This can be done in different ways.

For example, Higuera et al. (2018) used Lagrangian kinematics and surface elevation to specify the boundary condition for velocity on a front boundary of a rectangular CFD domain. Active dissipation was applied at the rear boundary and passive dissipation at the side boundaries of the domain. Another approach is using a cylindrical grid with a ring-shaped relaxation zone. This mesh type provides better resolution around a structure. The equations within the relaxation zone are modified to introduce a dissipation of disturbances of the incoming wave solution with a dissipation coefficient gradually increasing from zero at the inner edge of the relaxation zone to a high value at its outer edge. In this way, the incoming wave is generated at an outer boundary of the computational domain and propagates freely into the domain interior. At the same time, the waves reflected by the structure propagate freely in the field and dissipate inside the relaxation zone without being reflected back. The size of the computational domain is proportional to the wavelength of the incoming wave and the width of the relaxation zone to the length of the wave reflected or radiated by the structure. Normally, the reflected wave has the same length as the incoming wave, but in certain situations, e.g., for slender structures, the peak of the reflected spectrum is shifted towards higher frequencies. For such structures the width of the relaxation zone can be reduced. This is also the case for waves radiated by ringing structures. The optimal sizes of the main computation domain and the relaxation zone depend on a particular wave and structure. Practical recommendations on their selection for different types of structures and waves should be developed as a result of the convergence study with respect to these parameters.

Acknowledgments

The work on wave over sheared currents presented in this chapter is supported by EPSRC within the Supergen Marine Technology Challenge (Grant EP/J010316/1). The author also thanks Dr. Dimitris Stagonas for performing experiments on wave propagation over currents used in this work and for analysis of experimental data.

References

Bateman, W., Swan, C. and Taylor, P. (2001). On the efficient numerical simulation of directionally spread surface water waves. Journal of Computational Physics 174(1): 277–305.

Bitner-Gregersen, E. (May 2017). Hacking wave structure. Next generation wave-structure interaction codes. Feature article, DNV-GL.

Bonnefoy, F., Ducrozet, G., Touzé, D. L. and Ferrant, P. (2010). Time domain simulation of nonlinear water waves using spectral methods. pp. 129–164. *In*: Ma, Q. (ed.). Advances in Numerical Simulation of Nonlinear Water Waves. World Scientific.

Braess, H. and Wriggers, P. (2000). Arbitrary Lagrangian Eulerian finite element analysis of free surface flow. Computational Methods in Applied Mechanics and Engineering 190: 95–109.

Brennen, C. and Whitney, A. K. (1970). Unsteady, free surface flows; solutions employing the Lagrangian description of the motion. *In*: 8th Symposium on Naval Hydrodynamics. Office of Naval Research, pp. 117–145.

Buldakov, E. (2013). Tsunami generation by paddle motion and its interaction with a beach: Lagrangian modelling and experiment. Coastal Engineering 80: 83–94.

Buldakov, E. (2014). Lagrangian modelling of fluid sloshing in moving tanks. Journal of Fluids and Structures 45: 1–14.

Buldakov, E., Higuera, P. and Stagonas, D. (2019). Numerical models for evolution of extreme wave groups. Submitted to Applied Ocean Research.

Buldakov, E., Stagonas, D. and Simons, R. (2015). Langrangian numerical wave-current flume. *In*: Proceedings of 30th International Workshop on Water Waves and Floating Bodies. Bristol, UK, pp. 25–28.

Buldakov, E., Stagonas, D. and Simons, R. (2017). Extreme wave groups in a wave flume: Controlled generation and breaking onset. Coastal Engineering 128: 75–83.

Cai, X., Langtangen, H., Nielsen, B. and Tveito, A. (1998). A finite element method for fully nonlinear water waves. Journal of Computational Physics 143(2): 544–568.

Chan, R. K. -C. (1975). A generalized arbitrary Lagrangian-Eulerian method for incompressible flows with sharp interfaces. Journal of Computational Physics 17: 311–331.

Chen, L., Stagonas, D., Santo, H., Buldakov, E., Simons, R., Taylor, P. and Zang, J. (2019). Numerical modelling of interactions of waves and sheared currents with a surface piercing vertical cylinder. Coastal Engineering 145: 65–83.

Chen, L. -F., Ning, D. -Z., Teng, B. and Zhao, M. (2017). Numerical and experimental investigation of nonlinear wave-current propagation over a submerged breakwater. Journal of Engineering Mechanics 143(9): 04017061-1–04017061-14.

Craig, W. and Sulem, C. (1993). Numerical simulation of gravity waves. Journal of Computational Physics 108(1): 73–83.

Da Silva, A. and Peregrine, D. (1988). Steep, steady surface waves on water of finite depth with constant vorticity. Journal of Fluid Mechanics 195: 281–302.

Dias, F. and Bridges, T. J. (2006). The numerical computation of freely propagating time-dependent irrotational water waves. Fluid Dynamics Research 38(12): 803–830.

Dommermuth, D. and Yue, D. (1987a). A high-order spectral method for the study of nonlinear gravity waves. Journal of Fluid Mechanics 184: 267–288.

Dommermuth, D. and Yue, D. (1987b). Numerical simulations of nonlinear axisymmetric flows with a free surface. Journal of Fluid Mechanics 178: 195–219.

Dommermuth, D., Yue, D., Lin, W. and Rapp, R. (1988). Deep-water plunging breakers: A comparison between potential theory and experiments. Journal of Fluid Mechanics 189: 423–442.

Ducrozet, G., Bingham, H. B., Engsig-Karup, A. P., Bonnefoy, F. and Ferrant, P. (2012a). A comparative study of two fast nonlinear free-surface water wave models. International Journal for Numerical Methods in Fluids 69(11): 1818–1834.

Ducrozet, G., Bonnefoy, F., Le Touzé, D. and Ferrant, P. (2007). 3-D HOS simulations of extreme waves in open seas. Natural Hazards and Earth System Science 7(1): 109–122.

Ducrozet, G., Bonnefoy, F., Le Touzé, D. and Ferrant, P. (2012b). A modified high-order spectral method for wavemaker modeling in a numerical wave tank. European Journal of Mechanics-B/ Fluids 34: 19–34.

Eatock Taylor, R., Wu, G., Bai, W. and Hu, Z. (2008). Numerical wave tanks based on finite element and boundary element modeling. Journal of Offshore Mechanics and Arctic Engineering 130(3): 031001-1–031001-8.

Engsig-Karup, A., Eskilsson, C. and Bigoni, D. (2016). A stabilised nodal spectral element method for fully nonlinear water waves. Journal of Computational Physics 318: 1–21.

Fenton, J. and Rienecker, M. (1982). A Fourier method for solving nonlinear water-wave problems: Application to solitary-wave interactions. Journal of Fluid Mechanics 118: 411–443.

Fenton, J. D. (1999). Numerical methods for nonlinear waves. pp. 241–324. *In*: Liu, P. L. F. (ed.). Advances in Coastal and Ocean Engineering. Vol. 5. World Scientific.

Fernandez, H., Sriram, V., Schimmels, S. and Oumeraci, H. (2014). Extreme wave generation using self correcting method-revisited. Coastal Engineering 93: 15–31.

Fochesato, C. and Dias, F. (2006). A fast method for nonlinear three-dimensional free-surface waves. Proceedings of the Royal Society of London A: Mathematical, Physical and Engineering Sciences 462(2073): 2715–2735.

Fochesato, C., Grilli, S. and Dias, F. (2007). Numerical modeling of extreme rogue waves generated by directional energy focusing. Wave Motion 44(5): 395–416.

Fritts, M. and Boris, J. (1979). The Lagrangian solution of transient problems in hydrodynamics using a triangular mesh. Journal of Computational Physics 31(2): 173–215.

Gomez-Gesteira, M., Rogers, B. D., Dalrymple, R. A. and Crespo, A. J. (2010). State-of-the-art of classical SPH for free-surface flows. Journal of Hydraulic Research 48(sup1): 6–27.

Gouin, M., Ducrozet, G. and Ferrant, P. (2016). Development and validation of a non-linear spectral model for water waves over variable depth. European Journal of Mechanics-B/Fluids 57: 115–128.

Grilli, S., Guyenne, P. and Dias, F. (2001). A fully non-linear model for three-dimensional overturning waves over an arbitrary bottom. International Journal for Numerical Methods in Fluids 35(7): 829–867.

Grilli, S., Skourup, J. and Svendsen, I. (1989). An efficient boundary element method for nonlinear water waves. Engineering Analysis with Boundary Elements 6(2): 97–107.

Grilli, S. and Svendsen, I. (1990). Corner problems and global accuracy in the boundary element solution of nonlinear wave flows. Engineering Analysis with Boundary Elements 7(4): 178–195.

Grilli, S. T., Dias, F., Guyenne, P., Fochesato, C. and Enet, F. (2010). Progress in fully nonlinear potential flow modeling of 3D extreme ocean waves. pp. 75–128. *In*: Ma, Q. (ed.). Advances in Numerical Simulation of Nonlinear Water Waves. World Scientific.

Guignard, S. and Grilli, S. (2001). Modeling of wave shoaling in a 2D-NWT using a spilling breaker model. *In*: Proceedings of the Eleventh International Offshore and Polar Engineering Conference. Vol. 3. Stavanger, Norway, pp. 116–123.

Guo, L. -D., Sun, D. -P. and Wu, H. (2012). A new numerical wave flume combining the 0–1 type BEM and the VOF method. Journal of Hydrodynamics, Ser. B 24(4): 506–517.

Haussling, H. and Coleman, R. (1979). Nonlinear water waves generated by an accelerated circular cylinder. Journal of Fluid Mechanics 92(4): 767–781.

Herivel, J. W. (1955). The derivation of the equations of motion of an ideal fluid by Hamilton's principle. Proc. Cambridge Philos. Soc. 51: 344–349.

Higuera, P., Buldakov, E. and Stagonas, D. (2018). Numerical modelling of wave interaction with a FSPO using a combination of OpenFOAM and Lagrangian models. *In*: Proceedings of the Twenty Eighth International Offshore and Polar Engineering Conference. Sapporo, Japan, pp. 1486–1491.

Hirt, C., Cook, J. and Butler, T. (1970). A Lagrangian method for calculating the dynamics of an incompressible fluid with free surface. Journal of Computational Physics 5(1): 103–124.

Isaacson, M. (1982). Nonlinear-wave effects on fixed and floating bodies. Journal of Fluid Mechanics 120: 267–281.

Jiang, S. -C., Teng, B., Gou, Y. and Ning, D. -Z. (2012). A precorrected-fft higher-order boundary element method for wavebody problems. Engineering Analysis with Boundary Elements 36(3): 404–415.

Kawahara, M. and Anjyu, A. (1988). Lagrangian finite element method for solitary wave propagation. Computational Mechanics 3(5): 299–307.

Keller, H. B. (1971). A new difference scheme for parabolic problems. *In*: Numerical Solution of Partial Differential Equations, II (SYNSPADE 1970) (Proc. Sympos., Univ. of Maryland, College Park, Md., 1970). Academic Press, New York, pp. 327–350.

Kim, C., Clément, A. and Tanizawa, K. (1999). Recent research and development of numerical wave tanks—a review. International Journal of Offshore and Polar Engineering 9(4): 241–256.

Kim, S. -H., Yamashiro, M. and Yoshida, A. (2010). A simple two-way coupling method of BEM and VOF model for random wave calculations. Coastal Engineering 57(11): 1018–1028.

Lachaume, C., Biausser, B., Grilli, S., Fraunié, P. and Guignard, S. (2003). Modeling of breaking and post-breaking waves on slopes by coupling of BEM and VOF methods. *In*: Proceedings of the Thirteenth International Offshore and Polar Engineering Conference. Honolulu, Hawaii, USA, pp. 1698–1704.

Lamb, H. (1932). Hydrodynamics. Cambridge University Press.

Landrini, M., Colagrossi, A., Greco, M. and Tulin, M. (2012). The fluid mechanics of splashing bow waves on ships: A hybrid BEM-SPH analysis. Ocean Engineering 53: 111–127.

Lin, P. (2008). Numerical Modeling of Water Waves. Taylor & Francis, London.

Longuet-Higgins, M. S. and Cokelet, E. D. (1976). The deformation of steep surface waves on water. I. A numerical method of computation. Proceedings of the Royal Society of London A: Mathematical, Physical and Engineering Sciences 350(1660): 1–26.

Ma, Q. (ed.). (2010). Advances in numerical simulation of nonlinear water waves. Vol. 11 of Advances in Coastal and Ocean Engineering. World Scientific.

Ma, Q. and Yan, S. (2006). Quasi ALE finite element method for nonlinear water waves. Journal of Computational Physics 212(1): 52–72.

Ma, Q. W., Wu, G. X. and Eatock Taylor, R. (2001). Finite element simulation of fully non-linear interaction between vertical cylinders and steep waves. Part 1: methodology and numerical procedure. International Journal for Numerical Methods in Fluids 36(3): 265–285.

Ma, Q. W. and Yan, S. (2010). QALE-FEM method and its application to the simulation of free-responses of floating bodies and overturning waves. pp. 165–202. *In*: Ma, Q. (ed.). Advances in Numerical Simulation of Nonlinear Water Waves. World Scientific.

NAG. (2016). NAG Library Manual, Mark 26. The Numerical Algorithms Group Ltd, Oxford, UK.

New, A., McIver, P. and Peregrine, D. (1985). Computations of overturning waves. Journal of Fluid Mechanics 150: 233–251.

Nimmala, S., Yim, S. and Grilli, S. (2013). An efficient three-dimensional FNPF numerical wave tank for large-scale wave basin experiment simulation. Journal of Offshore Mechanics and Arctic Engineering 135(2): 021104-1–021104-10.

Ning, D., Zang, J., Liu, S., Eatock Taylor, R., Teng, B. and Taylor, P. (2009). Free-surface evolution and wave kinematics for nonlinear uni-directional focused wave groups. Ocean Engineering 36(15): 1226–1243.

Nishimura, H. and Takewaka, S. (1988). Numerical analysis of two-dimensional wave motion using Lagrangian description. Doboku Gakkai Rombun-Hokokushu/Proceedings of the Japan Society of Civil Engineers 9(5): 1910–199, in Japanese.

Ohyama, T. and Nadaoka, K. (1991). Development of a numerical wave tank for analysis of nonlinear and irregular wave field. Fluid Dynamics Research 8(5-6): 231–251.

Ölmez, H. and Milgram, J. (1995). Numerical methods for nonlinear interactions between water waves. Journal of Computational Physics 118(1): 62–72.

Radovitzky, R. and Ortiz, M. (1998). Lagrangian finite element analysis of newtonian fluid flows. International Journal for Numerical Methods in Engineering 43(4): 607–619.

Ramaswamy, B. and Kawahara, M. (1987). Lagrangian finite element analysis applied to viscous free surface fluid flow. International Journal for Numerical Methods in Fluids 7(9): 953–984.

Robertson, I. and Sherwin, S. (1999). Free-surface flow simulation using hp/spectral elements. Journal of Computational Physics 155(1): 26–53.

Ryu, S., Kim, M. H. and Lynett, P. J. (2003). Fully nonlinear wave-current interactions and kinematics by a BEM-based numerical wave tank. Computational Mechanics 32(4): 336–346.

Schäffer, H. A. (2008). Comparison of Dirichlet-Neumann operator expansions for nonlinear surface gravity waves. Coastal Engineering 55(4): 288–294.

Seiffert, B. R. and Ducrozet, G. (2018). Simulation of breaking waves using the high-order spectral method with laboratory experiments: wave-breaking energy dissipation. Ocean Dynamics 68(1): 65–89.

Seiffert, B. R., Ducrozet, G. and Bonnefoy, F. (2017). Simulation of breaking waves using the high-order spectral method with laboratory experiments: Wave-breaking onset. Ocean Modelling 119: 94–104.

Sharma, G., Agarwala, A. and Bhattacharya, B. (2013). A fast parallel Gauss Jordan algorithm for matrix inversion using CUDA. Computers & Structures 128: 31–37.

Skyner, D. (1996). A comparison of numerical predictions and experimental measurements of the internal kinematics of a deep-water plunging wave. Journal of Fluid Mechanics 315: 51–64.

Souli, M. and Benson, D. J. (2013). Arbitrary Lagrangian Eulerian and Fluid-Structure Interaction: Numerical Simulation. Wiley-ISTE.

Sriram, V., Ma, Q. and Schlurmann, T. (2014). A hybrid method for modelling two dimensional non-breaking and breaking waves. Journal of Computational Physics 272: 429–454.

Stagonas, D., Buldakov, E. and Simons, R. (2018a). Experimental generation of focusing wave groups on following and adverse-sheared currents in a wave-current flume. Journal of Hydraulic Engineering 144(5): 04018016-1–04018016-11.

Stagonas, D., Higuera, P. and Buldakov, E. (2018b). Simulating breaking focused waves in CFD: Methodology for controlled generation of first and second order. Journal of Waterway, Port, Coastal and Ocean Engineering 144(2): 06017004-1–06017004-8.

Staroszczyk, R. (2009). A Lagrangian finite element analysis of gravity waves in water of variable depth. Archives of Hydroengineering and Environmental Mechanics 56(1-2): 43–61.

Subramani, A., Beck, R. and Schultz, W. (1998). Suppression of wave breaking in nonlinear water wave computations. *In*: Proceedings of 13th International Workshop on Water Waves and Floating Bodies. Alphen aan den Rijn, The Netherlands, pp. 139–142.

Tian, Z., Perlin, M. and Choi, W. (2012). An eddy viscosity model for two-dimensional breaking waves and its validation with laboratory experiments. Physics of Fluids 24(3): 036601-1–036601-23.

Tsai, W. and Yue, D. K. P. (1996). Computation of nonlinear free-surface flows. Annual Review of Fluid Mechanics 28: 249–278.

Vinje, T. and Brevig, P. (1981). Numerical simulation of breaking waves. Advances in Water Resources 4(2): 77–82.

Violeau, D. and Rogers, B. D. (2016). Smoothed particle hydrodynamics (SPH) for free-surface flows: past, present and future. Journal of Hydraulic Research 54(1): 1–26.

Wang, C. and Wu, G. (2011). A brief summary of finite element method applications to nonlinear wave-structure interactions. Journal of Marine Science and Application 10(2): 127–138.

Wang, P., Yao, Y. and Tulin, M. P. (1995). An efficient numerical tank for non-linear water waves, based on the multi-subdomain approach with BEM. International Journal for Numerical Methods in Fluids 20(12): 1315–1336.

West, B. J., Brueckner, K. A., Janda, R. S., Milder, D. M. and Milton, R. L. (1987). A new numerical method for surface hydrodynamics. Journal of Geophysical Research: Oceans 92(C11): 11803–11824.

Wu, G. and Eatock Taylor, R. (1994). Finite element analysis of two-dimensional non-linear transient water waves. Applied Ocean Research 16(6): 363–372.

Wu, G. and Eatock Taylor, R. (1995). Time stepping solutions of the two-dimensional nonlinear wave radiation problem. Ocean Engineering 22(8): 785–798.

Wu, G. and Eatock Taylor, R. (2003). The coupled finite element and boundary element analysis of nonlinear interactions between waves and bodies. Ocean Engineering 30(3): 387–400.

Wu, G. X. and Hu, Z. Z. (2004). Simulation of nonlinear interactions between waves and floating bodies through a finite-element-based numerical tank. Proceedings of the Royal Society of London A: Mathematical, Physical and Engineering Sciences 460(2050): 2797–2817.

Xue, M., Xü, H., Liu, Y. and Yue, D. (2001). Computations of fully nonlinear three-dimensional wave-wave and wave-body interactions. Part 1. Dynamics of steep three-dimensional waves. Journal of Fluid Mechanics 438: 11–39.

Yan, H. and Liu, Y. (2011). An efficient high-order boundary element method for nonlinear wave-wave and wave-body interactions. Journal of Computational Physics 230(2): 402–424.

Yan, S. and Ma, Q. W. (2010). QALE-FEM for modelling 3D overturning waves. International Journal for Numerical Methods in Fluids 63(6): 743–768.

Zienkiewicz, O., Taylor, R. and Nithiarasu, P. (2005). Finite Element Method for Fluid Dynamics, 6th Edition. Elsevier.

Chapter 3

Wave Breaking and Air Entrainment

Pierre Lubin

1 Introduction

Surface wave breaking, occurring from the ocean to the coastal zone, is a complex and challenging two-phase flow phenomenon, which plays an important role in numerous processes, including air–sea transfer of gas, momentum and energy, and in a number of technical applications, such as coastal and military applications. However, most recent modeling attempts are still struggling with the lack of physical understanding of the fine details of the breaking processes, which makes the task of parameterizing breaking effects very difficult since no universal scaling laws for physical variables have been found so far. The complex 3D structure of turbulent flows under breaking waves has not been properly addressed yet in the literature and remains poorly understood, therefore, a better knowledge of the temporal and spatial evolution of the aerated region under breaking waves is crucial.

Describing this flow process is still a key scientific challenge. Wave breaking has been shown to influence coastal erosion, climate and intensification of tropical cyclones (Véron et al., 2015), and cause ocean ambient noise (Prosperetti, 1988). The breakup and evolution of the entrained air into numerous bubbles is a source of acoustic noise, influencing acoustic underwater communications. Bubble clouds also modify the optical properties of the water column. The hydrodynamic performance of ships is influenced by the wake, which depends heavily on the air entrainment, and the sound generated by the bubble clouds make the ships visible to detection, for military applications. The wake starts at the stem of a ship's hull, where bow waves are observed to break, playing a role in the floating stability. In hydraulic engineering, large spillways are often protected from cavitation damage by controlling aeration. Tsunami waves and storm surges are also a great threat for coastlines, where civil nuclear facilities can often be found. Numerical methods for tsunami hazard assessment need an accurate modeling of the wave-breaking phenomenon, in order to provide guidance for risk assessment on the nuclear facilities and more broadly on the human occupation of the coastal zone. Offshore structures are also very often impacted by breaking waves during extreme weather conditions, and thus need to be designed properly, taking breaking waves into account. When large tankers travel in narrow rivers or channels, river banks are impacted by breaking waves due to the wakes, and can cause major corrosion. Due to climate change (Nicholls, 2004), coastal zones are more often subjected to extreme conditions where overtopping occurs, damaging coastal dunes and beaches, and affecting dikes.

Univ. Bordeaux, CNRS, Bordeaux INP, I2M, UMR 5295, F-33400, Talence, France.
Email: pierre.lubin@bordeaux-inp.fr

A popular method for simulating turbulence in breaking waves is the Large Eddy Simulation (LES) method. This method is now widely used to simulate academic test cases and highly complex industrial applications, but also to investigate the physical phenomena that occur in complex geophysical flows such as breaking waves. In LES, the contribution of the large, energy-carrying structures is exactly simulated, and the effect of the non-simulated smallest scales of turbulence has to be modeled. The scope of this chapter is to review the very recent progresses obtained so far, thanks to the LES technique, to discuss the limitations and identify the key issues which have to be overcome. This chapter is organised as follows. The first section is dedicated to the physical description of the wave breaking process in detail. The governing equations and associated numerical methods are introduced in the next section. Then the numerical results are analysed, leading to the investigation of a new three-dimensional type of vortical structure observed under breaking waves. The chapter is then concluded

2   Physics of breaking

Waves are generated in the ocean when wind blows and modifies the water surface. This results in an oscillatory movement of water masses, transmitted gradually and controlled by the force of gravity. Most often, when the wind blows with a certain fetch, the surface of the water oscillates randomly to form short-crested wind waves; whereas far from the windy zone, the wave motion is organized into fully developed and coherent waves to form the swell. Offshore, the waves will evolve according to their interaction with the wind and other wave trains they encounter as they propagate through the oceans. Near the coasts, wave propagation is directly influenced by variations in water depth: waves are directly modified due to the local bathymetry. Finally, waves break, air is entrained in vortical structures and turbulence is generated.

2.1   Wave breaker types

As the waves approach the beaches, they undergo heavy transformations, which occur in four distinctive zones:

1. the shoaling zone, where the waves slow down and their height begins to increase. The waves lose their symmetrical appearance, and the front of the wave begins to steepen.

2. the breaking zone, where the waves overturn and break. When the waves start to decelerate and the front face becomes almost vertical, the water particles at the crest of the waves will have a higher velocity than the local wave celerity, thus leading to irreversible breaking (Peregrine, 1983).

3. the surf zone, where we observe the transformation of the waves into a strongly aerated, highly unsteady and chaotic flow. The turbulence is generated at the surface and advected to bottom, potentially causing the suspension of the elements constituting the bottom (e.g., sediment) (Battjes, 1988; Svendsen and Putrevu, 1996).

4. the swash zone, where the broken waves finally dissipate a significant amount of energy until they reach the beach, and finally the breaking bores are observed to run-up and run-down the sloping beach (Mory et al., 2011).

Therefore, the surf process can be seen as an irreversible transition from a well-organized, two-dimensional wave motion, into a rotational, highly turbulent and three-dimensional motion.

When arriving on beaches under oblique attack, the waves produce a current parallel to the coastline (coastal drift current). In addition, water masses brought in by breaking waves on beaches then return to the open sea, sometimes generating strong cross-shore currents: rip-currents and undertow currents, channeled by bathymetry channels. However, since the wave breaking area is strongly related to the water depth, the tide also modulates the effects of the swell by varying the average level of the water body, between high and low tide. The trigger point of the surge, therefore, varies as the day progresses.

There are several types of breaking waves. The terminology associated with breaking waves dates back to the late 1940's, when the terms "plunging" and "spilling" can be found in a U.S. Navy Hydrographic Office military manual (Bigelow and Edmondson, 1947). The collapsing and surging breaker will be identified in later works in which definitions will be given to distinguish them (Galvin, 1968). This following terminology is conventionally adopted:

- *spilling breaker*, the front face of the wave turns into foam, and large rollers are observed made of a mixture of air and turbulent water propagating from the breaking point to the beach, the wave gradually losing its height and speed, as it dissipates its energy ;

- *plunging breaker*, which is the most spectacular form, where a very impressive overturning motion of the wave can be seen, with the formation of a jet ejected from the crest of the wave, which encloses a large tube of air before violently impacting the water to form large splashes of foam ;

- *collapsing breaker*, observed when the waves are propagating to the beach and deforming, but without breaking, to finally form an almost vertical wall of water that will fall in its lower part, violently throwing a mixture of water, air and sand up the beach ;

- *surging breaker*, is a form of breaking wave that is observed when the waves propagate to the beach, with very narrow surf zone, deform, and break through their base where a tongue of water is propelled upward from the beach, with almost no foam. This type of breaker is usually encountered in steep slopes or in the vicinity of man-made structures.

By observation, the Iribarren number or Iribarren parameter (also known as the surf similarity parameter or breaker parameter) is a dimensionless number used to describe the occurrence and the type of wave breaking on sloping beaches (Battjes, 1974). It has been established that there is a continuous spectrum ranging from spilling to surging breaking, varying according to three parameters: the slope of the beach (slope inclination from mild to strong), the steepness of the wave (the ratio between the height of the wave and its wavelength, which is an indicator of the degree of non-linearity of the wave) and the dispersion parameter (the ratio of the depth and wavelength). This last parameter, which reflects the influence of the bathymetry variations on the propagation of the wave, classifies if the water depth is shallow or deep. For an equivalent incoming wave condition, the breaking will be "softer" or "milder" for low beach slope, while the breaking will be more violent on steep slope (breaking plunging to frontal). Surging breaking is the limit of the spectrum, since we do not observe a large amount of foam nor a violent impact. The triggering of the breaking event occurs later when the beach goes from a gentle to a steep slope, when waves break closer and closer to the top of the beach. So, this confirms the intuition of the members the United States Army Corps of Engineers, who wanted to get information about the sea bed by observing breaking waves: when a wave breaks, we are often in the presence of a high point of bathymetry.

Moreover, according to the type of breaking, beaches will not undergo the same fate (Ting and Kirby, 1994). Surprisingly, beaches subjected to spilling waves are those that will erode the most (Ting and Kirby, 1995): indeed, the foam rollers propagating up the beaches will generate stronger return currents ("undertow") that will generate in the lower part of the water column, oriented from the beach to the open sea, transporting away the suspended sediment. On the contrary, the more violent breaking events, while occurring closer to the beach, will have a strong impact and put the sediment in motion mainly towards the beaches (Ting and Kirby, 1996).

2.2 Flow structure

During the last few decades, important contributions have focused on breaking waves. The mechanisms involved during the breaking process have been detailed (Peregrine, 1983), while the dynamics of the surf zone have been intensely studied (Battjes, 1988; Svendsen and Putrevu, 1996). Spilling breaking waves have been extensively studied, allowing a better understanding of the dynamics of surface tension effects, which can dramatically alter the deformation of the crest shape as the wave approaches breaking (Duncan et al., 1999; Duncan, 2001). The latest advances in numerical modeling and experimental measurement techniques used for the surf zone investigation have been discussed (Christensen et al., 2002). More recently, the most significant results have been presented and the scientific challenges and obstacles that remain to be addressed have been identified (Kiger and Duncan, 2012).

Experimental observations made over the last 30 years allowed significant progress to the understanding of flow physics and its consequences on the littoral zone, in terms of of induced currents, sediment suspension, and subsequent erosion or accretion phenomena. In the 1970's and 1980's, the first experimental works, complemented by the first numerical simulations, allowed scientists to detail the large structures of turbulent flow generated by the waves breaking on beaches. The wave breaking processes induce vortices that play an important role in the pick-up and suspension of sediments in the surf zone.

When a wave breaks, the wave that propagates to the shore loses its symmetrical shape and steepens. When the front of the wave becomes almost vertical, a jet of water is ejected from the crest (the lip of the wave, as surfers call it). This jet then free-falls down and will impact the front of the wave, enveloping more or less air (sometimes forming a tube, a cavity prized by surfers). The impacting jet is often followed by large series of splash-ups, projected upward and forward. According to the plunge point position on the front face of the wave, different types of breaking waves can be observed. If the plunge point is located near the crest of the wave, the subsequent splash-up will be directed downwards leading to spilling breaking. If the jet impact is further down the face of the wave, away from the crest down the front face of the wave, then a plunging wave is observed.

Among the most important works, experiments have been presented to discuss the internal structure of the flow under breaking waves (Miller, 1976), highlighting what is now coined the "breaker vortices", whose size and intensity depend on the type of breaking experienced. The process of successive splash-ups has been shown for the first time Miller (1976): once the impacting jet creates a first splash-up, a second smaller splash-up is then generated, and so on until the energy of the wave is dissipated. Pictures show the formation of large quantities of air bubbles during these splash-up cycles. Vortex structures generated by plunging breaking waves were proved to affect significantly the bottom (Miller et al.). Sand bars were suggested to be formed when the structures created by the plunging wave were present throughout the flow, while these bars were suppressed by spilling breaking when the structures were confined in the upper part of the flow, near the free surface.

This difference between spilling and breaking has also been confirmed by a series of experiments conducted to identify the internal structure of the velocity field in the surf zone (Nadaoka and Kondoh, 1982). Measurements clearly indicate that the velocity field is divided into two regions, the upper part of the water column, near the free surface, and the lower part, near the bottom. The high rates of turbulence associated with air entrainment were quantified and proved to be responsible for the attenuation of wave energy in the surf zone (Nadaoka and Kondoh, 1982). Air bubbles are mainly contained in large vortical structures and transported to the bottom by them. The upper part of the flow is characterized by the presence of large structures and a high aeration rate, while the bottom zone is where smaller structures coexist, coming from the upper part of the flow and those generated by friction at the bottom.

During all the stages of a breaking event, different vortices will appear (Zhang and Sunamura, 1994). Two major categories have been identified: structures with a horizontal axis of rotation and those with an oblique rotation (pointing from the surface towards the bottom). The first category, mixing co- and counter-rotating vortices (Sakai et al., 1986; Bonmarin, 1989; Kimmoun and Branger, 2007) due to successive splash-ups generated during plunging breaking, generates high levels of turbulence and consists of a large amount air. They are found in the large rollers of foam in the surf zone. Oblique structures are those that are most likely to reach the bottom and put sediment in suspension (Nadaoka et al., 1989). These structures appear more typically under spilling breakers, confirming observations of greater beach erosion for this type of breaking. It can be seen that the large horizontal structures are present in the bore front propagating towards the shore, while behind the crest of the breaking wave, the structure of the flow changes rapidly into a three-dimensional structure stretching along an oblique axis from the free surface towards the bottom. These structures are very intense and carry a great amount of turbulence from the surface, thus agitating the bottom. The schematic structure of the flow is classified and synthesized in (Zhang and Sunamura, 1990; Zhang and Sunamura, 1994).

More recently, (Kubo and Sunamura, 2001) identified a new type of large-scale turbulence, named "downbursts", which can be present in the breaker zone along with the previously observed oblique vortex. (Ting, 2006) also identified these downbursts of turbulence descending from the free-surface. A downburst is characterized by a rapidly descending water mass without marked rotational characteristics, impinging the bed and causing the sediment particles to move more vigorously than oblique vortices. Very few numerical works have been dedicated to the simulation of these three-dimensional structures observed under the breaking waves (Watanabe and Saeki, 1999; Christensen and Deigaard, 2001; Lubin, 2004; Watanabe et al., 2005; Christensen, 2006; Lubin et al., 2006; Iafrati, 2009; Lakehal and Liovic, 2011). In addition, a very limited number of these works have been devoted to smaller processes, such as the striation of free-falling jets or their dislocation into drops before impact (Longuet-Higgins, 1995; Narayanaswamy and Dalrymple, 2002; Watanabe et al., 2005; Saruwatari et al., 2009).

3   Numerical model

Performing numerical simulations of three-dimensional breaking waves requires a large number of mesh grid nodes and long CPU time calculations in order to compute the hydrodynamics from the largest to the smallest length and time scales (Lubin et al., 2011). Moreover, robust and accurate methods are needed for precise numerical simulations. Recent progress in computational power allowed us to run fine three-dimensional simulations which gave us the opportunity to study, for the first time, the fine vortex filaments generated during the early stage of the plunging wave breaking process, and the subsequent air entrainment.

In this section, we discuss the results from the numerical simulations performed to describe the whole breaking event (Lubin et al., 2006; Lubin and Glockner, 2015; Lubin et al., 2019). Therefore, the governing equations and numerical methods will only be briefly summarized here, further details can be found in (Lubin and Glockner, 2015).

We solve the Navier-Stokes equations in air and water, using the Large Eddy Simulation (LES) framework. The resulting set of equations describing the entire hydrodynamic and geometrical processes involved in the motion of non-miscible multiphase media is given by Equations 1–3 below:

$$\nabla \cdot \mathbf{u} = 0 \tag{1}$$

$$\rho \left(\frac{\partial \mathbf{u}}{\partial t} + \mathbf{u} \cdot \nabla \mathbf{u} \right) = -\nabla p + \rho \mathbf{g} + \nabla \cdot \left((\mu + \mu_t) \left[\nabla \mathbf{u} + \nabla^t \mathbf{u} \right] \right) \tag{2}$$

and

$$\frac{\partial C}{\partial t} + \mathbf{u} \cdot \nabla C = 0 \tag{3}$$

where $\mathbf{u}$ is the velocity, C the phase function used to locate the different fluids, t the time, p the pressure, $\mathbf{g}$ the gravity vector, ρ the density, μ the dynamic viscosity. The turbulent viscosity μ_t is calculated with the Mixed Scale model (Sagaut, 1998). x, z and y are respectively the horizontal, vertical and transverse coordinates. u_x, u_z and u_y are the corresponding velocity components of the velocity vector $\mathbf{u}$. The air and water physical properties were used (see Table 1). The main limitation was surface tension is not considered in this study, as discussed by (Lubin and Glockner, 2015) and in the discussion section of this chapter.

The interface tracking is achieved by a Volume Of Fluid method (VOF) and a Piecewise Linear Interface Calculation (PLIC) (Youngs et al., 1982; Scardovelli and Zaleski, 1998). This method has the advantage of building a sharp interface between air and water. A phase function C is used to locate the different fluids. The real air and water physical properties were used. The numerical tool has already been shown to give accurate results for environmental and industrial applications (Lubin et al., 2006; Lubin et al., 2010; Glockner et al., 2011; Brouilliot and Lubin, 2013). All the details of the numerical methods used to discretize and solve the equations are given in (Lubin and Glockner, 2015).

The main evolution between (Lubin et al., 2006) and (Lubin and Glockner, 2015; Lubin et al., 2019) was the use of a parallel version of the numerical tool, which allowed us to obtain much more detailed results as shown in the next section.

The numerical tool runs on parallel distributed memory architectures in production mode. The computational domain is partitioned at the beginning of the code execution in order to reduce the memory space required. The load balance is achieved with an accuracy of one mesh in each direction of the space. The equations modeling a physical problem are discretized implicitly, thus leading to the resolution of linear systems. To solve it, we use the libraries of open source solvers and preconditioners made available to the international scientific community by specialists in the field, namely HYPRE. The Navier-Stokes equations are discretized by the finite volume method on a Cartesian mesh. A pressure correction method (Goda) is used for the treatment of incompressibility and the coupling of velocity and pressure. The prediction step is solved by a BiCGStab preconditioned by the Jacobi method. The pressure increment correction step is solved by a GMRES associated with a PFMG multigrid preconditioning of the HYPRE library. This choice has proved to be the most efficient on simulations of multiphase problems.

Table 1: Physical parameters used for the numerical simulations.

Water density, ρ_w	1000 $kg.m^{-3}$	Water viscosity, μ_w	1×10^{-3} $kg.m^{-1}.s^{-1}$
Air density, ρ_a	1.1768 $kg.m^{-3}$	Air viscosity, μ_a	1.85×10^{-5} $kg.m^{-1}.s^{-1}$
Gravity, g	9.81 $m.s^{-2}$	Surface tension	Neglected

Table 2: Mesh grid densities used in the previous works and the actual time taken from the start of the simulation to the end, from 625 000 to 0.819 billion of mesh grid points.

Cases	Mesh grid densities	Mesh grid resolutions (m)	Processors	Wall-clock time
(Lubin et al., 2006)	250 x 100 x 25	$\Delta x \simeq \Delta z = \Delta y = 4.10^{-4}$	1	Several weeks
(Lubin and Glockner, 2015)	256 x 125 x 50	$\Delta x \simeq \Delta z = \Delta y = 4.10^{-4}$	8	1 hour
(Lubin and Glockner, 2015)	512 x 250 x 100	$\Delta x \simeq \Delta z = \Delta y = 2.10^{-4}$	256	Few hours
(Lubin and Glockner, 2013)	1024 x 400 x 200	$\Delta x \simeq \Delta z = \Delta y = 10^{-4}$	576	2 days
(Lubin and Glockner, 2015)	1024 x 500 x 200	$\Delta x \simeq \Delta z = \Delta y = 10^{-4}$	1024	1 day
(Lubin and Glockner, 2015)	2048 x 1000 x 400	$\Delta x \simeq \Delta z = \Delta y = 0.5 \times 10^{-4}$	8192	3 days

The performances of Thetis have been verified on different supercomputers. The scalability of the different steps of the code with $50x50x50$ cells per core, from 1 node to 128 nodes, has been tested. The total number of points was 256 million mesh cells on 2048 cores and corresponded to the largest simulations we could perform. The scalability proved to be in accordance with the theoretical scalability for the whole part of the calculation time spent in the numerical code (filling of the matrices, resolution of the advection equation). For the time spent in the solvers, it is in conformity with the one expected for this kind of exercise with a contained increase of the computation time.

More than 100 million grid points have been used to discretize the three-dimensional numerical domain ($1024 \times 500 \times 200$), with non-uniform mesh grid cells. The grid was evenly distributed in longitudinal and transverse directions ($\Delta x = \Delta y = 10^{-4}$ m). In the vertical direction, the grid was clustered with a constant grid size $\Delta z = 10^{-4}$ m in the free-surface zone. The three-dimensional numerical domain has been partitioned into 1024 subdomains (one processor per subdomain). The computing time was approximately 24 hours, with 1024 cores, for a simulated physical time of 0.88 s, which comprised the initial stage of wave breaking and the subsequent air entrainment. As presented in Table 2, the parallel version of the numerical code can give access to very fine length scales providing high performance computing facilities.

4 Wave breaking of unstable sinusoidal wave

4.1 Initial configuration

The initial conditions used by many authors correspond to a sine wave of large amplitude in a periodic domain (Cokelet, 1979; Vinje and Brevig, 1981; Abadie et al., 1998; Chen et al., 1991; Abadie, 2001; Lubin et al., 2006; Iafrati, 2009; Hu et al., 2012). This is an artificial condition, since breaking is not due to bathymetric variations but caused by the instability of the initial wave, which has already demonstrated its effectiveness in describing all kinds of wave breaking types. Initial quantities (free surface, velocity and pressure fields) are calculated from the linear theory (Airy wave) (Craik, 2004). The initial wave being of large amplitude, is outside the range of validity of the analytical solution, which causes the velocity field and the free surface to be unstable. Therefore, this wave is initially steep, so it can not propagate in the periodic domain and will break more or less rapidly.

The overturning motion is controlled by only two initial parameters, namely the steepness and dispersion parameters, which makes this approach very interesting to study all types of breaking.

The numerical domain is periodic in the direction of the flow and is one wavelength long. The initial wave will propagate from left to right, from where it will leave the domain to re-enter the left limit. The initial quantities are calculated from the linear solution of Airy. A free slip boundary condition is imposed at the lower and lateral limits, and an open boundary condition at the upper limit.

This initial condition allows the use of a numerical domain of reasonable size, since it is limited to one wavelength, which requires less discretization points. However, this approach has some limitations. Indeed, the wave breaks in a medium of constant depth, on a flat bottom, and without pre-existing turbulence or currents since no prior breaking waves exist. The use of a periodic domain simulates an infinite number of waves breaking at the same time, which does not happen in reality.

These early works were original because many authors treated the problem under conditions of a non-viscous fluid (Longuet-Higgins and Cokelet, 1976; Cokelet, 1979; Peregrine et al., 1980; Vinje and Brevig, 1981; Longuet-Higgins, 1982; New et al., 1985; Skyner, 1996). In addition, many works using Navier-Stokes equations considered ratios of densities and viscosities not taking into account the characteristics of air and water (Abadie, 1998; Abadie, 1998; Chen et al., 1999; Yasuda et al., 1999; Abadie, 2001; Guignard, 2001; Guignard et al., 2001; Biausser and Suivi, 2003). For example, (Abadie, 1998) showed that it was necessary to add an artificial viscosity to eliminate parasitic disturbances observed near the free surface. In other works (Sakai et al., 1986; Takikawa et al., 1997; Lin and Liu, 1998; Lin and Liu, 1998; Bradford, 2000; Iafrati and Mascio, 2001; Watanabe and Saeki, 2002; Iafrati and Campana, 2003), the real air and water physical properties are used.

In addition, very few studies have considered three-dimensional numerical domains (Christensen, 1996; Watanabe and Saeki, 1999; Christensen and Deigaard, 2001; Lubin et al., 2002; Lubin et al., 2003; Watanabe et al., 2005; Christensen,2006).

The first step of the study was to vary the two initial parameters to check the validity ranges of the approach versus different types of breaking. For a set of steepness values, the whole spectrum of breaker types has been simulated (Lubin et al., 2006). What has been called a "weak plunging breaker", or weakly breaking, is actually the limit between plunging waves and spilling waves. A small jet was observed to be ejected from the crest of the wave and impacted at the top of the wave, very close to the crest. The spilling breaking wave usually involves a much more complex process of vorticity generation, due to disturbances appearing on the front side of the wave that steepens (Duncan et al., 1999). However, the whole process of plunging breaking has been precisely simulated, with a particular focus on the generation of splash-up and air entrainment (Lubin et al., 2006; Lubin and Glockner, 2015; Lubin et al., 2019).

4.2 Splash-up and large vortical structures

Many authors (Kiger and Duncan, 2012) have analyzed in detail the generation of the splash-up and the entry of the impacting free-falling jet in the front face of the wave. (Peregrine, 1981) discussed "splashes" in waterfalls and breaking waves and (Peregrine, 1983) then presented three possible cases of splash-up generation.

To discuss this point, (Lubin, 2004; Lubin et al., 2006) have investigated the splash-up generation process. Splash-ups contain water from both the impacting jet and the front face of the wave. In all the simulations performed for this study, it has been found that in the very first moments of the impact, the jet bounces, regardless of the position of the plunge point or the angle between the jet and the front face of the wave. Then a surprising fact has been noticed: the jet does not enter the wave, regardless of the breaking force. The water of the impacting jet separates into two parts, one feeding the upper part of the splash-up, the other wrapping the tube create by the overturning motion of the breaking wave (Figure 1). It has been observed that the jet transmits its momentum by percussion, while the air cavity enveloped by the free-falling jet is entrained and pushed into the water by the weight of the crest of the falling wave. This result was confirmed by experimental Bonmarin (1989) and numerical studies (Abadie et al., 1998; Yasuda et al., 1999; Lubin et al., 2006; Narayanaswamy and Dalrymple, 2002; Dalrymple and Rogers, 2006) which also highlighted this phenomenon of quasi-total reflection of water jet impacting the front face of the wave (Kiger and Duncan, 2012).

The impacting jet pushes the water from the front face of the wave, making the splash up grow in size and rise above the crest of the incident wave. A large amount of vorticity is then generated by rotating the pockets entrained during the process of successive splash-ups generation, as observed during some experiments (Miller, 1976; Bonmarin, 1989). Depending on the configuration of the splash-up, several vortical structures will then be created: the co- and counter-rotating vortices. Their number and their size are important parameters, as the dissipation of energy is related to their behavior during wave breaking.

As detailed previously (Section 2.2) (Zhang and Sunamura, 1990) classified the generation of oblique and horizontal structures. Oblique structures were not observed in the numerical simulations, this being probably due to the use of a periodic domain, which did not allow enough time for this phenomenon to appear. Indeed, as shown by (Nadaoka et al., 1989), horizontal structures are mainly present in the breaking bore, whereas oblique structures are observed only well after the crest of the breaking wave.

Once the splash-up reaches its highest point, the splash-up falls back partly onto the impacting jet, and partly towards the front. generating some other successive splash-ups. The falling back part will be responsible for generating a dipole composed of two structures counter-rotating. This structure carries a large amount of momentum.

5 A new type of vortical structures under breaking waves

As described in the previous sections, depending on the type of breaker, different types of structures are observed under the breaking waves.

Thanks to the initial condition of an unstable sine wave already used previously (Lubin et al., 2006), the generation of a filament vortex could be observed. It can be clearly deduced from Table 3 that the vortex filaments are found only beneath plunging breaking waves.

We have observed that the generation of vortex filaments is an extremely rapid process (Lubin and Glockner, 2013; Lubin and Glockner, 2015). When the water jet collides with the front of the wave, we witness the tumultuous impact of a body of water. At the moment of the impact of the jet, a line of craters is created upstream of the point of impact, from where the splash-up develops. Craters do not penetrate deeply, but are deformed under pressure increasing between the mass of water of the impacting jet and the front face of the wave. The impacting jet splits into two parts, as we have already described (Lubin et al., 2006), creating a line of discontinuity. The craters are then stretched along this line, and rotate.

The rotating structures are then clearly identified as soon as the jet enters the front face of the wave. Vortex filaments are located at the exact boundary between the impacting jet water and the water of the front face of the wave, along the discontinuity line. The structures at the beginning of the surf are disorganized, while during the development phase of the splash-up a more regular organization can be observed with coherent structures.

To identify and visualize the vortex filaments, we use the criterion Q (which is the second invariant of the gradient tensor) which will enable us to analyze the evolution of structures in space over time (Jeong and Hussain, 1995). It can be seen that aerobic vortex structures coincide with the coherent structures identified by the Q criterion. We can see that the vortices are regularly spaced and seem regularly organized. These fine structures look like

Table 3: Vortex filaments observed as a function of the breaker type. WP: weak plunging; PL: plunging breaker.

d/L	H/L	Breaker type	Vortex occurrence
	0.08	PL	Yes
0.10	0.10	PL	Yes
	0.12	PL	Yes
	0.09	PL	Yes
0.13	0.11	PL	Yes
	0.13	PL	Yes
	0.10	WP	No
0.17	0.12	PL	Yes
	0.14	PL	Yes
	0.11	WP	No
0.20	0.13	PL	Yes
	0.15	PL	Yes

ribs, connecting the splash-up and the air tube formed during the overturning motion of the wave. The air entrainment coincides with the structures identified by the Q criterion, while it seems sucked into the vortex core. Figure 2 shows the spiral movement of the water, confirming the presence of these vortex filaments under the breaking waves. It can also be observed that the spiraling structures are areas of low pressure. Once the vortex is formed, the air is entrained and the pressure at the heart of the vortices is weaker than that of the surrounding water. The spiral motion of the filament shell is illustrated in Figure 2, the current lines wrapping around structures. Figure 2 confirms the observation that vortices do not have a preferential direction of rotation.

6   Discussion and future work

Previous works dedicated to the exact same configuration (Lubin et al., 2006), where the vortex filaments have never been observed, used a sequential version of the numerical tool. The simulations were then limited in mesh grid resolution in order to have an affordable computation time. However, the computational time usually took more than several weeks. In order to discuss this point, a mesh grid sensitivity investigation was proposed, considering four mesh grid densities (Lubin and Glockner, 2015). This numerical study confirmed that the vortex filaments could not be simulated using a coarse mesh grid density. As it could be expected, much less detail could be seen for the air entrainment, and no vortex filaments were detected. Then, the finest mesh grid density allowed a better description of the aeration inside the vortex filaments. The structures could also be better identified and aeration could be observed to last longer in the core of the filaments, due to a more accurate free-surface description. One of the most interesting results was to conclude that the coarse grid was sufficient to describe the largest eddies in the flow and account for the correct decay rate of the total energy of the wave (Lubin and Glockner, 2015).

Callaghan et al., 2016 indicated that estimating individual breaking wave energy dissipation in the field remains a fundamental problem. At the same time, it remains a very challenging issue. Thus, laboratory experiments and numerical simulations are still mandatory to fill the gaps in our knowledge.

To date, there are a very limited number of studies on air entrainment induced by breaking waves. Recent studies on air entrainment induced by breaking waves detailed the

Figure 1: Sketch of the colliding jet with the flow separation: downstream the plunge point, the main tube of air is entrapped, while the splash-up is growing upstream. The arrows represent the opposing flows meeting and separating, creating a line of discontinuity. Taken from (Lubin and Glockner, 2015).

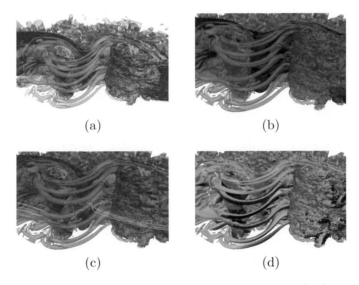

(a)

(b)

(c)

(d)

Figure 2: Evolution of the coherent vortical structures underneath the plunging breaking wave at $t = 0.17$ s (left column) and $t = 0.24$ s (right column), for $H/L = 0.13$ and $d/L = 0.13$ (L1 configuration). Taken from (Lubin and Glockner, 2015). The coherent vortex filaments are educed using the Q-criterion (Hunt et al., 1988). The vortex envelopes are visualised with the positive $Q = 1$ isosurfaces (in green). (a): the free-surface is identified with the isocontour of the phase function (in blue), showing the air entrainment; (b): pressure isocontour ($p = 0.7$ $N.m^{-2}$, with the reference $p = 0$ located at the free-surface) illustrating the low pressure inside the vortex filaments; (c): streamlines showing spiralling flow around and inside the fine elongated vortex filaments; (d): the colour scale on the isosurfaces of $Q = 1$ refers to the local value of the spanwise velocity. Regions associated with positive spanwise velocity are in red, negative in blue.

relation between underwater bubbles generation, turbulence production and energy dissipation. Plunging waves tend to carry large amounts of air at greater depths than other types of breaker type. The dynamics of a bubble under unsteady perturbations are affected by the balance between forces acting on the bubble; the residence time of the bubble in the turbulent flow field also needs to be taken into account (Chanson and Lee, 1997; Chanson et al., 2002). A large number of bubbles are generated and broken in the shear zones, where velocity gradients play a great role in the breakup process. This aspect of the flow remains a technical challenge for both the numerical models and experimental techniques. Our current understanding indicates that some research effort needs to be done to elucidate the physics of the unsteady motion of a breaking roller (Lubin and Chanson, 2017). Practically, there is still a need to evaluate:

- a Froude number (Fr) characteristic of wave breaking

- the roller height

- the roller length

- the roller celerity

- the mean front slope angle of the roller

- the energy dissipation in the roller region

- the bubble size distributions.

The Hinze scale is classically admitted to determine the size of the smallest bubbles affected by turbulence (Hinze, 1949; Hinze, 1955; Hinze, 1975). (Ho-Joon et al., 2007) and (Byoungjoon et al., 2016) studied experimentally the relation between air entrainment, vortical structures and energy dissipation, highlighting the difficulty to perform measurements in highly aerated areas. Using particle image velocimetry (PIV) and bubble image velocimetry (BIV) techniques, they carefully detailed the air entrainment at every stage of splash-ups and vortical structures generation during the breaking event. Void fraction has proved to be an important quantity to consider, as well as bubble size distribution (Callaghan et al., 2016). (Wang et al., 2016) and (Deike et al., 2016) numerically investigated 3D breaking waves and subsequent air entrainment. These works proved the great progress made by numerical tools, but also reported the great costs in terms of computational times required by such simulations. They also showed that a major parameter for accurate numerical simulations is the choice of the mesh size, which will determine the size of the smallest resolved structures of the flow. Usually, the experimental findings indicated the targeted length scales which are to be resolved.

Moreover, the bubble cloud physics has been shown to be also affected by other parameters like surfactant, temperature or salinity (Callaghan et al., 2014; Callaghan et al., 2017; Anguelova and Huq, 2018), which are out of reach for present Navier-Stokes numerical simulations. Consequently, more detailed numerical simulations and experiments are needed to investigate and understand the details of the flow. The metrology is clearly an issue to get accurate and reliable physical measurements in this rapidly-varied aerated turbulent flow. Moreover, in the breaking process even without wind, the role of air is crucial. Recent numerical studies showed that an important part of breaking dissipation and turbulence generation comes from the air entrainment and the presence of vortices above the free surface (Iafrati et al., 2013).

It has been discussed that instabilities (Rayleigh-Taylor and Kelvin-Helmholtz) can be possibly observed at different stages of the wave breaking event (Lubin et al., 2019): plunging jet ejection at the wave crest, plunging jet growth and disintegration in droplets, plunging jet impact and splash-up occurrence. This, in turn, leads to droplets and bubbles generation. Understanding which physical parameters control the instabilities will allow to understand and predict many aspects of wave breaking. Taking surface tension into account in future simulations is thus required to go further into the analysis of plausible instability mechanisms. The challenge is to deduce a physically based model as time advances during a single breaking event. (Deike et al., 2017) indeed showed that modelling air entrainment induced by a single wave breaker (Deike et al., 2016) was a necessary step towards an averaged estimation of the amount of air entrained by breaking waves in the open ocean.

The aim of this chapter was to present the first study of these coherent eddy structures. 3D wave breaking processes, including wind and current, are thus still poorly understood, yet essential in a number of open ocean and nearshore processes. Therefore, the understanding of the complete physical processes induced by breaking waves and the subsequent flow structure is still lacking in its finest details. (Zhou et al., 2017) investigated the interaction between wave-breaking induced turbulent coherent structures and suspended sediment transport, indicating that the understanding of the whole 3D flow structures in the surf zone (including turbulent eddies, aeration, currents and turbulence near the bed) is clearly needed to be able to evaluate the erosion budget for beaches. But studying the characteristics of a propagating breaking roller is still a major challenge (Leng and Chanson, 2019).

References

Abadie, S. (1988). Modélisation numérique du déferlement plongeant par méthode VOF. Ph.D. thesis, Université Bordeaux I.

Abadie, S. (2001). Numerical modelling of the flow generated by plunging breakers. *In*: Proceedings Coastal Dynamics, pp. 202–211.

Abadie, S., Caltagirone, J. -P. and Watremez, P. (1998). Mécanisme de génération du jet secondaire ascendant dans un déferlement plongeant. Comptes Rendus de l'Acadmie des Sciences 326: 553–559.

Anguelova, M. D. and Huq, P. (2018). Effects of salinity on bubble cloud characteristics. Journal of Marine Science and Engineering 6.

Battjes, J. A. (1988). Surf-zone dynamics. Annual Review of Fluid Mechanics 20: 257–293.

Battjes, J. (1974). Surf Similarity, pp. 466–480.

Biausser, B. (2003). Suivi d'interface tridimensionnel de type Volume of Fluid: application au déferlement. Ph.D. thesis, Université de Toulon et du Var.

Bigelow, H. B. and Edmondson, W. T. (1947). Wind waves at sea, breakers and surf. U.S. Govt. Print. Off., 1947.

Bonmarin, P. (1989). Geometric properties of deep-water breaking waves. Journal of Fluid Mechanics 209: 405–433.

Bradford, S. F. (2000). Numerical simulation of surf zone dynamics. Journal of Waterway, Port, Coastal, and Ocean Engineering 126: 1–13.

Brouilliot, D. and Lubin, P. (2013). Numerical simulations of air entrainment in a plunging jet. Journal of Fluids and Structures 43: 428–440.

Byoungjoon, N., Kuang-An, C., Zhi-Cheng, H. and Ho-Joon, L. (2016). Turbulent flow field and air entrainment in laboratory plunging breaking waves. Journal of Geophysical Research: Oceans 121: 2980–3009.

Callaghan, A. H., Deane, G. B. and Stokes, M. D. (2016). Laboratory air-entraining breaking waves: Imaging visible foam signatures to estimate energy dissipation. Geophysical Research Letters 43: 11320–11328.

Callaghan, A. H., Deane, G. B. and Stokes, M. D. (2017). On the imprint of surfactant-driven stabilization of laboratory breaking wave foam with comparison to oceanic whitecaps. Journal of Geophysical Research: Oceans 122: 6110–6128.

Callaghan, A. H., Stokes, M. D. and Deane, G. B. (2014). The effect of water temperature on air entrainment, bubble plumes, and surface foam in a laboratory breaking-wave analog. Journal of Geophysical Research: Oceans 119: 7463–7482.

Chanson, H. and Lee, J. -F. (1997). Plunging jet characteristics of plunging breakers. Coastal Engineering 31: 125–141.

Chanson, H., Aoki, S. and Maruyama, M. (2002). Unsteady air bubble entrainment and detrainment at a plunging breaker: dominant time scales and similarity of water level variations. Coastal Engineering 46: 139–157.

Chen, G., Kharif, C., Zaleski, S. and Li, J. J. (1999). Two-dimensional navier-stokes simulation of breaking waves. Physics of Fluids 11: 121–133.

Christensen, E. D. (1996). Large eddy simulation of breaking waves. Ph.D. thesis, Technical University of Denmark.

Christensen, E. D. (2006). Large eddy simulation of spilling and plunging breakers. Coastal Engineering 53: 463–485.

Christensen, E. D. and Deigaard, R. (2001). Large eddy simulation of breaking waves. Coastal Engineering 42: 53–86.

Christensen, E. D., Walstra, D. -J. and Emerat, N. (2002). Vertical variation of the flow across the surf zone. Coastal Engineering 45: 169–198.

Cokelet, E. D. (1979). Mechanics of wave-induced forces on cylinders, Ed. T.L. Shaw, pp. 287–301.

Craik, A. D. (2004). The origins of water wave theory. Annual Review of Fluid Mechanics 36: 1–28.

Dalrymple, R. A. and Rogers, B. D. (2006). Numerical modeling of water waves with the sph method. Coastal Engineering 53: 141–147.

Deike, L., Lenain, L. and Melville, W. K. (2017). Air entrainment by breaking waves. Geophysical Research Letters 44: 3779–3787.

Deike, L., Melville, W. K. and Popinet, S. (2016). Air entrainment and bubble statistics in breaking waves. Journal of Fluid Mechanics 801: 91129.

Dommermuth, D. G., Yue, D. K. P., Lin, W. M., Rapp, R. J., Chan, E. S. and Melville, W. K. (1988). Deep-water plunging breakers: a comparison between potential theory and experiments. Journal of Fluid Mechanics 189: 423–442.

Duncan, J. H. (2001). Spilling breakers. Annual Review of Fluid Mechanics 33: 519–547.

Duncan, J. H., Qiao, H., Philomin, V. and Wenz, A. (1999). Gentle spilling breakers: crest profile evolution. Journal of Fluid Mechanics 39: 191–222.

Galvin, C. J. (1968). Breaker type classification on three laboratory beaches. Journal of Geophysical Research 73: 3651–3659.

Guignard, S. (2001). Suivi d'interface de type VOF—application au déferlement des ondes de gravité d aux variations bathymétriques. Ph.D. thesis, Université de Toulon et du Var.

Guignard, S., Marcer, R., Rey, V., Kharif, C. and Fraunié, P. (2001). Solitary wave breaking on sloping beaches: 2-d two phase flow numerical simulation by sl-vof method. European Journal of Mechanics-B/Fluids 20: 57–74.

Hinze, J. O. (1949). Critical speeds and sizes of liquid globules. App. Sci. Res. A1: 273–288.

Hinze, J. O. (1955). Fundamentals of the hydrodynamic mechanism of splitting in dispersion processes. A.I.Ch.E.J. 1: 289–295.

Hinze, J. O. (1975). Turbulence Second Edition, McGraw-Hill.

Ho-Joon, L., Kuang-An, C., Zhi-Cheng, H. and Byoungjoon, N. (2015). Experimental study on plunging breaking waves in deep water. Journal of Geophysical Research: Oceans 120: 2007–2049.

Hu, Y., Guo, X., Lu, X., Liu, Y., Dalrymple, R. A. and Shen, L. (2012). Idealized numerical simulation of breaking water wave propagating over a viscous mud layer. Physics of Fluids 24: 112104.

Hunt, J. C. R., Wray, A. A. and Moin, P. Eddies, streams, and convergence zones in turbulent flows. *In*: Proceedings of the Summer Program 1988, Center for Turbulence Research, pp. 193–208.

Iafrati, A. (2009). Numerical study of the effects of the breaking intensity on wave breaking flows. Journal of Fluid Mechanics 622: 371–411.

Iafrati, A. and Campana, E. F. (2003). A domain decomposition approach to compute wave breaking (wave breaking flows). International Journal for Numerical Methods in Fluids 41: 419–445.

Iafrati, A., Babanin, A. and Onorato, M. (2013). Modulational instability, wave breaking, and formation of large-scale dipoles in the atmosphere. Phys. Rev. Lett. 110: 184504.

Iafrati, A., Mascio, A. D. and Campana, E. F. (2001). A level set technique applied to unsteady free surface flows. International Journal for Numerical Methods in Fluids 35: 281–297.

Jeong, J. and Hussain, F. (1995). Coherent structures near the wall in a turbulent channel flow. Journal of Fluid Mechanics 285: 69–94.

Kiger, K. T. and Duncan, J. H. (2012). Air-entrainment mechanisms in plunging jets and breaking waves. Annual Review of Fluid Mechanics 44: 563–596.

Kimmoun, O. and Branger, H. (2007). A piv investigation on laboratory surf-zone breaking waves over a sloping beach. Journal of Fluid Mechanics 588: 353–397.

Kubo, H. and Sunamura, T. (2001). Large-scale turbulence to facilitate sediment motion under spilling breakers. pp. 212–221. *In*: Hanson, H. and Larson, M. (eds.). Proceedings of the ASCE 4th International Conference on Coastal Dynamics.

Lakehal, D. and Liovic, P. (2011). Turbulence structure and interaction with steep breaking waves. Journal of Fluid Mechanics 674: 1–56.

Leng, X. and Chanson, H. (2019). Two-phase flow measurements of an unsteady breaking bore. Experiments in Fluids 60: 42.

Lin, P. and Liu, P. L.-F. (1998). A numerical study of breaking waves in the surf zone. Journal of Fluid Mechanics 359: 239–264.

Lin, P. and Liu, P. L.-F. (1998). Turbulence transport, vorticity dynamics, and solute mixing under plunging breaking waves in surf zone. Journal of Geophysical Research 103: 15677–15694.

Longuet-Higgins, M. S. (1982). Parametric solutions for breaking waves. Journal of Fluid Mechanics 121: 403–424.

Longuet-Higgins, M. S. (1995). On the disintegration of the jet in a plunging breaker. Journal of Physical Oceanography 25: 2458–2462.

Longuet-Higgins, M. S. and Cokelet, E. D. (1976). The deformation of steep surface waves on water a numerical method of computation. Proc. R. Soc. Lond. A 350: 1–26.

Lubin, P. (2004). Large Eddy Simulation of Plunging Breaking Waves. Ph.D. thesis, Universite Bordeaux I, 2004. In English.

Lubin, P. and Glockner, S. (2013). Detailed numerical investigation of the three-dimensional flow structures under breaking waves. *In*: Proceedings of the 7th International Conference on Coastal Dynamics Conference, pp. 1127–1136.

Lubin, P. and Glockner, S. (2015). Numerical simulations of three-dimensional plunging breaking waves: generation and evolution of aerated vortex laments. Journal of Fluid Mechanics 767: 364393.

Lubin, P. and Chanson, H. (2017). Are breaking waves, bores, surges and jumps the same flow? Environmental Fluid Mechanics 17: 47–77.

Lubin, P., Abadie, S., Vincent, S. and Caltagirone, J. -P. (2002). Etude du déferlement par modélisation numérique 2d et 3d. pp. 85–92. *In*: du Littoral, C. F. (ed.). Génie Civil-Génie Cotier 7émes Journées Nationales, volume 1.

Lubin, P., Chanson, H. and Glockner, S. (2010). Large eddy simulation of turbulence generated by a weak breaking tidal bore. Environmental Fluid Mechanics 10: 587–602.

Lubin, P., Glockner, S., Kimmoun, O. and Branger, H. (2011). Numerical study of the hydrodynamics of regular waves breaking over a sloping beach. European Journal of Mechanics-B/Fluids 30: 552–564.

Lubin, P., Kimmoun, O., Véron, F. and Glockner, S. (2019). Discussion on instabilities in breaking waves: Vortices, air-entrainment and droplet generation. European Journal of Mechanics-B/Fluids 73: 144–156.

Lubin, P., Vincent, S., Abadie, S. and Caltagirone, J. -P. (2006). Three-dimensional large eddy simulation of air entrainment under plunging breaking waves. Coastal Engineering 53: 631–655.

Lubin, P., Vincent, S., Caltagirone, J. -P. and Abadie, S. (2003). Fully three-dimensional direct simulation of a plunging breaker. Comptes Rendus de l'Acadmie des Sciences 331: 495–501.

Miller, R. L. (1976). Role of vortices in surf zone predictions: sedimentation and wave forces. Soc. Econ. Paleontol. Mineral. Spec. Publ., R. A. Davis and R. L. Ethington, pp. 92–114.

Mory, M., Abadie, S., Mauriet, S. and Lubin, P. (2011). Run-up flows of collapsing bores over a beach. European Journal of Mechanics-B/Fluids 30: 565–576.

Nadaoka, K. and Kondoh, T. (1982). Laboratory measurements of velocity field structure in the surf zone by ldv. Coastal Engineering in Japan 25: 125–145.

Nadaoka, K., Hino, M. and Koyano, Y. (1989). Structure of the turbulent flow field under breaking waves in the surf zone. Journal of Fluid Mechanics 204: 359–387.

Narayanaswamy, M. and Dalrymple, R. A. (2002). An experimental study of surface instabilities during wave breaking. pp. 344–355. *In*: Smith, J. M. (ed.). Proceedings of the ASCE 28th International Conference on Coastal Engineering, volume 1, World Scientic.

New, A. L., McIver, P. and Peregrine, D. H. (1985). Computations of overturnings waves. Journal of Fluid Mechanics 150: 233–251.

Nicholls, R. J. (2004). Coastal flooding and wetland loss in the 21st century: Changes under the sres climate and socio-economic scenarios. Global Environmental Change 14: 69–86.

Peregrine, D. H. (1981). The fascination of fluid mechanics. Journal of Fluid Mechanics 106: 59–80.

Peregrine, D. H. (1983). Breaking waves on beaches. Annual Review of Fluid Mechanics 15: 149–178.

Peregrine, D. H., Cokelet, E. D. and McIver, P. (1980). The fluid mechanics of waves approaching breaking. *In*: Proc. 17th Int. Conf. Coastal Eng., pp. 512–528.

Prosperetti, A. (1988). Bubble-related ambient noise in the ocean. Journal of the Acoustical Society of America 84: 1042–1054.

Sagaut, P. (1998). Large Eddy Simulation for Incompressible Flows—An introduction, Springer Verlag.

Sakai, T., Mizutani, T. Tanaka, H. and Tada, Y. (1986). Vortex formation in plunging breaker. *In*: Proc. ICCE, pp. 711–723.

Saruwatari, A., Watanabe, Y. and Ingram, D. M. (2009). Scarifying and fingering surfaces of plunging jets. Coastal Engineering 56: 1109–1122.

Scardovelli, R. and Zaleski, S. (1999). Direct numerical simulation of free-surface and interfacial flow. Annual Review of Fluid Mechanics 31: 567–603.

Skyner, D. (1996). A comparison of numerical predictions and experimental measurements of the internal kinematics of a deep-water plunging wave. Journal of Fluid Mechanics 315: 51–64.

Svendsen, I. A. and Putrevu, U. (1996). Surf-zone hydrodynamics, volume 2 of Advances in Coastal and Ocean Eng., World Scientic, pp. 1–78.

Takikawa, K., Yamada, F. and Matsumoto, K. (1997). Internal characteristics and numerical analysis of plunging breaker on a slope. Coastal Engineering 31: 143–161.

Ting, F. C. K. (2006). Large-scale turbulence under a solitary wave. Coastal Engineering 53: 441–462.

Ting, F. C. K. and Kirby, J. T. (1994). Observation of undertow and turbulence in a laboratory surf zone. Coastal Engineering 24: 51–80.

Ting, F. C. K. and Kirby, J. T. (1995). Dynamics of surf-zone turbulence in a strong plunging breaker. Coastal Engineering 24: 177–204.

Ting, F. C. K. and Kirby, J. T. (1996). Dynamics of surf-zone turbulence in a spilling breaker. Coastal Engineering 27: 131–160.

Véron, F. (2015). Ocean spray. Annu. Rev. Fluid Mech. 47: 507– 538.

Vinje, T. and Brevig, P. (1981). Numerical simulation of breaking waves. J. Adv. Water Resour. 4: 77–82.

Wang, Z., Yang, J. and Stern, F. (2016). High-delity simulations of bubble, droplet and spray formation in breaking waves. Journal of Computational Physics 792: 307–327.

Watanabe, Y. and Saeki, H. (1999). Three-dimensional large eddy simulation of breaking waves. Coastal Engineering in Japan 41: 281–301.

Watanabe, Y. and Saeki, H. (2002). Velocity field after wave breaking. International Journal for Numerical Methods in Fluids 39: 607–637.

Watanabe, Y., Saeki, H. and Hosking, R. J. (2005). Three-dimensional vortex structures under breaking waves. Journal of Fluid Mechanics 545: 291–328.

Yasuda, T., Mutsuda, H., Mizutani, N. and Matsuda, H. (1999). Relationships of plunging jet size to kinematics of breaking waves with spray and entrained air bubbles. Coastal Engineering in Japan 41: 269–280.

Youngs, D. L. (1982). Time-dependent multimaterial flow with large fluid distortion, K.W. Morton and M.J. Baines, Numerical Methods for Fluid Dynamics, Academic Press, New York.

Zhang, D. and Sunamura, T. (1990). Conditions for the occurence of vortices induced by breaking waves. Coastal Engineering in Japan 33: 145–155.

Zhang, D. and Sunamura, T. (1994). Multiple bar formation by breaker-induced vortices: a laboratory approach. pp. 2856–2870. *In*: Edge B. L. (ed.). Proceedings of the ASCE 24th International Conference on Coastal Engineering.

Zhou, Z., Hsu, T. -J., Cox, D. and Liu, X. (2017). Large-eddy simulation of wave-breaking induced turbulent coherent structures and suspended sediment transport on a barred beach. Journal of Geophysical Research: Oceans 122: 207–235.

Chapter 4

Air Compressibility and Aeration Effects in Coastal Flows

Zhihua Ma, * *Ling Qian* and *Derek Causon*

1   Introduction

Multiphase flow impact problems are frequently encountered in natural events and industrial applications. Violent water waves may cause severe damage to ships, offshore platforms and coastal defences. Recent examples are the winter storms that occurred in the United Kingdom from December 2013 to February 2014, during which huge waves destroyed railway lines and coastal walls (Kay, 2014; Rodgers and Bryson, 2014). These disasters have resulted in huge economic loss. Scientific investigations from engineering, environmental and other perspectives are a necessity to understand and mitigate these naturally occurring events.

In the past several decades, great efforts have been made to study wave impacts on breakwaters, sea walls and liquid storage tanks, etc. through carefully controlled experiments (Bullock et al., 2007; Lugni et al., 2010), field measurements (Crawford, 1999) and theoretical analysis (Peregrine, 2003; Korobkin, 2006). The extreme impulsive pressures recorded in violent wave impact events can be tens or even hundreds of times those of impacts induced by ordinary non-breaking waves (Lugni et al., 2010). Intentions to classify the impact events into different types can be found in the works of Kirkgoz (1982), Schmidt et al. (1992) and Oumeraci et al. (1993). Here, we follow the ideas of Lugni et al. (2010) to divide these into three modes including (a) impact of an incipient breaking wave, (b) impact of a broken wave with an air pocket and (c) impact of a broken wave with water-air mixing. The second impact mode of an overturning wave enclosing an air cavity is of particular interest in the present study.

Traditionally the influence of trapped air pockets on plunging waves was ignored due to the small density of air compared to water. However, laboratory and field observations disclose that air may play an important role in the impact process (Bullock et al., 2007; Lugni et al., 2010; Peregrine, 2003; Chan et al., 1988; Lugni et al., 2010). During the transition of a plunging wave, an air pocket (or pockets) may be trapped in the body of the wave and compressed by the water mass; thus a portion of the wave energy will be transferred to the pocket. Once the wave front impacts the surface of the structure, the air pocket starts expanding to release the stored energy. The strongest pressure peak in the form of a "cathedral-roof" shape and subsequent pressure oscillations will be experienced by the structure. This distinct phenomenon has been discovered in experiments (Bullock et al., 2007; Chan et al., 1988; Lugni et al., 2010). In addition, the slamming of blunt bodies could

Centre for Mathematical Modelling and Flow Analysis, The Manchester Metropolitan University, Manchester M1 5GD, UK.
* Corresponding author: z.ma@mmu.ac.uk

also trap air into water beneath the lower surface of the structures. The air is extensively compressed and forms a very thin layer as the structure approaches the water surface. It may also expand afterwards when the pressure drops. The trapped air layer may repeat the expansion and contraction cycle leading to pulsating loads on the structure.

The problem becomes even more complicated when air bubbles are present in the water. In the ocean, bubbles are entrained/generated in the water by a number of different processes including biological production, entrapment of air by capillary waves, white capping and wave breaking (Crawford, 1999). These bubbles generally persist for many wave periods in seawater and do not tend to coalesce and hence remain small, rising slowly through the water (Scott et al., 1975). The peak pressure and impact duration are strongly influenced by trapped large air pockets as well as entrained small air bubbles. This might be expected to be closely related to the compressibility of the air and water-air mixture (Bredmose et al., 2009; Peregrine et al., 2005; Plumerault, 2009). Furthermore, negative pressures essentially gauge values below the atmospheric pressure have been recorded in field measurements (Crawford, 1999) and laboratory experiments (Bullock et al., 2007; Lugni et al., 2010; Oumeraci et al., 1993). Bullock et al. stated that negative pressures have the potential to induce large seaward forces resulting in the removal of blocks from masonry structures (Bullock et al., 2007). Medina-Lopez et al. suggested that cavitation may have made a significant contribution to the failure of the Mutriku Breakwater Wave Plant, particularly causing localised damages at joints between the staked cell units (Medina-Lopez, 2015). Crawford pointed out that such large forces could produce a sufficient overturning moment to cause overall failure of important structures like breakwaters (Crawford, 1999). Additionally, Lugni et al. indicated that even small pressure fluctuations might induce local flow cavitation for conditions close to the cavitation threshold (Lugni et al., 2010). Therefore, it is necessary to be aware of these issues and the need to include all relevant physics when theoretical analysis, experiments or numerical computations are used to investigate wave impact problems.

Compared to experimental investigations of plunging wave impacts, which have made significant progress regarding measurement of peak pressures and forces (Bullock et al., 2007, Lugni et al., 2010; Chan et al., 1988; Lugni et al., 2010), numerical simulations are not yet adequate to fulfil industrial and academic requirements due to the extreme complexity of these problems. Not to mention the challenge of resolving the free surface, which might overturn, break and experience further strong deformations, the compressibility of air and the water-air mixture and possible flow cavitation make the problem much more difficult. Earlier efforts for developing 3D numerical wave tanks (NWTs) were mostly based on single-fluid incompressible potential flow theory (De Divitiis and De Socio, 2002; Riccardi and Iafrati, 2004; Semenov and Iafrati, 2006; Xu et al., 2010; Zhang et al., 1996). Since air is not explicitly considered in the mathematical model, computation of entrapped air pockets and/or entrained air bubbles in waves cannot directly be achieved with these single-fluid NWTs. Two-fluid NWTs based on the incompressible Navier-Stokes equations have been proposed to simulate both the liquid and gas phases for violent wave breaking problems (Christensen and Deigaard, 2001; Greaves, 2006; Greaves, 2004; Lubin et al., 2006; Qian et al., 2006; Ma et al., 2011). However, these treat both the water and air as incompressible fluids, which means the density of each fluid remains constant throughout the process. Unfortunately, compressibility effects in the air pocket and water-air mixture cannot be handled properly by these models, nor, importantly, cavitation effects.

More recently, researchers have started to explore the importance of the compressibility of air and the water-air mixture for wave impact problems. Peregrine et al. (2005) and Bredmose et al. (2009) proposed a weakly compressible flow model combined with a single-phase potential flow solver to compute wave impact events. A fully conservative flow model based on the compressible Euler equations was adopted in the impact zone to describe the water wave with entrapped air pocket or wave with entrained air bubbles. In the energy equation, only the compressibility of the gas phase was included without considering compressibility in the liquid phase. Systematic numerical analysis of wave impacts on vertical walls were conducted and promising results were presented. However, model tests of a one-dimensional shock wave passing through a water-air interface exhibited strong nonphysical pressure oscillations at the material interface. This Gibbs phenomenon is a well known numerical artefact when using a conservative variable scheme and measures should be taken to preclude them. Plumerault et al. (2012) also carried out studies of aerated-water wave problems. They described the flow with a three-fluid model for gas, liquid and gas-liquid components in the flow. Energy conservation was not enforced explicitly with the corresponding equation removed from the equation set in their mathematical model. Instead, a pressure relaxation method (Saurel and Abgrall, 1999) was employed to solve the three-fluid model. Strong non-physical oscillations also arose at the material interface in their water-air shock tube results (Plumerault et al., 2012). They presented results for a deep water breaking Stokes wave in the incompressible limit. We observe that these works still have deficiencies in treating the material interface between water and air. In addition, it is unclear whether they have the capability explicitly to deal with local flow cavitation, as pointed out by Lugni et al. (2010).

The compressibility of air is also a crucial factor influencing the performance and resonance characteristics of coastal oscillating water columns (OWCs) even for non-violent flows (Dimakopoulos et al., 2017). Due to scaling effects, data obtained laboratory experiments cannot be properly extrapolated to full scale prototypes, where significant discrepancy in phase between air pressure and hydrodynamic pressure could change the peak load and performance of OWCs (Weber et al., 2007).

It is of great importance to develop an appropriate wave modelling tool, which could deal with trapped air pockets and a water-air mixture and to produce physical solutions in both high and low pressure regions fully accounting for compressibility and cavitation effects. This chapter summarises the recently developed compressible multiphase hydrocode AMAZON-CW. The compressible wave model is presented in Section 2. The numerical method, which utilises a finite volume approach, a third-order MUSCL reconstruction and the HLLC approximate Riemann solver to compute the convective fluxes in the governing equations, is presented in Section 3. Numerical results for benchmark test cases are presented in Section 4. Final conclusions are drawn in Section 5.

2 Flow model for dispersed water waves

In hydrodynamics, ordinary (non-breaking) water waves are usually considered as separated-phase flows, in which the free-surface separates the gas- and liquid-components completely. Neither the water sprays ejected from the wave front nor the air bubbles entrained into the water are considered in the separated flow model. However for breaking waves as mentioned in the above section, the water mass may trap large air pockets and/or small air bubbles when the crest starts to curl. At the same time, water droplets may also be ejected into the air as illustrated in Figure 1. In this case, the flow is of dispersed gas- and liquid-phases. In order to construct the mathematical model for wave impact problems, we adopt the homogeneous equilibrium approach to make the following assumptions:

1. The bubbly fluid is assumed to be a homogeneous mixture of air and water.
2. Each component obeys the conservation laws of mass, momentum and energy.
3. The mixture obeys the conservation laws of mass, momentum and energy.

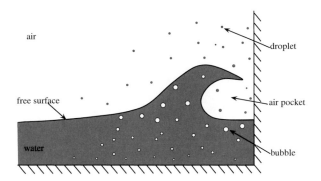

Figure 1: The inherent multi-phase nature of an overturning wave.

4. The pressure and velocity of all the phases and components are identical within a small fluid particle.
5. The fluid undergoes an adiabatic thermodynamic process during the compression and expansion of the trapped air pockets.

These assumptions are based on the belief that differences in the thermodynamic and mechanical variables will promote momentum, energy and mass transfer between the phases rapidly enough so that equilibrium is reached (Corradini, 1997; Bernard-Champmartin, 2014). The equilibrium model is usually considered an appropriate approach to treat free-surface flows (Saurel and Lemetayer, 2001).

It is important to pay attention to the thermodynamics of air enclosed wave impact events. Simply assuming the fluids (especially the gas phase) to be isothermal could be problematic, because an isothermal process usually should occur slowly enough to allow the system to continually adjust its temperature through heat exchange with the surrounding environment or an outside thermal reservoir. However, a wave impact event usually lasts for only a few seconds or less. Within such a short time, it is difficult for the whole system to adjust its temperature to reach a uniform and constant state. The recent work of Abrahamsen and Faltinsen (2011) reveals that the gas phase undergoes an adiabatic process during air enclosed plunging wave impacts.

2.1 Mathematical model

For each individual fluid component i ($i = 1$ for air, $i = 2$ for water), its basic material properties can be described as follows

1. Density $\rho_i = m_i/\Omega_i$, where m_i is mass and Ω_i volume.
2. Pressure $p = p_1 = p_2$, both components have the same pressure.
3. Velocity $\vec{V} = \vec{V}_1 = \vec{V}_2$, both components have the same velocity.
4. Internal energy $\rho_i e_i^{\mathrm{I}}$.
5. Kinetic energy $\rho_i e_i^{\mathrm{K}} = \frac{1}{2}\rho_i \vec{V}^2$.
6. Total energy $\rho e_i^{\mathrm{T}} = \rho_i e_i^{\mathrm{I}} + \rho_i e_i^{\mathrm{K}}$.

To determine the internal energy of fluid component i, an appropriate equation of state should be adopted. Since the ideal gas equation of state is not suitable for liquid, the stiffened-gas equation of state is utilised for both fluid components in the present study.

Therefore, the internal energy of component i can be described as

$$\rho_i e_i^{\text{I}} = \frac{p + \gamma_i p_{\text{c},i}}{\gamma_i - 1} = \Gamma_i p + \Pi_i \tag{1}$$

and the total energy is

$$\rho_i e_i^{\text{T}} = \frac{p + \gamma_i p_{\text{c},i}}{\gamma_i - 1} + \frac{1}{2}\rho_i \vec{V}^2 = \Gamma_i p + \Pi_i + \frac{1}{2}\rho_i \vec{V}^2 \tag{2}$$

where γ_i is a polytropic constant and $p_{\text{c},i}$ is a pressure constant. The parameters Γ_i and Π_i are defined as

$$\Gamma_i = \frac{1}{\gamma_i - 1}, \quad \Pi_i = \frac{\gamma_i p_{\text{c},i}}{\gamma_i - 1} \tag{3}$$

Additionally, the speed of sound for each component can be calculated as

$$c_i = \sqrt{\frac{\gamma_i}{\rho_i}(p + p_{\text{c},i})} \tag{4}$$

The formulae (1), (2) and (4) given to calculate the internal/total energy and speed of sound comply with the adiabatic assumption.

In order to describe the material properties of the homogeneous mixture, we introduce the volume fraction function α_1 for air, defined as

$$\alpha_1 = \frac{\Omega_1}{\Omega_1 + \Omega_2} \tag{5}$$

Accordingly, we have $\alpha_2 = 1 - \alpha_1$ for water. Based on these values, the material properties of the water-air mixture can be expressed by the following

1. Density $\rho = \sum_{i=1}^{N} \alpha_i \rho_i$
2. Momentum $\rho\vec{V} = \sum_{i=1}^{N} \alpha_i(\rho_i\vec{V})$
3. Kinematic energy $\rho e^{\text{K}} = \sum_{i=1}^{N} \alpha_i(\rho_i e_i^{\text{K}})$
4. Internal energy $\rho e^{\text{I}} = \sum_{i=1}^{N} \alpha_i(\rho_i e_i^{\text{I}})$
5. Total energy $\rho e^{\text{T}} = \rho e^{\text{I}} + \rho e^{\text{K}} = \sum_{i=1}^{N} \alpha_i(\rho_i e_i^{\text{T}})$

in which $N = 2$. Substituting Equation (2) into the formulation of mixture total energy, the pressure can be computed as

$$p = \frac{\rho e^{\text{T}} - \frac{1}{2}\rho\vec{V}^2 - \sum_{i=1}^{N}\alpha_i\Pi_i}{\sum_{i=1}^{N}\alpha_i\Gamma_i} \tag{6}$$

The speed of sound for the bubbly water-air mixture can be estimated by Wood's formula (Wood, 1941)

$$\frac{1}{\rho c^2} = \sum_{i=1}^{N} \frac{\alpha_i}{\rho_i c_i^2} \tag{7}$$

The mathematical model used here for the flow of the water-air mixture consists of the mass, momentum and energy conservation laws for the mixture. A conservation law of mass for each component is also included. In particular, gravitational effects should be considered and included for water wave problems. Consequently, the underlying conservative part of the flow model can be expressed in the following form

$$\frac{\partial \tilde{\mathbf{U}}}{\partial t} + \frac{\partial \tilde{\mathbf{F}}_1}{\partial x} + \frac{\partial \tilde{\mathbf{F}}_2}{\partial y} + \frac{\partial \tilde{\mathbf{F}}_3}{\partial z} = \tilde{\mathbf{G}} \tag{8}$$

in which $\tilde{\mathbf{U}}$ is the vector of conservative variables, $\tilde{\mathbf{F}}$ is the flux function, $\tilde{\mathbf{G}}$ are the source terms and these are defined as

$$
\tilde{\mathbf{U}} = \begin{bmatrix} \alpha_1\rho_1 \\ \alpha_2\rho_2 \\ \rho u \\ \rho v \\ \rho w \\ \rho e^{\mathrm{T}} \end{bmatrix}, \tilde{\mathbf{F}}_1 = \begin{bmatrix} \alpha_1\rho_1 u \\ \alpha_2\rho_2 u \\ \rho u^2 + p \\ \rho vu \\ \rho wu \\ \rho hu \end{bmatrix}, \tilde{\mathbf{F}}_2 = \begin{bmatrix} \alpha_1\rho_1 v \\ \alpha_2\rho_2 v \\ \rho uv \\ \rho v^2 + p \\ \rho wv \\ \rho hv \end{bmatrix}, \tilde{\mathbf{F}}_3 = \begin{bmatrix} \alpha_1\rho_1 w \\ \alpha_2\rho_2 w \\ \rho uw \\ \rho vw \\ \rho w^2 + p \\ \rho hw \end{bmatrix}, \tilde{\mathbf{G}} = \begin{bmatrix} 0 \\ 0 \\ 0 \\ -\rho g \\ 0 \\ -\rho g v \end{bmatrix} \tag{9}
$$

where u, v and w are the velocity components along x, y and z axes; g is the gravitational acceleration; h is the enthalpy given by

$$
h = (\rho e^{\mathrm{T}} + p)/\rho \tag{10}
$$

In addition to the conservative part, the advection of volume fraction function $\mathrm{D}\alpha_1/\mathrm{D}t$ also needs to be considered. Here, we adopt Kapila et al.'s one-dimensional advection equation (Kapila et al., 2001)

$$
\frac{\partial \alpha_1}{\partial t} + u\frac{\partial \alpha_1}{\partial x} = K\frac{\partial u}{\partial x} \tag{11}
$$

extended to three dimensions

$$
\frac{\partial \alpha_1}{\partial t} + u\frac{\partial \alpha_1}{\partial x} + v\frac{\partial \alpha_1}{\partial y} + w\frac{\partial \alpha_1}{\partial z} = K\left(\frac{\partial u}{\partial x} + \frac{\partial v}{\partial x} + \frac{\partial w}{\partial z}\right) \tag{12}
$$

where K is a function of the volume fraction and sound speed given by

$$
K = \alpha_1\alpha_2\left(\frac{1}{\rho_1 c_1^2} - \frac{1}{\rho_2 c_2^2}\right)\rho c^2 \tag{13}
$$

Equation (11) is derived from the pressure equilibrium assumption, and its right hand side term assures that the material derivatives of the phase entropy are zero in the absence of shock waves. If we neglect the right hand side of Equation (11), then this is a standard transport equation for α_1 as pointed out by Murrone and Guillard (2005).

The overall flow model includes Equations (8) and (13) and we write it in the following form

$$
\frac{\partial \alpha_1}{\partial t} + \vec{V}\cdot\nabla\alpha_1 = K\nabla\cdot\vec{V} \tag{14a}
$$

$$
\frac{\partial \tilde{\mathbf{U}}}{\partial t} + \frac{\partial \tilde{\mathbf{F}}_1}{\partial x} + \frac{\partial \tilde{\mathbf{F}}_2}{\partial y} + \frac{\partial \tilde{\mathbf{F}}_3}{\partial z} = \tilde{\mathbf{G}} \tag{14b}
$$

The underlying reason we do not choose a fully conservative or primitive model but a quasi-conservative model is due to the following factors. Fully conservative flow models have the aforementioned difficulties at material interfaces where nonphysical oscillations inevitably occur even for first order schemes (Saurel and Abgrall, 1999) due to a nonphysical pressure update (Ivings et al., 1998; Johnsen and Colonius, 2006) or negative volume fraction (Abgrall and Karni, 2001) during numerical computations. Although a primitive variable flow model can avoid these oscillations, difficulties arise when resolving strong shock waves to maintain the correct shock speed (Ivings et al., 1998). For complicated problems consisting of both material interfaces and shock waves, combining the fully conservative and primitive variable model formulations has previously been found to be an effective strategy (Ivings et al., 1997). However this method is quite intricate as switching is required between the two models for the different regions (Ivings et al., 1998). Quasi-conservative flow models, which combine the conservation laws with a non-conservative scalar (volume fraction or other material property) advection equation, have proved proficient and much simpler in the past (Saurel and Abgrall, 1999; Johnsen and Colonius, 2006).

If gravitational effects are excluded and only the x direction is considered, Equation (14) will reduce to a five-equation system which can be named the five-equation reduced model Murrone and Guillard (2005) or Kapila et al. model (Saurel et al., 2009). For the one-dimensional five-equation reduced model, Murrone and Guillard proved that it can be derived from the two-pressure and two-velocity Baer-Nunziato equations in the limit of zero relaxation time (Murrone and Guillard, 2005). They also proved that the reduced system has five real eigenvalues (three eigenvalues are equal) and is strictly hyperbolic with five linearly independent eigenvectors. Important information about other mathematical properties of the one-dimensional system, which include the structure of the waves, expressions for the Riemann invariants and the existence of a mathematical entropy, can also be found in their work. The three-dimensional two-pressure two-velocity Baer-Nunziato model has eleven equations in total (Tokareva and Toro, 2010). The model (14) is computationally less expensive as it deals with only seven equations.

3 Numerical Method

3.1 Treatment of the advection equation

System (14) has a volume fraction transport equation which is not conservative. As indicated by Johnsen and Colonius (2006), directly using the non-conservative formula will present difficulties when dealing with the material interface due to an inconsistency between the wave speeds (shock wave, rarefaction and contact discontinuity) and the velocity vector $\vec{V}$. They advise transforming the advection equation into a conservative formulation to overcome this obstacle for one-dimensional multi-fluid problems. Based on their work, here, we move forward to re-construct the volume fraction equation appropriately for three-dimensional multiphase flow problems.

Introducing a vector $\vec{f}$ defined as

$$\vec{f} = \alpha_1 \vec{V} = [\alpha_1 u, \alpha_1 v, \alpha_1 w]^{\mathrm{T}} \tag{15}$$

then the divergence of $\vec{f}$ is given by

$$\nabla \cdot \vec{f} = \frac{\partial}{\partial x}(\alpha_1 u) + \frac{\partial}{\partial y}(\alpha_1 v) + \frac{\partial}{\partial z}(\alpha_1 w) \tag{16a}$$

$$= \alpha_1 \left(\frac{\partial u}{\partial x} + \frac{\partial v}{\partial y} + \frac{\partial w}{\partial z} \right) + \left(u \frac{\partial \alpha_1}{\partial x} + v \frac{\partial \alpha_1}{\partial y} + w \frac{\partial \alpha_1}{\partial z} \right) \tag{16b}$$

$$= \alpha_1 \nabla \cdot \vec{V} + \vec{V} \cdot \nabla \alpha_1 \tag{16c}$$

Consequently, we can obtain

$$\vec{V} \cdot \nabla \alpha_1 = \nabla \cdot \vec{f} - \alpha_1 \nabla \cdot \vec{V} \tag{17}$$

Substituting Equation (17) into Equation (12) to obtain the quasi-conservative form

$$\frac{\partial \alpha_1}{\partial t} + \nabla \cdot \vec{f} = (\alpha_1 + K)\nabla \cdot \vec{V} \tag{18}$$

then replacing the non-conservative equation in system (14) by Equation (18), the integral formulation of the overall system can now be expressed as

$$\frac{\partial}{\partial t} \int_{\Omega} \mathbf{U} \, \mathrm{d}\Omega + \oint_{\partial\Omega} \mathbf{F}(\mathbf{U}) \, \mathrm{d}S = \int_{\Omega} \mathbf{G} \, \mathrm{d}\Omega \qquad (19)$$

in which Ω represents the flow domain, $\partial\Omega$ is its boundary; the vectors $\mathbf{U}$, $\mathbf{F}$ and $\mathbf{G}$ are given by

$$\mathbf{U} = \begin{bmatrix} \alpha_1 \\ \alpha_1\rho_1 \\ \alpha_2\rho_2 \\ \rho u \\ \rho v \\ \rho w \\ \rho e^{\mathrm{T}} \end{bmatrix}, \quad \mathbf{F} = \begin{bmatrix} \alpha_1 q \\ \alpha_1\rho_1 q \\ \alpha_2\rho_2 q \\ \rho u q + p n_x \\ \rho v q + p n_y \\ \rho w q + p n_z \\ \rho e^{\mathrm{T}} q + p q \end{bmatrix} = q\,\mathbf{U} + p\,\mathbf{N}_q, \quad \mathbf{G} = \begin{bmatrix} (\alpha_1 + K)\nabla \cdot \vec{V} \\ 0 \\ 0 \\ 0 \\ -\rho g \\ 0 \\ -\rho g v \end{bmatrix} \qquad (20)$$

where (n_x, n_y, n_z) is a unit normal vector across the boundary $\partial\Omega$, $q = un_x + vn_y + wn_z$ and $\mathbf{N}_q = [0, 0, 0, n_x, n_y, n_z, q]^{\mathrm{T}}$.

3.2   Spatial discretisation

For three-dimensional problems, hexahedrons are used to fill the domain Ω. Without loss of generality, if Ω is a cuboid, we may simply partition it using a Cartesian mesh with $I \times J \times K$ cells. Within the mesh, any cell can be indicated by the subscripts i, j and k. As shown in Figure 2, a hexahedral volume cell is closed by six quadrilateral faces. For this cell, six neighbouring cells are selected to form a computational stencil.

Supposing the mesh does not vary with time, the discrete form of Equation (19) for a mesh cell $m_{i,j,k}$ can be written as

$$\Omega_m \frac{\mathbf{U}_m^{n+1} - \mathbf{U}_m^n}{\Delta t} + \sum_{f=1}^{6} \mathbf{F}_f A_f = \Omega_m\,\mathbf{G}_m \qquad (21)$$

where the subscript m stands for the mesh cell itself, the subscript f represents a mesh cell face and A is the area of the face. Introducing the residual defined as

$$\mathbf{R}_m = -\Omega_m\,\mathbf{G}_m + \sum_{f=1}^{6} \mathbf{F}_f A_f \qquad (22)$$

Equation (21) can be written as

$$\Omega_m \frac{\mathbf{U}_m^{n+1} - \mathbf{U}_m^n}{\Delta t} + \mathbf{R}_m = 0 \qquad (23)$$

As flow discontinuities (shock waves and material interfaces, etc.) may be expected to occur in multi-phase multi-component compressible flows, attention should be paid to the computation of surface flux terms. In the present work, the surface flux term $\mathbf{F}$ across any mesh cell face f is evaluated by an approximate Riemann solution based on the numerical flux function

$$\mathbf{F}_f = \mathbf{F}(\mathbf{U}_f^+, \mathbf{U}_f^-) \qquad (24)$$

where the symbols $^+$ and $^-$ indicate the left and right sides of the face considering a normal vector n_f. The conservative variables at neighbouring cell centres may be assumed piecewise

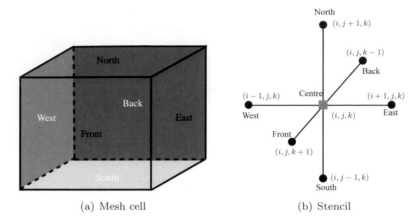

(a) Mesh cell   (b) Stencil

Figure 2: A mesh cell and its computational stencil.

constant and assigned directly to $\mathbf{U}_f^+$ and $\mathbf{U}_f^-$ respectively, and this very simple but diffusive treatment is first order accurate in space. To improve the accuracy, the solution data is formally reconstructed using a third order MUSCL scheme (monotone upstream-centred schemes for conservation laws Bram van Leer, 1979) based on the primitive variables $\mathbf{W} = (\alpha_1, \rho_1, \rho_2, u, v, w, p)^{\mathrm{T}}$, written as

$$\mathbf{W}_f^+ = \mathbf{W}_{\mathrm{L}} + \frac{1}{4}\left[(1 - \kappa)\,\mathbf{\Delta}_{\mathrm{L}} + (1 + \kappa)\,\mathbf{\Delta}\right] \tag{25a}$$

$$\mathbf{W}_f^- = \mathbf{W}_{\mathrm{R}} + \frac{1}{4}\left[(1 - \kappa)\,\mathbf{\Delta}_{\mathrm{R}} + (1 + \kappa)\,\mathbf{\Delta}\right] \tag{25b}$$

where the subscripts L and R represent the left and right neighbouring mesh cells respectively,

$$\begin{cases} \mathbf{\Delta}_{\mathrm{L}} &= \mathbf{W}_{\mathrm{L}} - \mathbf{W}_{\mathrm{LL}} \\ \mathbf{\Delta}_{\mathrm{R}} &= \mathbf{W}_{\mathrm{RR}} - \mathbf{W}_{\mathrm{R}} \\ \mathbf{\Delta} &= \mathbf{W}_{\mathrm{R}} - \mathbf{W}_{\mathrm{L}} \end{cases} \tag{26}$$

in which the subscript LL indicates the left neighbour of the left cell, and RR represents the right neighbour of the right cell. The parameter κ provides different options of upwind or centred schemes

$$\kappa = \begin{cases} -1 & \text{second order upwind} \\ 0 & \text{second order centred} \\ 1/3 & \text{third order semi-upwind} \end{cases} \tag{27}$$

In the present work, $\kappa = 1/3$ is adopted. To prohibit spurious oscillations introduced by the high order interpolation near discontinuities, a slope limiter function ϕ is utilised to modify the reconstruction procedure as follows

$$\mathbf{W}_f^+ = \mathbf{W}_{\mathrm{L}} + \frac{\phi_{\mathrm{L}}}{4}\left[(1 - \kappa\phi_{\mathrm{L}})\,\mathbf{\Delta}_{\mathrm{L}} + (1 + \kappa\phi_{\mathrm{L}})\,\mathbf{\Delta}\right] \tag{28a}$$

$$\mathbf{W}_f^- = \mathbf{W}_{\mathrm{R}} + \frac{\phi_{\mathrm{R}}}{4}\left[(1 - \kappa\phi_{\mathrm{R}})\,\mathbf{\Delta}_{\mathrm{R}} + (1 + \kappa\phi_{\mathrm{R}})\,\mathbf{\Delta}\right] \tag{28b}$$

In this study, we apply van Albada's limiter (van Albada, 1997) defined as

$$\phi_{\mathrm{L,R}} = \max\left(0, \frac{2r_{\mathrm{L,R}}}{r_{\mathrm{L,R}}^2 + 1}\right) \tag{29}$$

where

$$r_{\rm L} = \frac{\boldsymbol{\Delta}_{\rm L}}{\boldsymbol{\Delta}}, \quad r_{\rm R} = \frac{\boldsymbol{\Delta}}{\boldsymbol{\Delta}_{\rm R}} \tag{30}$$

In smooth regions, the slope limiter doesn't operate on spatial derivatives and the MUSCL scheme provides high resolution. The slope limiter operates in critical regions of flow discontinuities to prevent extrema and ensure stability and monotonicity of the solution (van Albada, 1997). The reconstructed conservative variables at the left and right sides of a mesh cell face can be easily obtained as

$$\mathbf{U}_f^+ = \mathbf{U}(\mathbf{W}_f^+), \quad \mathbf{U}_f^- = \mathbf{U}(\mathbf{W}_f^-) \tag{31}$$

The numerical flux term represented by Equation (24) is calculated by employing the HLLC approximate Riemann solver (ARS) for the homogeneous mixture.

3.3 The HLLC Riemann solver

The HLLC ARS originates from the work of Harten et al. (1983) and Toro et al. (1992). Its applications for separated multiphase multicomponent compressible flows can be found in the work of Hu et al. (2009) and Johnsen and Colonius (2006), etc. However, to the authors' knowledge, no work has been reported thus far that solves dispersed multi-phase multi-component compressible flows governed by the one-dimensional Kapila model or three-dimensional system (14) with the HLLC ARS. The numerical flux term represented by Equation (24) is calculated by the HLLC ARS defined as

$$\mathbf{F}_f = \begin{cases} \mathbf{F}_{\rm L} & 0 \le S_{\rm L} \\ \mathbf{F}_{\rm L} + S_{\rm L}^*(\mathbf{U}_{\rm L}^* - \mathbf{U}_{\rm L}) & S_{\rm L} < 0 \le S_{\rm M} \\ \mathbf{F}_{\rm R} + S_{\rm R}^*(\mathbf{U}_{\rm R}^* - \mathbf{U}_{\rm R}) & S_{\rm M} < 0 \le S_{\rm R} \\ \mathbf{F}_{\rm R} & 0 \ge S_{\rm R} \end{cases} \tag{32}$$

in which the middle left and right states are evaluated by

$$\mathbf{U}_{\rm L,R}^* = \frac{(q_{\rm L,R} - S_{\rm L,R})\mathbf{U}_{\rm L,R} + (p_{\rm L,R}\mathbf{N}_{q_{\rm L,R}} - p_{\rm L,R}^*\mathbf{N}_{q_{\rm L,R}^*})}{q_{\rm L,R}^* - S_{\rm L,R}} \tag{33}$$

the intermediate wave speed can be calculated by

$$S_{\rm M} = q_{\rm L}^* = q_{\rm R}^* = \frac{\rho_{\rm R}q_{\rm R}(S_{\rm R} - q_{\rm R}) - \rho_{\rm L}q_{\rm L}(S_{\rm L} - q_{\rm L}) + p_{\rm L} - p_{\rm R}}{\rho_{\rm r}(S_{\rm R} - q_{\rm R}) - \rho_{\rm L}(S_{\rm L} - q_{\rm L})} \tag{34}$$

and the intermediate pressure may be estimated as

$$p_{\rm M} = \rho_{\rm L}(q_{\rm L} - S_{\rm L})(q_{\rm L} - S_{\rm M}) + p_{\rm L} = \rho_{\rm R}(q_{\rm R} - S_{\rm R})(q_{\rm R} - S_{\rm M}) + p_{\rm R} \tag{35}$$

The left and right state wave speeds are computed by

$$S_{\rm L} = \min(q_{\rm L} - c_{\rm L}, \tilde{q} - \tilde{c}) \tag{36a}$$
$$S_{\rm R} = \max(q_{\rm R} + c_{\rm R}, \tilde{q} + \tilde{c}) \tag{36b}$$

where $\tilde{q}$ is an averaged velocity component evaluated by the component Roe-averages (Batten et al., 1997). The averaged speed of sound $\tilde{c}$ is calculated by the formulation proposed by Hu et al. (2009)

$$\tilde{c}^2 = \widetilde{\Psi} + \widetilde{\Gamma}\left(\widetilde{\frac{p}{\rho}}\right) \tag{37}$$

More detail of Equation (37) can be found in reference (Hu et al., 2009).

3.4 Temporal discretisation

In the present work, a third order total variation diminishing Runge-Kutta scheme (Gottlieb and Shu, 1998) is applied to update the numerical solution from time level n to $n+1$

$$\mathbf{U}^{(1)} = \mathbf{U}^n + \Delta t \mathbf{L}(\mathbf{U}^n) \tag{38a}$$

$$\mathbf{U}^{(2)} = \frac{3}{4}\mathbf{U}^n + \frac{1}{4}\mathbf{U}^{(1)} + \frac{1}{4}\Delta t \mathbf{L}(\mathbf{U}^{(1)}) \tag{38b}$$

$$\mathbf{U}^{n+1} = \frac{1}{3}\mathbf{U}^n + \frac{2}{3}\mathbf{U}^{(1)} + \frac{2}{3}\Delta t \mathbf{L}(\mathbf{U}^{(2)}) \tag{38c}$$

For a mesh cell m, the partial differential operator $\mathbf{L}$ is defined as

$$\mathbf{L}(\mathbf{U}_m) = -\frac{\mathbf{R}(\mathbf{U}_m)}{\Omega_m} \tag{39}$$

4 Results

The AMAZON-CW code has been verified through a series of benchmark test cases including 1D gravity-induced liquid piston, water-air shock tube, free drop of a water column in a closed 2D tank, slamming of a 2D flat plate into pure water and plunging wave impact at a vertical wall (Ma et al., 2014). It has also been applied to study intensive underwater explosions in close proximity to structures (Ma et al., 2015) and to investigate aeration effects on the slamming of blunt bodies (Ma et al., 2016). A few representative benchmark test cases are selected here to demonstrate the hydro-code's capability properly to handle the compressibility of air and water-air mixture as well as liquid cavitation in both and high pressure regimes.

4.1 1D problems

In order to verify the capability of the developed code in dealing with fluid compressibility and cavitation, we firstly exam it with 1D water-air shock tube and water cavitation problems.

4.1.1 Water-air shock tubes

Here, we consider two shock tube problems involving water-air mixtures. The shock tube initially has an imaginary membrane in the middle of the tube, which separates it into left and right parts filled with fluids at different thermodynamic states as shown in Figure 3. The initial conditions, stopping times and number of mesh cells used for the two shock tubes are listed in Table 1 (cases 1 and 2). Gravitational effects are not included for these problems.

Computed results for these two cases are shown in Figure 4 and 5. Computations of water-air shock tube problems are known to be challenging for numerical methods, particularly around the material interface where spurious nonphysical oscillations may occur. A

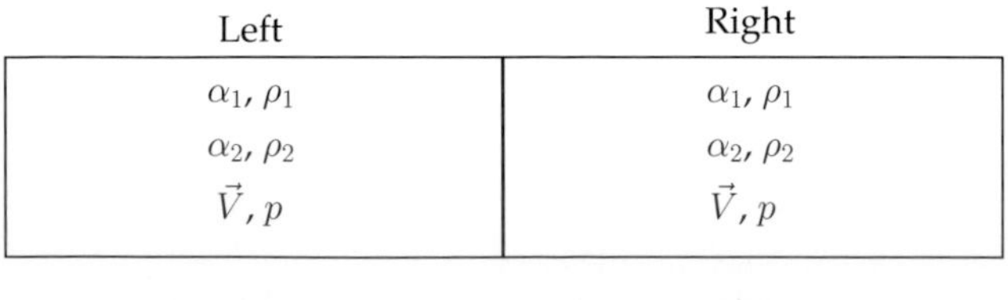

Left	Right
α_1, ρ_1	α_1, ρ_1
α_2, ρ_2	α_2, ρ_2
$\vec{V}, p$	$\vec{V}, p$

Figure 3: Setup of the water-air shock tube.

Table 1: Initial conditions for water-air shock tubes (cases 1 and 2) and 1D cavitation test (case 3). ρ: density (units in kg/m^3); γ: polytropic constant; P_c: pressure constant (units in MPa); p: pressure (units in bar); u: velocity (units in m/s); T: stop time (units in μs); N: number of mesh cells.

Case	T	N	$\alpha_1(\%)$	ρ_1	γ_1	P_{c1}	ρ_2	γ_2	P_{c2}	u	p	Side
1	200	1000	0.5	50	1.4	0	1000	4.4	600	0	10000	Left
			0.5	50	1.4	0	1000	4.4	600	0	1	Right
2	551.37	800	0.00195	6.908	1.4	0	1027	4.4	600	0	10	Left
			0.01	1.33	1.4	0	1027	4.4	600	0	1	Right
3	1850	1000	0.01	1	1.4	0	1000	4.4	600	-100	1	Left
			0.01	1	1.4	0	1000	4.4	600	100	1	Right

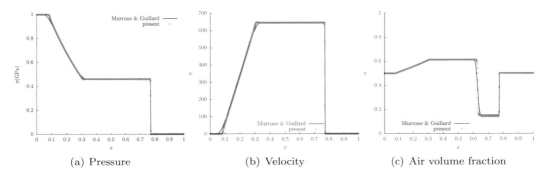

(a) Pressure	(b) Velocity	(c) Air volume fraction

Figure 4: Comparison of the present computation (red curve) and Murrone and Guillard's results (Murrone and Guillard, 2005) (black line) for water-air shock tube 1.

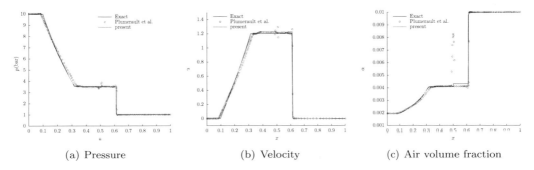

(a) Pressure	(b) Velocity	(c) Air volume fraction

Figure 5: Comparison of the present computation (red curve), exact solution (black curve) and Plumerault et al. 's results (Plumerault et al., 2012) (blue dots) for water-air shock tube 2.

close examination of these results shows generally good agreement with exact solutions and published independent numerical predictions with other flow codes. This verifies that the present method resolves wave speeds correctly and does not produce spurious nonphysical oscillations near shocks or at material interfaces.

In particular in Figure 5, it can be seen, in contrast, that the solution produced by Plumerault et al. exhibits strong nonphysical numerical oscillations around the shock and material interface and the present method has superior performance. Agreement with the exact solution is good apart from a slight underprediction of the discontinuity in density in the present results at the material interface in the region around $x = 0.5$.

4.1.2 Cavitation test

This case is used to test the capability of a numerical method to deal with liquid flow cavitation. Figure 6 shows the setup of the water expansion tube. The imaginary membrane is placed at the middle of the tube at $t = 0$. The initial conditions for this case are listed in Table 1 (case 3). The obtained solution is depicted in Figure 7. Rarefaction waves are observed to propagate outwards from the centre of the tube and the pressure decreases symmetrically from the middle in both left and right directions. Responding to the reduction of pressure, air entrained in the water begins to expand and its volume fraction increases. Accordingly, the volume fraction of the liquid phase falls. This creates a cavitation pocket in the middle of the tube and results in the dynamic appearance of two interfaces that were not present initially (Saurel et al., 2009). These results indicate that the present method can deal with low pressures quite well neither violating the physics nor producing a nonphysical negative value of absolute pressure that would cause the solver to diverge numerically. The model can also handle the creation, during the computation, of interfaces not present initially. Excellent agreement with the exact solution and Saurel et al. 's calculated results is observed in Figure 7.

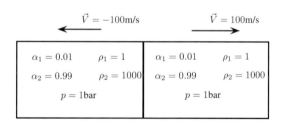

Figure 6: Setup of the water cavitation tube.

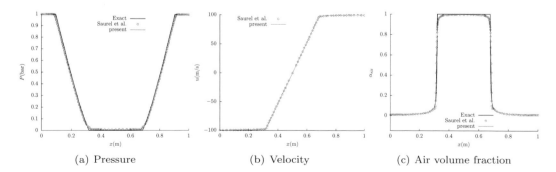

(a) Pressure   (b) Velocity   (c) Air volume fraction

Figure 7: Liquid expansion tube with a cavitation pocket. The present results (dashed line) are compared to the exact solution (black line) and Saurel et al. 's independent computation (Saurel et al., 2009) (blue dots).

4.2 Free drop of a water column in a closed tank

This benchmark test was proposed by the liquid sloshing community at the ISOPE 2010 conference to exam numerical codes' capability of handling multidimensional water impact problems involving air entrapment (Dias and Brosset, 2010). Figure 8 shows the setup for this problem. A rectangular water column ($\rho_2 = 1000$ kg/m^3) with width 10 m and height 8 m is initially at rest in the closed tank and air ($\rho_1 = 1$ kg/m^3) fills the remainder of the tank. The initial pressure is $p = 1$ bar in the tank. Under gravity, the water column will drop and impact upon the bottom of the tank at around $t = 0.64 \sim 0.65$ s. The impact pressure at this moment is of particular interest to ship structural engineers for this type of problem, since it is fundamentally a key issue for the safety of liquefied natural gas carriers.

To obtain an accurate prediction of the pressure peak, use of a fine mesh is recommended by other researchers (Braeunig et al., 2009; Guilcher et al., 2010). Therefore, we equally distribute 1500 mesh cells in the vertical direction and 200 mesh cells in the horizontal direction. Figure 9 gives two snapshots of the pressure field in the tank just before impact and at the impact time $t = 0.65$ s. It is clearly shown that the highest pressure occurs at the bottom centre of the tank and the distribution of pressure is symmetric about the central section (y axis). In Figure 10, we present several snapshots of the volume fraction contours in the liquid tank. An interesting finding is that a small amount of air is trapped between the water body and the bottom surface of the tank. This body of air undergoes not only compression due to the liquid impact (at $t = 0.65$ s) but also expansion when the gas

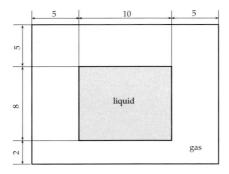

Figure 8: Initial setup for free drop of a water column in a closed 2D tank (Units in metres).

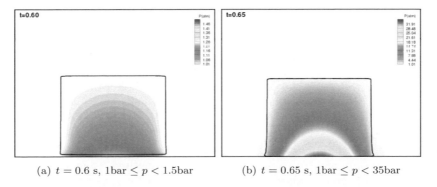

(a) $t = 0.6$ s, 1bar $\leq p < 1.5$bar   (b) $t = 0.65$ s, 1bar $\leq p < 35$bar

Figure 9: Snapshots of the pressure contours in the closed tank. The black curve is the free surface ($\alpha_1 = 0.5$).

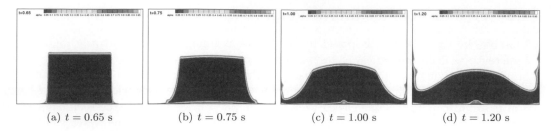

(a) $t = 0.65$ s　　(b) $t = 0.75$ s　　(c) $t = 1.00$ s　　(d) $t = 1.20$ s

Figure 10: Snapshots of the volume fraction for the gas phase in the closed tank. The water column starts to deform upon impacting the bottom of the tank. A small amount of air is trapped between the free surface and the bottom of the tank, and the air undergoes compression and expansion.

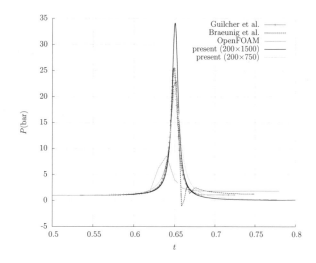

Figure 11: Time history of the absolute pressure at the bottom centre of the tank. There is good agreement with the impact time $t = 0.65$ predicted by Guilcher et al. (2010), Braeunig et al. (2009) and the present work, while incompressible OpenFOAM gives an earlier estimate at $t = 0.64$ and a lower pressure peak.

phase pressure exceeds the environmental liquid phase pressure. At $t = 0.75$ s, the portion of trapped air pocket appears to be a very thin layer, then has a cylindrical shape at $t = 1$ s and a half-cylindrical shape at $t = 1.2$ s.

The time history of the absolute pressure at the bottom centre of the tank is plotted in Figure 11. Guilcher et al.'s results computed independently using an SPH code (Guilcher et al., 2010) are represented by the red curve with "+" symbol. Braeunig et al.'s computation (Braeunig et al., 2009) is represented by the blue dashed line. We use a black curve to illustrate the results obtained by the present compressible method. The green curve indicates the result obtained on a coarse mesh (200×700 cells) with the present method. We have also used the *interFoam* module from the open-source software OpenFOAM®2.1.1, which is based on an incompressible two-fluid finite volume method, to compute this problem on the same mesh and present these results with a green line. Comparing the impact time, the present compressible results agree well with Guilcher et al.'s solution and Braeunig et al.'s work ($t = 0.65$ s), while *interFoam* gives a slightly early prediction at $t = 0.64$ s and the lowest pressure peak at 8.7 bar. The others produce much higher predictions at over 20 bar. After the peak, the pressure begins to fall. The minimum pressure following the peak obtained by *interFoam* is about 2 bar; Braeunig et al. produced a non-physical negative pressure with some oscillations (Braeunig et al., 2009); Guilcher et al. gave a value of 1 bar (Braeunig et al., 2009); the present compressible method obtains much lower but positive values for the minimum pressure after the pressure peak. Only the present method permits the fluid to expand sufficiently far to be in tension.

4.3   Underwater explosion near a planar rigid wall

This case is chosen to investigate a shock wave interaction with a planar rigid wall in water with possible hull cavitation loading/re-loading on the wall. Figure 12 depicts the set up for this problem. At the initial state, an explosive gas bubble of unit radius is located at the origin $(0,0)$ in water. The density of the air bubble is $\rho_g = 1270$ kg/m^3 , the pressure is $p_g = 8290$ bar and the ratio of specific heats is $\gamma_g = 2$. For the water, the density is $\rho_l = 1000$ kg/m^3, the pressure is $p_l = 1$ bar, the polytropic constant is $\gamma_l = 7.15$ and the pressure constant is $p_c = 3 \times 10^8$.

The computational domain is a rectangular region of $[-6, 6] \times [-6, 3]$. The horizontal rigid planar wall is located at $y = 3$. The domain is uniformly covered by 360×270 mesh cells which is the same as the grid used in Xie's work (Xie, 2005). The problem is also solved on a fine mesh with 720×540 cells to ensure the convergence of the numerical solution. A numerical pressure gauge is placed at the centre of the planar wall (point P) to monitor the impact pressure. Around this gauge point, we also integrate the pressure in a square region of 4×4 to obtain the impact force or loading. The length of this region is along the planar wall, and the width is perpendicular to the $x - y$ plane. The boundary conditions for this problem correspond a rigid upper wall and all other boundaries are transmissive (see Sections 4.2 and 5.2 of Xie (2005)).

Figure 13 shows several snapshots of the pressure contours in the domain computed on the coarse mesh. At $t = 1.5$ ms, the explosion generated main shock has been reflected from the planar wall. The reflected shock interacts with the expanding gas bubble undergoing a refraction process that results in a strong rarefaction wave propagating back towards the planar wall at $t = 2$ ms and a weaker reflected shock traversing the gas bubble. After the rarefaction wave impacts the wall and is reflected at $t = 3$ ms, a low pressure region forms near the planar wall causing a cavitation region to appear in this low pressure region at $t = 4$ ms. Simultaneously, the main shock continues to propagate through the computational domain.

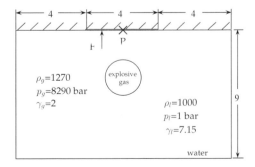

Figure 12: Setup of an underwater explosion near a planar rigid wall problem. The top boundary is a solid wall, all the others are open boundaries. The radius of the gas bubble is 1 m. The domain is discretised by a uniform mesh with 360×270 cells.

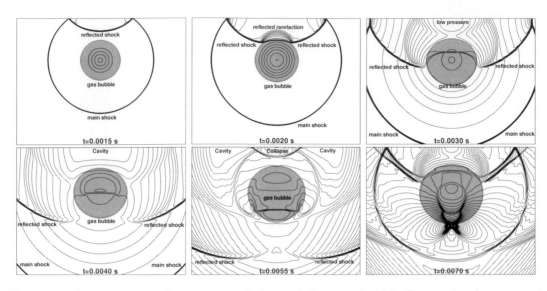

Figure 13: Line contours of pressure and the exploding gas bubble (grey colour) computed on the mesh with 360×270 cells.

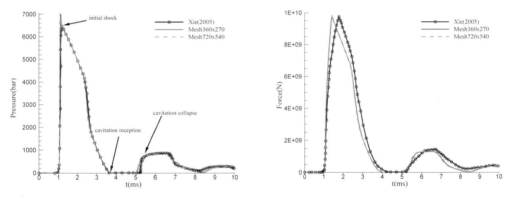

Figure 14: Comparison of the impact loading on the solid wall. Left: pressure time series at point P; Right: force acting on a 4m $\times$ 4m area of the solid wall. Red and green lines are the present results computed on the coarse (360×270) and fine mesh (720×540) respectively, black lines are Xie's computation on a mesh with 360×270 cells (Xie, 2005).

At $t = 5.5$ ms, the cavitation pocket collapses from the centre of the cavitation zone, resulting in a water-jet forming directed towards the planar wall and a relatively strong secondary compression wave propagating towards the gas bubble. The secondary compression wave impacts the gas bubble and leads to a second rarefaction wave propagating back towards the planar wall at $t = 7$ ms in a cyclic process that subsequently weakens with time.

Figure 14 shows the pressure (at point P) and force (covering the blue region around point P) acting upon the planar rigid wall. Numerical solutions computed on the coarse and fine meshes are almost identical and this confirms the convergence of the results. It can clearly be seen from this figure that the structural loading consists of shock loading and subsequent cavitation collapse reloading. The peak pressure of the shock is much higher than the peak pressure of cavitation collapse, whilst the duration of cavitation collapse reloading is longer than that of shock.

The present work gives about 7% higher prediction of the shock wave peak pressure (7000 bar) compared to Xie's numerical solution (6500 bar) (Xie, 2005). The start time of the cavitation collapse predicted by the present work is $t = 5.05$ ms, while Xie's one-fluid solver prediction is $t = 5.25$ ms (Xie, 2005). The amplitude of the pressure at cavitation collapse computed by the present method is very close to Xie's calculation. Considering the impact force, the present work and Xie's solution give almost the same amplitude of the peak force at about 9.8×10^{10} N. The rise time of the peak force calculated by Xie is slightly higher than our prediction. This can be attributed to small differences in the numerical dissipation present in the respective flow solvers.

4.4 Water entry of a rigid plate

4.4.1 Experiment and computation setup

The experimental study was carried out in the Ocean Basin at Plymouth University's COAST Laboratory. The ocean basin is 35 m long by 15.5 m wide and has an adjustable floor allowing operation at different water depths up to 3 m. The falling block includes a rigid flat impact plate connected to two driver plates and its total mass can be varied from 32 kg (block 1) to 52 kg (block 2). The impact plate is 0.25 m long and 0.25 m wide with a thickness of 0.012 m. The impact velocity can vary between 4 m/s and 7 m/s by adjusting the initial position of the plate. Pressures were measured during impact by five miniature pressure transducers installed on the impact plate as illustrated in Figure 15. The actual impact velocity was integrated from the measured data recorded by an accelerometer mounted on the top of the flat plate.

In the laboratory, the drop height of the plate is calibrated to obtain impact velocity 5.5 m/s while the present numerical model is constructed in the Eulerian frame of reference consistent with a finite volume method. To solve the water entry problem, we fix the flat plate in the numerical mesh and cause the water to move upward with the prescribed impact velocity. This strategy is appropriate for a short-duration impact process (Ng and Kot, 1992). Since violent hydrodynamic impacts are usually inertia dominant, viscous effects are currently ignored in the numerical computation. A compatible no-penetration velocity boundary condition therefore should be applied at the flat plate.

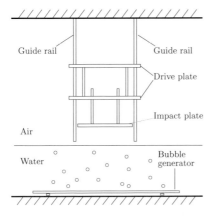 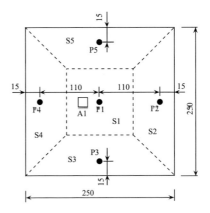

Figure 15: Experiment setup. Left: a sketch of the facility; right: configuration of the instrumentation on the impact plate (Units in mm). P1–P5 are pressure transducers, S1–S5 are the influence areas of the pressure transducers and A1 is the accelerometer.

Before the trapped air layer breaks into small bubbles, turbulence does not play a significant role in the impact process, thus the flow could reasonably be assumed to be symmetric about the two central sections of the plate. Consequently, we compute only a quarter of the whole flow field chosen as a cubic region $[0, 0.4] \times [0, 0.4] \times [0, 0.4]$ ($x - y - z$, units in m) to cover a quarter of the plate located in the region $[0, 0.125] \times [0.2, 0.212] \times [0, 0.125]$ ($x - y - z$, units in m). The thickness of the plate is 0.012 m. The initial calm water surface is set at $y = 0.1$ m and so the distance to the plate is 0.1 m in the hydro-code. Symmetry conditions are applied at the boundaries corresponding to the central sections of the plate. The numerical domain is discretised by $80 \times 1200 \times 80$ mesh cells. The purpose of using large number of cells in y-direction is to capture as accurately as possible the peak loadings. The numerical results and experimental measurements for pure and aerated water entries are presented in the following, respectively.

4.4.2 Pure water entry

Figure 16 presents the time series of gauge pressures and total slamming force on the plate for an impact velocity $v = 5.5$ m/s. The phases of these data have been properly shifted to correlate the first pressure peak at P1 to time $t = 0$. It is clearly shown that the evolution of the impact pressure loading at the plate centre consists of distinct stages including shock loading, fluid expansion (low pressure) loading and secondary re-loading.

When the plate approaches the water surface, the pressure rises very sharply to the first and highest peak value in less than 1 ms and then drops very quickly to zero. The first and highest pressure spike could be considered as a shock loading (Mitsuyasu, 1966). The duration of the shock loading is less than 2 ms. Table 2 lists the peak loadings on the plate. At P1, the maximum value of the shock load is 21.57 bar for the numerical simulation and is within 22.16 bar and 23.66 bar for the laboratory measurements. At the other four symmetric gauge points P2, 3, 4 and 5, the computed peak values are 10.44 bar; the laboratory measurement shows that the data varies from 6.53 bar to 11.42 bar for block 1 and it varies from 7.77 bar to 10.71 bar for block 2. The discrepancy between these pressure gauge measurements taken at symmetric positions on the plate may be due to plate flexure or minor variations in plate impact angle. Any of these factors may cause some transducers to contact the water slightly earlier than others. Regarding the total impact loading, the slamming force is positive (upward direction) with a peak value around 68.35 kN to 69.57 kN during this stage (see Table 2). Just after the first peak, the measurement data for P1 also shows a pressure spike up to 4.6 bar in the time series at around $t=0.6$ ms. This spike is not captured in numerical simulation. The underlying reason that caused this small secondary pressure spike might be due to the reflection of shock (pressure wave) from the bottom of the basin.

The impulses of shock loading on the plate obtained in experiments and numerical simulations are presented in Table 3. These values are calculated as temporal integrals of local pressures or overall forces through the interval $[t_b, t_a]$, which covers the lifespan of the shock load. The computed peak pressures at P1 and overall forces listed in Table 2 are close to the measurement for entry velocity $v = 5.5$ m/s , but the rise time is longer compared to experiment. Therefore the computed impulses are greater than experiment results as shown in Table 3.

After the shock load, the extensively compressed air layer trapped beneath the plate expands and induces the pressure to further decrease towards a vacuum. The lowest absolute pressure obtained in the numerical computation is 0.20 bar at P1 and 0.36 bar at P2 (see Table 4). The measured lowest absolute pressures show a variety of values including negative and positive readings. These negative values may be due to the acoustic noise generated by the shock in the ocean basin. However the observed trend in the measured data is close to the

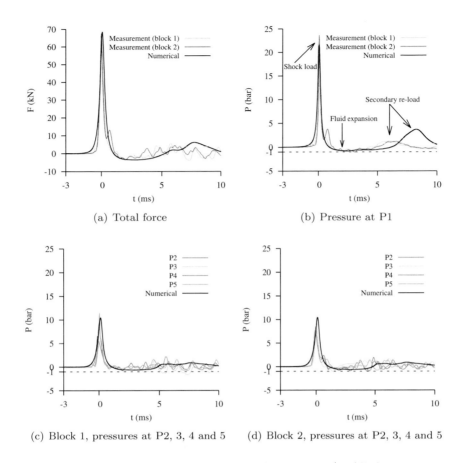

(a) Total force   (b) Pressure at P1

(c) Block 1, pressures at P2, 3, 4 and 5   (d) Block 2, pressures at P2, 3, 4 and 5

Figure 16: Pure water slamming loads on the plate for v=5.5 m/s. All the pressures presented here are gauge values. The horizontal dashed line represents a perfect vacuum pressure. Phases of all the results are adjusted to correlate the first pressure peak to time t=0. The masses of blocks 1 and 2 are 32 kg and 52 kg respectively.

Table 2: Peak gauge pressures and forces on the plate for pure water entry.

Instrument	P1(bar)	P2(bar)	P3(bar)	P4(bar)	P5(bar)	F(kN)
Block 1	22.16	6.53	8.46	9.43	11.42	68.35
Block 2	23.66	8.31	10.71	7.77	8.63	69.57
Numerical	21.57	10.44	10.44	10.44	10.44	68.41

numerical simulation results. After the lowest value, the pressure starts to increase towards zero with a second contraction of the trapped air. At P1, the duration of the fluid expansion load is 5.2 ms for the numerical simulation and 3.5 ms ∼ 3.7 ms for the experiment. At the four symmetric gauge points, the duration of the fluid expansion load is 3.7 ms for the numerical simulation and around 3.5 ms for experiment. The sub-atmospheric pressure on the lower surface causes the plate to experience negative (downward in direction) impact force for around 4 ms as shown in Figure 16. The computed minimum impact force on the plate is –3.54 kN, the measured data is –3.92 kN for block 1 and is –3.29 kN for block 2.

Table 3: Impulses of shock loading on the plate for pure water entry. Pressure and force impulses are calculated as $P^I = \int_{t_b}^{t_a} p\,dt$ and $F^I = \int_{t_b}^{t_a} f\,dt$ respectively, where t_b and t_a are the times immediately before and after the shock loading.

Instrument	P_1^I(Pa·s)	P_2^I(Pa·s)	P_3^I(Pa·s)	P_4^I(Pa·s)	P_5^I(Pa·s)	F^I(N·s)
Block 1	652.46	335.21	327.90	355.95	359.73	25.27
Block 2	719.71	381.59	359.25	388.14	402.53	28.55
Numerical	849.10	614.33	614.33	614.33	614.33	35.02

Table 4: Minimum absolute pressures and forces on the plate for pure water entry, $t \in [0, 5\text{ms}]$.

Instrument	P1(kPa)	P2(kPa)	P3(kPa)	P4(kPa)	P5(kPa)	F(kN)
Block 1	0.59	41.74	-62.19	27.86	24.05	-3.92
Block 2	-2.67	8.81	-6.37	3.86	22.69	-3.29
Numerical	20.03	36.15	36.15	36.15	36.15	-3.54

When the pressure recovers from sub-atmospheric to atmospheric values, the air layer does not stop contracting but is further compressed and leads to secondary re-loading of the plate. This re-loading is less severe compared to the shock loading. At P1, the numerical simulation produces stronger re-loading (3.71 bar) compared to the laboratory measurement (1.5 bar ~ 2 bar). This may be due to the inviscid assumption adopted in the numerical simulation, which currently ignores the fact that part of the fluid kinematic energy would be transformed into turbulence and/or dissipated by viscosity. The amplitude of the entire re-loading (force) obtained from numerical simulation is slightly higher than experiments.

Snapshots of the loadings on the plate obtained by numerical simulation and/or experiment are given in Figure 17. The left side represents the solution at t=−0.035 ms, just before the occurrence of shock load peak. The right side indicates the result at t=2.365 ms, which is at a time very close to the trough of the fluid expansion load. The time t=0 corresponds to the first pressure peak at P1. The top row illustrates contours of the computed pressure distribution on the lower surface of the plate. The middle row depicts the pressure distribution along the horizontal central section of the plate (lower surface). The bottom row shows the computed velocity field at centres of the mesh cells that are just beneath the lower surface of the flat plate. At t=−0.035 ms, the pressure is nearly axisymmetric on the plate apart from the four corners. The computed highest absolute pressure (21.30 bar) is found to be at the plate centre and decreases outwards. The measured pressure here is 21.54 bar for block 1 and 21.95 bar for block 2. The computed lowest pressures (5.78 bar) are at the four corners of the plate. The velocity vectors are directed outward from the plate centre. This means the fluid beneath the plate is being expelled away from the plate. At t=2.365 ms, the plate is experiencing non-axisymmetric sub-atmospheric pressure loading. The computed lowest pressure (0.20 bar) is at the plate centre and it increases outwards. Meanwhile, the measured pressure at the plate centre is 0.33 bar for block 1 and 0.21 bar for block 2. The computed highest pressures (1.04 bar) are at the four corners of the plate. The velocity vectors are directed inwards towards the plate centre. This indicates that ventilation may happen as the fluid is being drawn in under the plate.

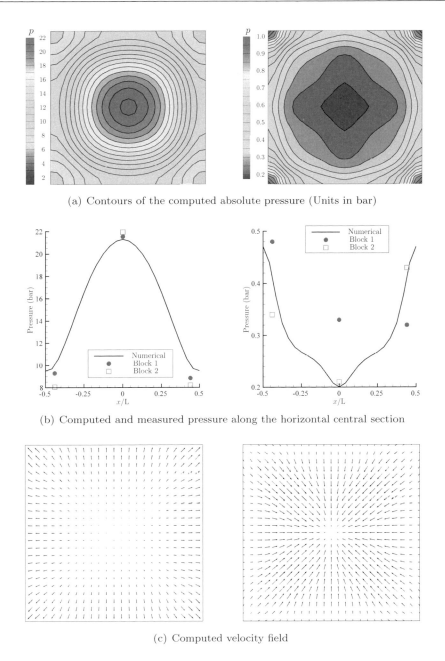

(a) Contours of the computed absolute pressure (Units in bar)

(b) Computed and measured pressure along the horizontal central section

(c) Computed velocity field

Figure 17: Impact loadings on the flat plate at T=-0.035 ms (left) and T=2.365 ms (right) for v=5.5 m/s, pure water entry.

Some researchers have claimed that cavitation occurs in hydrodynamic slamming events as the pressure descends dramatically after the shock load (Faltinsen, 2000). In the present experimental study, some transducers provide negative minimum absolute pressures, the measured positive minimum absolute pressures are within the range of 0.59 kPa to 47.14 kPa. The present minimum pressures obtained in the numerical simulation at P1 is 20.03 kPa. Compared to the water saturation pressure of 1.70 kPa at 15°C, most of the valid/positive experimental data and all the numerical results are still above the cavitation condition except the lowest pressures 0.59 kPa measured at P1 of block 1. Consequently, we are currently not able to conclude that cavitation certainly occurs in the water entry of the square plate. However, it is obvious that the computed and measured low pressures are very close to the cavitation condition, and so the surrounding water is very likely to be in tension.

4.4.3 Aerated water entry

As the shock load is the most severe and dangerous force acting on the flat plate structure, here we focus on investigating effect of aeration on the first peak loading. Before presenting the results for aerated water entry, it is necessary to clarify the differences in introducing air bubbles into the water between the experiment and the numerical computations.

In the laboratory, the bubble generation system was placed at the bottom of ocean basin, which is filled with fresh water for a depth of 1 m. The air bubbles rise from the bottom of the ocean basin to the water surface and break up. The size of the bubbles significantly increase as they ascend. Therefore the aeration level near the water surface is notably higher than at the bottom of basin. Nevertheless the bubble aeration was measured at one water depth only (0.25 m away from the water surface) and assumed uniform throughout depth.

The numerical model does not treat the air bubbles explicitly in a one-by-one manner, but instead assumes the water to be uniformly mixed with air bubbles at a specified concentration level (and thus implied speed of sound). Thus, the aeration level is represented by the air volume fraction α_1 in Equation (14). As the change of aeration level with water depth $\alpha_1(y)$ is currently not available, we apply the average representative value measured in laboratory to set up the initial conditions for the numerical computations and assume that $\alpha_1(y)$ does not vary with the water depth. Obviously, this is a simplified and conservative treatment as the real aeration level near the water surface is notably under-estimated in the numerical model's setup.

Figure 18 shows the time series of the gauge pressures on the plate under different aeration levels. The left column represents the pressures measured for block 1 (32 kg), the middle column indicates the data for block 2 (52 kg) and the right column illustrates the numerical results. From the top to bottom rows are pressures gauged/computed at locations P1, 2, 3, 4 and 5. Note that all the time series have been shifted to correlate the first peak of P1 to time $t = 0$.

Close inspection of these figure shows that the peak impact pressures on the plate are significantly reduced by aeration. In the numerical computation, at all gauge points, the pressures are continuously reduced by increased aeration levels. The highest aeration level $\alpha_1 = 1.6\%$ reduces all the pressures by $49.4\% \sim 50.7\%$. Looking at the measurement data, the reduction in pressure is greater when the same level of aeration is used. Here, we would like to emphasise again that the aeration level near the water surface is noticeably higher than the basin bottom. In the experiment, the aeration levels were measured at only one depth with a distance of 0.25 m away from the water surface. Since the real aeration near the water surface is higher than the measurement point, we only implement the significant average measurement value, which is rather conservative/under-estimated, in the numerical simulation.

Looking at the measured pressures at P2, 3, 4 and 5 shown in Figure 18, it is easy to spot that the occurrences of the pressure peaks are notably non-synchronous with P1 by phase shifts up to 4.2 ms. In the experiment, the water surface is not flat for the aerated cases due to the disturbance caused by rising bubbles, and the surface disturbance is greater for higher aeration levels. Therefore the pressure transducers installed on the plate are not in contact with the water surface at the same time. While in the numerical simulation, the surface is assumed to be flat. This leads to the significant difference in phase and amplitude of the pressure peaks. The details of peak pressures as well as impact forces measured/computed

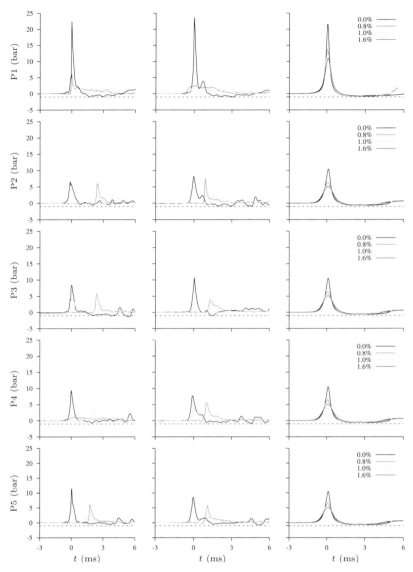

Figure 18: Impact pressures on the plate in water with different aeration levels ($\alpha_1 = 0 \sim 1.6\%$) for $v = 5.5$ m/s. Left: block 1 (32 kg); Middle: block 2 (52 kg); Right: numerical computation. From the top to bottom rows are pressures at gauge points P1, 2, 3, 4 and 5. The computed results at P2, 3, 4 and 5 are identical.

under different aeration levels and impact velocities are listed in Table 5. The computed minimum absolute pressures and forces under different aeration levels are listed in Table 6.

Figure 19 shows the computed pressures on the plate near the occurrence of peak loading and trough loading in 0.8% aerated water. Similar to the pure water entry cases, at the time near the occurrence of peak loading, the pressure distribution on the plate is nearly axisymmetric apart from the four corners. At the time near the occurrence of trough loading, the pressure distribution is non-axisymmetric. The trapped air undergoes intensive compression and expansion. The pressure distribution for cases with higher aeration levels are similar to these two figures (except the amplitude of pressure), therefore they are not included here.

Table 5: Maximum gauge pressures and forces on the plate for aerated water entry.

Instrument	$\alpha_1(\%)$	P1(bar)	P2(bar)	P3(bar)	P4(bar)	P5(bar)	F(kN)
Block 1	0.0	22.16	6.53	8.46	9.43	11.42	68.35
	0.8	6.35	6.40	6.14	2.47	6.23	18.55
	1.0	3.77	3.86	3.87	4.82	1.93	12.11
	1.6	–	–	–	–	–	–
Block 2	0.0	23.66	8.31	10.71	7.77	8.63	69.57
	0.8	2.89	7.50	3.94	6.08	6.27	21.69
	1.0	3.67	3.97	4.03	4.12	2.92	15.87
	1.6	–	–	–	–	–	–
Numerical	0.0	21.57	10.44	10.44	10.44	10.44	68.41
	0.8	13.75	6.41	6.41	6.41	6.41	46.65
	1.0	12.81	5.97	5.97	5.97	5.97	43.72
	1.6	10.92	5.14	5.14	5.14	5.14	37.76

Table 6: Computed minimum absolute pressures and forces on the plate for aerated water entry.

$\alpha_1(\%)$	P1(kPa)	P2(kPa)	P3(kPa)	P4(kPa)	P5(kPa)	F(kN)
0.0	20.03	36.15	36.15	36.15	36.15	-3.54
0.8	8.67	28.79	28.79	28.79	28.79	-4.34
1.0	9.56	32.47	32.47	32.47	32.47	-4.21
1.6	14.61	44.72	44.72	44.72	44.72	-3.65

Figure 20 shows the peak pressures at P1 and total impact forces on the plate for pure and aerated water entries obtained in experiment and numerical simulation. All the numerical results indicate that both the pressure and force decline with increased aeration levels. When bubbles are generated to aerate the fresh water in laboratory, the water surface is greatly disturbed by the quickly rising and breaking bubbles. This causes some difficulty in measuring impact loadings and leads to obvious phase shifts of pressure peaks. Nevertheless the experimental data illustrates that aeration can dramatically reduce the peak loadings.

Figure 21 shows the impulses of shock loading on the plate for entry into pure and aerated water. These impulses are calculated as time integrals of pressure or force in the interval $[t_b, t_a]$, which covers the lifespan/duration of the shock load. Looking at the pressure impulse at P1, the numerical simulation shows a declining trend with increased aeration level (but the reduction is mild compared to the reduction of peak pressures); the experiment does not clearly show this declining trend. Looking at the force impulse on the plate, neither numerical computation nor experiment shows a clear declining or rising trend: the force impulse experiences both rise and decline. The difference in shock load impulse between pure water and aerated water with highest aeration level is presented in Table 7. For pressure impulse at P1, the maximum aeration level seems to be able to reduce it by 11.9%~19.1%. Meanwhile the maximum aeration level causes up to 17.2% variation of force impulse.

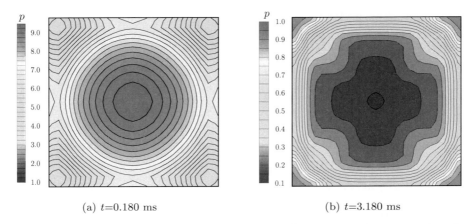

(a) t=0.180 ms   (b) t=3.180 ms

Figure 19: Computed contours of absolute pressures (Units in bar) on the plate for v=5.5 m/s and aeration level 0.8%. Left: t =0.18 ms, just after the occurrence of peak loading at P1; Right: t=3.18 ms, near the occurrence of trough loading at P1.

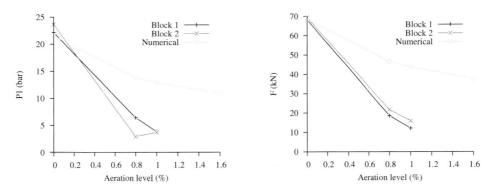

Figure 20: Aeration effects on the peak impact loadings for v=5.5 m/s. Left: Peak gauge pressure at P1; Right: Total impact force on the plate.

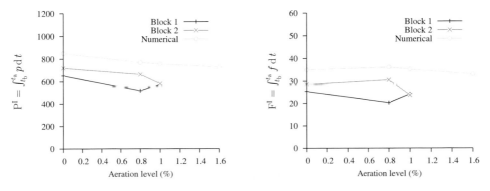

Figure 21: Aeration effects on the impulse of shock loadings for v=5.5 m/s. Left: pressure impulse at P1 (Units in Pa·s); Right: Total force impulse on the plate (Units in N·s).

Table 7: Variation of shock load impulse between pure water impact and aerated water (with highest aeration level) impact.

Instrument	P_1^I (Pa·s)			F^I (N·s)		
	$\alpha_1 = 0$	$\max(\alpha_1)$	ΔP_1^I (%)	$\alpha_1 = 0$	$\max(\alpha_1)$	ΔF^I (%)
Block 1	652.46	574.85	-11.9	25.27	24.13	-4.5
Block 2	719.71	582.58	-19.1	28.55	23.65	-17.2
Numerical	849.10	728.90	-14.2	35.02	33.08	-5.5

4.5   Plunging wave impact at a vertical wall

Here we perform an exploratory calculation to establish the viability and promise of the method for violent wave impact simulations involving air pockets and aeration. Figure 22 shows the setup for a plunging wave impact event. The length of the wave tank is 3 m and the height is 0.8 m. A step of 0.2 m height is placed in the bottom right part of the wave tank starting at $x = 1.75$ m with a 45° slope to cause the wave to steepen and break. A piston type wave maker is placed at the left boundary of the domain to generate waves. The still water depth is $d = 0.3$ m. The whole NWT is divided into two sub-domains. A two-fluid NWT based on the incompressible Navier-Stokes equations developed in our previous work (Qian et al., 2006) is used to deal with the left sub-domain. This incompressible solver, which is named AMAZON-SC, adopts an interface-capturing method to treat the free surface as a discontinuity in density. We utilise the present compressible flow model (14) to handle the right-sub domain, where an air pocket will be trapped or enclosed by the water body. The dashed line in Figure 22 indicates the interface between the incompressible and compressible flow solvers. Buffer zones are used near the interface to exchange flow information between the two solvers. Within the buffer zones, one or two layers of mesh cells for each component solver as required are placed on the opposite side of the interface. The flow information including density, velocity and pressure at these mesh cells is obtained from the companion solver domain through interpolation. More details of the coupling of component flow solvers will be reported separately in future work. A background uniform Cartesian mesh is used to overlay the flow domain, and the basic mesh cell is a square with size of $h = 0.01$ m. Solid boundaries not aligned with the Cartesian mesh in the left sub-domain are treated using the cut-cell method (Qian et al., 2006).

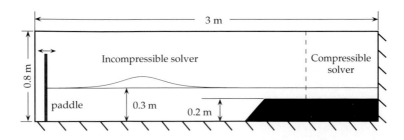

Figure 22: Setup of the plunging wave impact problem. The numerical wave tank is divided into two sub-domains occupied by the incompressible flow solver (Qian et al., 2006) and the present two-phase compressible flow solver.

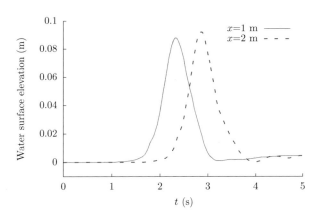

Figure 23: Water surface elevations in the numerical wave tank (without structure) at $x = 1$ m and $x = 2$ m.

Before computing the plunging breaker impact problem, we first conduct a simple test to generate a solitary wave using the incompressible solver. The solid step is removed from the NWT and the right boundary of the domain is treated as an open boundary. The amplitude of the solitary wave is $a = 0.09$ m. The wave is generated by prescribing the paddle movement according to Rayleigh's solitary wave theory (see Katell and Eric, 2002). Figure 23 shows the water surface elevations at $x = 1$ m and $x = 2$ m. The wave crest takes 0.52 s to travel between these two locations. Obviously, the phase speed of the solitary wave is equal to $c^* = 1.92$ m/s. The theoretical phase speed for solitary waves can be calculated as $c = \sqrt{g(d+a)}$. We obtain $c = 1.95$ m/s for this test case so the relative error between the computed and theoretical wave phase speeds is less than 1.5%.

The integrated numerical wave tank is now used to solve the plunging wave impact problem. The paddle is used again to generate a solitary wave with height $a = 0.2$ m. In addition to the integrated NWT, we also use the established standalone in-house incompressible two-fluid NWT AMAZON-SC to solve this problem for the purposes of comparison. According to field measurements and laboratory experiments, the first pressure spike in the form of a "church-roof" shape is a key to the safety to structures. Therefore, we focus on this phase of the impact in the current discussion. Computation of the subsequent wave evolution will be considered in future work.

Figure 24 shows the profiles of the free surface at different times. The red solid line represents the results with AMAZON-SC, the stand-alone incompressible solver, and the blue dashed line indicates the solution obtained with the compressible solver. At $t = 2.13$ s, the two solvers give almost the same profiles. We notice that an obvious discrepancy of the free surface profiles appears at $t = 2.15$ s. Although the wave crests are almost the same distance away from the vertical wall in the horizontal direction, the wave crest obtained by the incompressible solver is higher than the compressible solver. The wave trough obtained by the compressible solver moves upward along the wall faster than the incompressible solver. The free surface beneath the wave crest obtained by the compressible solver is closer to the vertical wall than the incompressible solver. At $t = 2.16$ s, we notice that the wave crest obtained by the incompressible solver moves upward significantly higher than the compressible solver, and this trend continues to $t = 2.17$ s when the waves almost impact the wall. The water continuously moves upward after the wave impacts the wall. From the figures, it is not difficult to observe that the trapped air pocket predicted by the compressible solver is much smaller than the incompressible solver. It would seem that compressibility effects play an important role in changing the shape of the air pocket and the

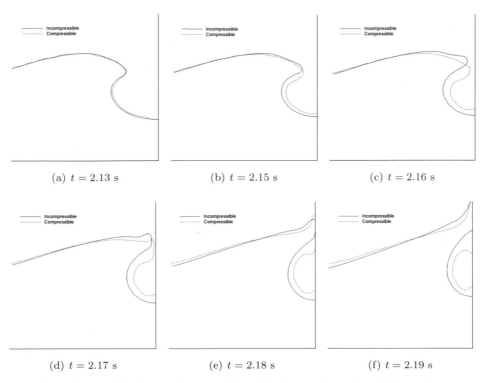

(a) $t = 2.13$ s   (b) $t = 2.15$ s   (c) $t = 2.16$ s

(d) $t = 2.17$ s   (e) $t = 2.18$ s   (f) $t = 2.19$ s

Figure 24: Snapshots of the free surface profiles. Mesh step size is $h = 0.01$ m. Incompressible and compressible solvers produce almost the same wave crest velocities in the x-direction. Comparison of the time and spatial evolution of the air pockets between the two solvers show discrepancies due to compressibility effects. The volume of the air pocket predicted by the compressible solver is much smaller than the incompressible solver.

free surface. The incompressible assumption appears to lead an overestimate of the volume of the air pocket for this type of problem.

In Figure 25, we present several snapshots of the pressure distribution in the wave field at different times. The first row illustrates the results with the stand-alone incompressible solver AMAZON-SC and the second row corresponds to computations with the compressible solver. For the compressible solver, we can clearly see that the pressure in the air pocket increases dramatically, and the pressure rise travels downstream along the vertical wall and tank bottom. At $t = 1.95$ s, a second compression (pressure increase) in the air pocket is captured by the compressible solver. These phenomena are much less apparent in the predictions with the incompressible solver.

Quantitative comparison of pressures is made and presented in Figure 26. We gauge the pressures in the air pocket. For this problem, the size of the air pocket is relatively large as its diameter is around 10 cm. A significant amount of the wave energy is stored in the pocket through compression. Consequently, the air pressure rises significantly to about 114000 Pa as shown in the figure. Under the constant density assumption, the incompressible solver cannot deal with compressibility effects and only predicts a peak pressure of 104000 Pa. The rise time of the pressure spike for the incompressible solver is almost ten times that of the compressible solver. After the first peak, the air expands to a low pressure. The compressible solver predicts a fall to around 9500 Pa, whilst the incompressible solver predicts about 101000 Pa.

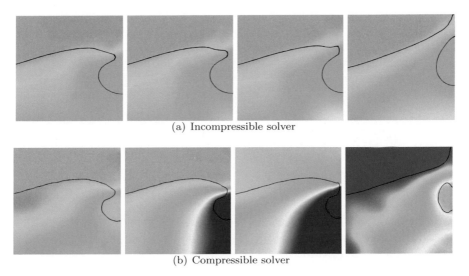

(a) Incompressible solver

(b) Compressible solver

Figure 25: Snapshots of the pressure distributions in the numerical wave tank. From left to right, $t = 2.155$, 2.16, 2.165 and 2.195 s. The black curve represents the free surface. Top: incompressible solver; Bottom: compressible solver. Pressure contour range: $p_{\min} = 100000$ Pa, $p_{\max} = 112000$ Pa, $\Delta p = 400$ Pa.

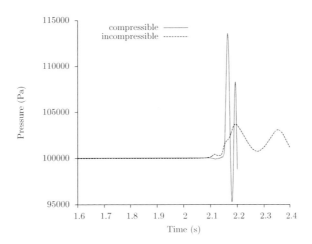

Figure 26: Time evolution of the pressure in the air pocket. Sub-atmospheric pressures are captured by the compressible solver (solid line) but not the incompressible solver (dashed line). There is a significant decrease from the first pressure peak to the second peak with the compressible solver. Rise times and pressures obtained by the compressible solver are markedly different to those of the incompressible solver.

The pressure distributions along the vertical wall at different times computed by the compressible solver are shown in Figure 27. The pressure in the region ($0.3 \leq y \leq 0.4$) is strongly influenced by the trapped air pocket. The air pocket is continuously compressed from $t = 2.156$ to 2.167 s. It is noted that at $t = 2.184$ s the air pocket undergoes expansion and the pressure reduces accordingly to sub-atmospheric values. Thus, this local region is experiencing seaward (suction) force. These numerical findings confirm Bullock et al.'s work (Bullock et al., 2007) that negative gauge pressures indeed occur in violent wave impact

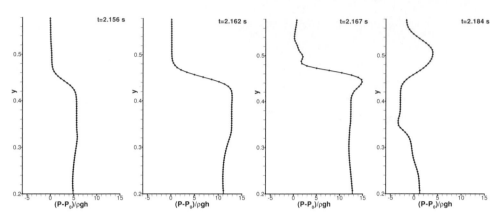

Figure 27: Pressure distribution on the vertical wall at $t = 2.156$, 2.162, 2.167 and 2.184 s (computed by the compressible solver). P_0 is the atmospheric pressure, $h = 0.1$ is the initial water depth before the vertical wall.

events and the resultant seaward force has the potential to cause the removal of blocks from masonry structures.

In their laboratory experiments of overturning wave breaking on structures, Lugni et al. observed that after the strongest first impact pressure peak, the pressure decreases to a value lower than atmospheric pressure and a subsequent second pressure peak is observed much lower than the first one (see Figures 4 and 5 of Lugni et al., 2010). The numerical findings produced by the present compressible solver of a steep pressure spike followed by a negative gauge pressure and subsequent lower second pressure peak, etc. agree qualitatively well with Lugni et al.'s experiments (Lugni et al., 2010), although the wave conditions are not exactly the same.

5   Conclusions

The development of AMAZON-CW, a compressible multiphase hydrocode for free surface flows, has been reviewed in this chapter. The model is based on the conservation laws of mass, momentum and energy in addition to a non-conservative advection equation for volume fraction. Detailed derivation of the model as well as the treatment of the non-conservative term are provided. A third-order finite volume scheme based on a MUSCL reconstruction and an HLLC approximate Riemann solver for compressible water-air mixtures is used to discretise the integral form of the mathematical equations. A number of test cases including a liquid cavitation tube, two water-air shock tubes, a free-falling water column in a closed tank, an underwater explosion near a planar rigid wall, the entry of a flat plate into pure and aerated water and an air enclosed plunging wave impacting at a vertical wall have been used to examine the code. The obtained results are reasonable and generally agree well with experiments, exact solutions and other numerical computations. These results verify the present method and clearly show the advantages of including air compressibility effects over other numerical methods based on single-fluid or two-fluid incompressible flow models for complex violent wave impact problems.

The developed code can reasonably resolve the free surface and large air pockets trapped in breaking waves. However, attention should be paid to the post-breaking process in order to fully understand the air entrainment process and to provide accurate aeration level and distribution in the water, which will significantly influence the subsequent wave impacts at structures in respect of intensity and duration.

Our future work will focus on extending the code's capability to resolve fine-scale flow features, e.g., bubble entrained in water and droplets ejected from wave fronts. This will be achieved through the further development of the hydrocode by including a hybrid turbulence modelling approach (DES) and surface tension as well as adaptive high order (5th/7th) low-dissipation numerical discretisation schemes. The code will also be parallelised on massive heterogeneous HPC facilities to solve complex 3D problems in coastal and offshore engineering.

For realistic industrial applications, running a compressible multiphase free surface code in the whole domain will be very expensive in terms of computer time and memory. To reduce the excessive amount of computation, effort needs to be made to enable the AMAZON-CW code be called and executed on the fly when compressibility and aeration have significant influence on waves. This will request a smooth linkage of the code with other free surface solvers, e.g., the overset mesh based incompressible multiphase flow code *ogFoam* (Ma et al., 2018) and multiphase hydro-elasticity codes, e.g., the *wsiFoam* code (Martnez Ferrer et al., 2016; Martnez Ferrer et al., 2018; Martnez Ferrer et al. 2018), which are currently under development.

Acknowledgements

This work was partially funded by the Engineering and Physical Sciences Research Council (EPSRC, UK) projects: Fundamentals and Reliability of Offshore Structure Hydrodynamics (EP/J012793/1), Modelling Marine Renewable Energy Devices; Designing for Survivability (EP/J010197/1), Virtual Wave Structure Interaction Simulation Environment (EP/K037889/1), A Zonal CFD Approach for Fully Nonlinear Simulations of Two Vessels in Launch and Recovery Operations (EP/N008839/1), A CCP on Wave Structure Interaction: CCP-WSI (EP/M022382/1) and High-fidelity Simulation of Air Entrainment in Breaking Wave Impacts (EP/S011862/1). The authors would also like to thank Dr. Tri Mai, Profs. Deborah Greaves and Alison Raby of Plymouth University for providing the experimental data for the rigid flat plate slamming problem.

References

Abgrall, R. and Karni, S. (2001). Computations of compressible multifluids. J. Comput. Phys. 169(2): 594–623.

Aggelos S. Dimakopoulos, Mark J. Cooker and Tom Bruce. (2017). The influence of scale on the air flow and pressure in the modelling of oscillating water column wave energy converters. International Journal of Marine Energy 19: 272–291.

Angelo Murrone and Herve Guillard. (2005). A five equation reduced model for compressible two phase flow problems. J. Comput. Phys. 202(2): 664–698.

Batten, P., Clarke, N., Lambert, C. and Causon, D. M. (1997). On the choice of wavespeeds for the HLLC Riemann solver. SIAM J. Sci. Comput. 18(6): 1553–1570.

Bernard-Champmartin, A., Poujade, O., Mathiaud, J. and Ghidaglia, J. -M. (2014). Modelling of an homogeneous equilibrium mixture model (HEM). Acta Applicandae Mathematicae 129(1): 1–21.

Bjorn C. Abrahamsen and Odd M. Faltinsen. (2011). The effect of air leakage and heat exchange on the decay of entrapped air pocket slamming oscillations. Phys. Fluids 23(10): 102107.

Braeunig, J. -P., Brosset, L., Dias, F. and Ghidaglia, J. -M. (2009). Phenomenological study of liquid impacts through 2d compressible two-fluid numerical simulations. In Proceedings of the 19th International Offshore and Polar Engineering Conference (ISOPE), Osaka, Japan, volume 7.

Bram van Leer. (1979). Towards the ultimate conservative difference scheme. V. A second-order sequel to Godunov's method. J. Comput. Phys. 32(1): 101–136.

Bredmose, H., Peregrine, D. H. and Bullock, G. N. (2009). Violent breaking wave impacts. Part 2: Modelling the effect of air. J. Fluid Mech. 641(1): 389–430.

Bullock, G. N., Obhrai, C., Peregrine, D. H. and Bredmose. H. (2007). Violent breaking wave impacts. Part 1: Results from large-scale regular wave tests on vertical and sloping walls. Coastal Eng. 54(8): 602–617.

Chan, E. S., Melville, W. K., Chan, E. S. and Melville, W. K. (1988). Deep-water plunging wave pressures on a vertical plane wall. Proc. R. Soc. Lond. A 417(1852): 95–131.

Christensen, E. D. and Deigaard, R. (2001). Large eddy simulation of breaking waves. Coastal Eng. 42: 53–86.

Corradini, M. L. (1997). Fundamentals of multiphase flow. Technical report, University of Wisconsin.

Crawford, A. R. (1999). Measurement and analysis of wave loading on a full scale coastal structure. Ph.D. thesis, Plymouth University.

De Divitiis, N. and De Socio, L. M. (2002). Impact of floats on water. J. Fluid Mech. 471: 365–379.

Dias, F. and Brosset, L. (2010). Comparative numerical study: description of the calculation case. In 20th Int. Oshore and Polar Eng. Conf., volume 3, Beijing, China.

Encarnacion Medina-Lopez, William Allsop, Aggelos Dimakopoulos and Tom Bruce. (2015). Conjectures on the failure of the OWC breakwater at Mutriku. In Proceedings of Coastal Structures and Solutions to Coastal Disasters Joint Conference, Boston, Massachusetts.

Eric Johnsen and Tim Colonius. (2006). Implementation of weno schemes in compressible multicomponent flow problems. J. Comput. Phys. 219(2): 715–732.

Faltinsen, O. M. (2000). Hydroelastic slamming. Journal of Marine Science and Technology 5(2): 49–65, cited By (since 1996)79.

Greaves, D. (2004). A quadtree adaptive method for simulating fluid flows with moving interfaces. Journal of Computational Physics 194(1): 35–56.

Greaves, D. (2006). Simulation of viscous water column collapse using adapting hierarchical grids. International Journal for Numerical Methods in Fluids 50(6): 693–711.

Guilcher, P. M., Oger, G., Brosset, L., Jacquin, E., Grenier, N. and Le Touze, D. (2010). Simulation of liquid impacts with a two-phase parallel SPH model. In Proceedings of 20th International Offshore and Polar Engineering Conference, June 20–26, Bejing, China, volume 3.

Harten, A., Lax, P. D. and van Leer, B. (1983). On upstream differencing and Godunov-type schemes for hyperbolic conservation laws. SIAM Rev. 25: 35–61.

Hisashi Mitsuyasu. (1966). Shock pressure of breaking wave. Coastal Engineering Proceedings 1(10): 268–283.

Hu, X. Y., Adams, N. A. and Iaccarino, G. (2009). On the HLLC Riemann solver for interface interaction in compressible multi-fluid flow. J. Comput. Phys. 228(17): 6572–6589.

Ivings, M. J., Causon, D. M. and Toro. E. F. (1997). On hybrid high resolution upwind methods for multicomponent flows. ZAMM. Z. angew. Math. Mech. 77(9): 645–668.

Ivings, M. J., Causon, D. M. and Toro, E. F. (1998). On Riemann solvers for compressible liquids. Int. J. Numer. Methods Fluids 28(3): 395–418.

John C. Scott. (1975). The role of salt in whitecap persistence. Deep Sea Research and Oceanographic Abstracts 22(10): 653–657.

Jon Kay. (2014). UK storms destroy railway line and leave thousands without power. BBC News http://www.bbc.co.uk/news/uk-26042990, February 2014.

Kapila, A. K., Meniko, R., Bdzil, J. B., Son, S. B. and Stewart, D.S. (2001). Two-phase modeling of deflagration-to-detonation transition in granular materials: Reduced equations. Phys. Fluids 13: 3002–3024.

Katell, G. N. and Eric, B. (2002). Accuracy of solitary wave generation by a piston wave maker. Journal of Hydraulic Research 40(3): 321–331.

Kirkgoz, M.S. (1982). Shock pressure of breaking waves on vertical walls. Journal of the Waterway Port Coastal and Ocean Division 108(1): 81–95.

Korobkin, A. A. (2006). Two-dimensional problem of the impact of a vertical wall on a layer of a partially aerated liquid. J. Appl. Mech. Tech. Phys. 47(5): 643–653.

Lubin, P., Vincent, S., Abadie, S. and Caltagirone, J. P. (2006). Three-dimensional large eddy simulation of air entrainment under plunging breaking waves. Coastal Eng. 53(8): 631–655.

Lucy Rodgers and Mark Bryson. (2014). 10 key moments of the UK winter storms. BBC News http://www.bbc.co.uk/news/uk-26170904, February 2014.

Lugni, C., Brocchini, M. and Faltinsen, O. M. (2010). Evolution of the air cavity during a depressurized wave impact. II. The Dynamic Field Phys. Fluids 22(5): 056102.

Lugni, C., Miozzi, M., Brocchini, M. and Faltinsen, O. M. (2010). Evolution of the air cavity during a depressurized wave impact. I. The kinematic flow field. Phys. Fluids 22: 056101.

Ma, Z. H., Qian, L., Causon, D. M. and Mingham, C. G. (2011). Simulation of solitary breaking waves using a two-fluid hybrid turbulence approach. In Proceedings of the Twenty-first (2011) International Offshore and Polar Engineering Conference.

Ma, Z. H., Causon, D. M., Qian, L., Mingham, C. G., Gu, H. B. and Martnez Ferrer, P. (2014). A compressible multiphase flow model for violent aerated wave impact problems. Proceedings of the Royal Society A 470(2172): 20140542.

Ma, Z. H., Causon, D. M., Qian, L., Gu, H. B., Mingham, C. G. and Martnez Ferrer, P. (2015). A GPU based compressible multiphase hydrocode for modelling violent hydrodynamic impact problems. Computers and Fluids 120: 1–23.

Ma, Z. H., Causon, D. M., Qian, L., Mingham, C. G., Mai, T., Greaves, D. and Raby, A. (2016). Pure and aerated water entry of a flat plate. Physics of Fluids 28: 016104.

Ma, Z. H., Qian, L., Martnez-Ferrer, P. J., Causon, D. M., Mingham, C. G. and Bai, W. (2018). An overset mesh based multiphase flow solver for water entry problems. Computers & Fluids 172: 689–705.

Martínez Ferrer, P. J., Causon, D. M., Qian, L., Mingham, C. G. and Ma, Z. H. (2016). A multi-region coupling scheme for compressible and incompressible flow solvers for two-phase flow in a numerical wave tank. Computers and Fluids 125: 116–129.

Martínez Ferrer, P. J., Ling Qian, Zhihua Ma, Derek M. Causon and Clive G. Mingham. (2018). An efficient finite-volume method to study the interaction of two-phase fluid flows with elastic structures. Journal of Fluids and Structures 83: 54–71.

Martínez Ferrer, P. J., Qian, L., Ma, Z. H., Causon, D. M. and Mingham, C. G. (2018). Improved numerical wave generation for modelling ocean and coastal engineering problems. Ocean Engineering, 257–272.

Ng, C. O. and Kot, S. C. (1992). Computations of a water impact on a two-dimensional flat-bottomed body with a volume-of-fluid method. Ocean Eng. 19: 377–393.

Oumeraci, H., Klammer, P. and Partenscky, H. W. (1993). Classification of breaking wave loads on vertical structures. Journal of Waterway, Port, Coastal, and Ocean Engineering 119(4): 381–397.

Peregrine, D. H. (2003). Water-wave impact on walls. Annu. Rev. Fluid Mech. 35(1): 23–43.

Peregrine, D. H., Bredmose, H., Bullock, G., Obhrai, C., Müller, G. and Wolters, G. (2005). Violent water wave impact on a wall. In Rogue Waves, Proceedings of the Aha Huliko a Hawaiian Winter Workshop, pp. 155–159.

Plumerault, L. -R. (2009). Numerical Modelling of Aerated-water Wave Impacts on a Coastal Structure. Ph.D. thesis, University of Pau and Pays de l'Adour.

Plumerault, L. -R., Astruc, D., Villedieu, P. and Maron, P. (2012). A numerical model for aerated-water wave breaking. Int. J. Numer. Methods Fluids 69(12): 1851–1871.

Qian, L., Causon, D. M., Mingham, C. G. and Ingram, D. M. (2006). A free-surface capturing method for two fluid flows with moving bodies. Proc. R. Soc. Lond. A 462: 21–42.

Riccardi, G. and Iafrati, A. (2004). Water impact of an asymmetric floating wedge. J. Engrg. Math. 49(1): 19–39.

Richard Saurel and Olivier Lemetayer. (2001). A multiphase model for compressible flows with interfaces, shocks, detonation waves and cavitation. J. Fluid Mech. 431: 239–271.

Saurel, R. and Abgrall, R. (1999). A multiphase godunov method for compressible multifluid and multiphase flows. J. Comput. Phys. 150(2): 425–467.

Saurel, R. and Abgrall, R. (1999). A simple method for compressible multifluid flows. SIAM J. Sci. Comput. 21(3): 1115–1145.

Saurel, R., Petitpas, F. and Berry, R. A. (2009). Simple and efficient relaxation methods for interfaces separating compressible fluids, cavitating flows and shocks in multiphase mixtures. Journal of Computational Physics 228(5): 1678–1712.

Schmidt, R., Oumeraci, H. and Partenscky, H. W. (1992). Impact loads induced by plunging breakers on vertical structures. In Proc. 23rd Int. Conf. Coastal Engng., 1992.

Semenov, Y. A. and Iafrati, A. (2006). On the nonlinear water entry problem of asymmetric wedges. J. Fluid Mech. 547: 231–256.

Sigal Gottlieb and Chi-Wang Shu. (1998). Total variation diminishing runge-kutta schemes. Mathematics of Computation of the American Mathematical Society 67(221): 73–85.

Tokareva, S. A. and Toro, E. F. (2010). HLLC-type Riemann solver for the Baer-Nunziato equations of compressible two-phase flow. J. Comput. Phys. 229(10): 3573–3604.

Toro, E. F., Spruce, M. and Speares, W. (1992). Restoration of the contact surface in the HLLC Riemann solver. Technical report, Craneld Institute of Technology.

van Albada, G. D., van Leer, B. and Roberts, W. W. (1997). A comparative study of computational methods in cosmic gas dynamics. In Upwind and High-Resolution Schemes, Springer, pp. 95–103.

Weber, J. (2007). Representation of non-linear aero-thermodynamic effects during small scale physical modelling of owc wecs. In Proceedings of the 7th European Wave and Tidal Energy Conference, volume 895.

Wengfeng Xie. (2005). A Numerical Simulation of Underwater Shock-avitation-structure Interaction. Ph.D. thesis, National University of Singapore.

Wood, A. B. (1941). A Textbook of Sound. G. Bell and Sons, London.

Xu, G. D., Duan, W. Y. and Wu, G. X. (2010). Simulation of water entry of a wedge through free fall in three degrees of freedom. Proc. R. Soc. Lond. A 466(2120): 2219–2239.

Zhang, S., Yue, D. K. P. and Tanizawa, K. (1996). Simulation of plunging wave impact on a vertical wall. J. Fluid Mech. 327: 221–254.

Chapter 5

Violent Wave Impacts and Loadings using the δ-SPH Method

Matteo Antuono, * *Salvatore Marrone* and *Andrea Colagrossi*

1   Introduction

Modelling of physical problems with violent impacts and strong fluid-structure interaction often represents a demanding challenge for many numerical schemes. The main difficulties arise from the occurrence of large deformations of the air-water interface (due to wave breaking events) and from the imposition of the boundary conditions along it. Differently from mesh-based schemes (which need specific techniques to model the free-surface evolution), particle methods can directly track the air-water interface thanks to their Lagrangian structure. In fact, numerical particles can be regarded as movable computational nodes in the bulk fluid. A further advantage of such schemes is the fact that it is possible to model solid bodies and boundaries through particles as well. This allows for a straightforward description of moving solid profiles in comparison to many grid-based schemes which generally use immersed boundaries (see, e.g., Peskin,1972; Mittal and Iaccarino, 2005), overset grids (Carrica et al., 2007; Muscari et al., 2011) or need to implement remeshing algorithms during the evolution (e.g., Löhner et al., 2008).

A promising particle method which has proved to be accurate and robust in a wide variety of engineering applications is the Smoothed Particle Hydrodynamics (SPH) scheme. This is a meshless Lagrangian scheme whose formulation relies on two main steps: (i) the representation of the continuous differential operators through convolution integrals with a compact-support kernel function, (ii) the discretization of such convolution integrals into a finite set of elementary fluid particles. SPH was initially conceived in astrophysics for the evolution of gaseous clouds (see Gingold and Monaghan, 1977; Lattanzio et al., 1989) and, later, was applied to hydrodynamics (Monaghan, 1994; Monaghan, 1996; Colagrossi and Landrini, 2003). Because of this historical path, the first SPH models were derived under the assumption that the fluid is weakly-compressible, i.e., that the density variation remains below 1% during the flow evolution. The density field is then used to compute the pressure field through the state equation for barotropic fluids. Through the years, however, several variants of SPH have been defined on the hypothesis that the fluid is incompressible, this being a more natural assumption in hydrodynamics (though not the most convenient for SPH, as we will discuss later). To date, the distinction between weakly-compressible and incompressible models still represents the main classification of the SPH schemes.

CNR-INM, Institute of Marine engineering, Via di Vallerano 139, Rome, 00128, Italy.
* Corresponding author: matteo.antuono@cnr.it

A further classification for applications in the hydrodynamic context can be achieved with respect to the modelling of the air phase. The air-water interface can be modelled in SPH either by a single-phase approximation (i.e., only the water phase is discretized) or by a two-phase approach (both air and water domains are discretized). The former allows to reduce the CPU costs of the simulation whereas the latter is needed whenever air-cushioning effects are of interest in the simulation.

In the present chapter we will only deal with single-phase SPH schemes; in particular, we will focus on weakly-compressible variants. One of the main advantages in this case is that the dynamic boundary condition along the free surface is implicitly satisfied during the flow evolution (see Colagross et al., 2009; Colagrossi et al., 2011). This implies that it is not necessary to detect the free surface, leading to a considerable reduction in computational cost. A further fundamental difference between compressible and incompressible schemes is that the former are explicit in time while the latter are implicit and rely on the solution of a Poisson equation for the pressure field. The linear system that stems from the above equation requires a large computational cost which essentially cancels out the advantage for a larger time step during the numerical simulations in comparison to the weakly-compressible variants. In turn, being explicit in time, the latter schemes are more easily parallelized and show good scalability properties (see Crespo et al., 2011; Oger et al., 2016).

The main drawback of the weakly-compressible models is represented by the generation of a spurious numerical acoustic noise that affects both the pressure and density fields. In the SPH literature, several attempts have been made to overcome this issue. In Colagrossi et al. (2003) a filtering of the density field was suggested through the use of a Moving Least Squares (MLS) integral interpolation. This approach gave good results for confined flows but, in the presence of a free surface and for long-time simulations, it did not conserve the total volume of the fluid system, since the hydrostatic component of the pressure field was filtered improperly (see example, Sibilla, 2007). An alternative approach was proposed by Vila, 1999 and Moussa and Vila, 2000, who studied the convergence of SPH schemes using approximate Riemann solvers to model the particle interactions. From their works, a family of SPH schemes based on Riemann solvers was developed (see Marongiu et al., 2010; Koukouvinis et al., 2013; Oger et al., 2016). These models generally provide accurate results but, unfortunately, do not allow an exact quantification of the dissipation caused by the numerical scheme that is generally quite large.

Exploiting the theory of Riemann solvers, Ferrari et al., 2009 defined a numerical diffusive term based on the use of a Rusanov flux. This term was added in the continuity equation to reduce the numerical noise affecting the density field. A similar approach was followed in Molteni and Colagrossi, 2009 where the diffusive term was modelled as the Laplacian of the density field. It is worth noting that a similar term was obtained in Clausen, 2013 by rewriting the continuity equation in terms of the pressure field. In particular, this was achieved by imposing the entropy variation to minimize the density fluctuations, this corresponding to an approximation of an incompressible flow. Unfortunately both the Ferrari et al., 2009 and Molteni and Colagrossi, 2009 schemes proved incompatible with the hydrostatic solution in the presence of a free surface, since the adopted diffusive terms introduced non-zero contributions close to the interface. To circumvent this issue and, at the same time, maintain the positive features of the above-mentioned diffusive schemes, Antuono et al., 2010 proposed a correction to the diffusive term of Molteni and Colagrossi, 2009. The corrected term proved to be compatible with the hydrostatic solution and to properly smooth out the numerical spurious oscillations from the pressure and density fields.

The scheme described in Antuono et al., 2010, which is now known as the δ-SPH scheme, has been widely inspected from both a theoretical (see, for example, Antuono et al., 2012; Antuono et al., 2015) and a numerical point of view (Marrone et al., 2011; Antuono et al.,

2015; Bouscasse et al., 2013; Bouscasse et al., 2013), proving to be an accurate, robust and reliable model for a broad variety of applications in fluid dynamics (see, e.g., Canelas et al., 2016; Meringolo et al., 2015; Crespo et al., 2017; Altomare et al., 2017). For these reasons, in the present chapter we will exclusively deal with the δ-SPH scheme, providing details on the implementation of the model, the procedure to assign boundary conditions along solid profiles and, finally, showing some relevant applications.

2 Governing equations

The governing equations at the basis of a generic weakly-compressible SPH scheme are the Euler equations for barotropic and compressible fluids. In a Lagrangian formalism, these read:

$$
\begin{cases}
\dfrac{d\rho}{dt} = -\rho\,\nabla\cdot\boldsymbol{u}\,, \\[2mm]
\dfrac{d\boldsymbol{u}}{dt} = -\dfrac{\nabla p}{\rho} + \boldsymbol{g}\,, \\[2mm]
\dfrac{d\boldsymbol{r}}{dt} = \boldsymbol{u}\,, \qquad p = F(\rho)\,,
\end{cases}
\tag{1}
$$

where $d/dt = \partial/\partial t + \boldsymbol{u}\cdot\nabla$ indicates the Lagrangian derivative. Here $\boldsymbol{u}$ is the flow velocity, $\boldsymbol{g}$ is gravity acceleration, p and ρ denote the pressure and density fields respectively and F represents the state equation. The hypothesis that the fluid is weakly-compressible corresponds to the fulfilment of the following inequality:

$$
\frac{dp}{d\rho} = c^2 \gg \max\left(\|\boldsymbol{u}\|^2, \frac{\Delta p}{\rho}\right),
\tag{2}
$$

where $c = c(\rho)$ is the sound speed and Δp is the maximum pressure variation (see, e.g., Monaghan et al., 1994; Marrone et al., 2015).

The SPH model at the continuum is obtained from system (1) by substituting the spatial differential operators by their mollified (smoothed) counterparts. Below, we give some details about such a filtering (or, equivalently, *smoothing*) procedure. Let us consider a generic flow field f and its convolution integral with a weight function W (also known as kernel) over the fluid domain (hereinafter denoted by Ω):

$$
\langle f\rangle(\boldsymbol{r}) = \int_\Omega f(\boldsymbol{r}')\,W(\boldsymbol{r}'-\boldsymbol{r};h)\,dV'\,.
\tag{3}
$$

In particular $W(\boldsymbol{r}'-\boldsymbol{r};h)$ is a positive, radial, strictly decreasing function of $\|\boldsymbol{r}'-\boldsymbol{r}\|$ and has a compact support. The symbol h (usually referred to as the *smoothing length*) indicates the characteristic length of such a support which, from a physical perspective, represents the domain of influence of a fluid point-like particle at the position $\boldsymbol{r}$. A very comprehensive review of the SPH framework can be found in reference Monaghan, 2005. Hereinafter, the notation $W(\boldsymbol{r}'-\boldsymbol{r};h)$ will be shortened as $W(\boldsymbol{r}'-\boldsymbol{r})$ and the dependence on h will be implicitly assumed. In the limit as the smoothing length h goes to zero, the original field of the convolution integral (3) should be recovered. In order to satisfy this requirement, the kernel W must integrate to one.

$$
\int_\Omega W(\boldsymbol{r}'-\boldsymbol{r};h)\,dV' = 1\,.
\tag{4}
$$

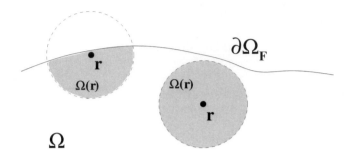

Figure 1: Configurations of the kernel support $\Omega(\boldsymbol{r})$ with respect to the fluid domain boundary $\partial \Omega_F$.

As extensively discussed in Colagrossi et al., 2009, such a property is not satisfied when the kernel support $\Omega(\boldsymbol{r})$ is not completely immersed inside the fluid domain which is a common configuration for particles close to the domain boundary $\partial \Omega$ (see Figure 1). The incompleteness of the kernel support close to the domain boundary plays a major role in the definition of the SPH differential operators at the continuum. In any case, the theoretical analysis of the SPH is not of concern in the present chapter. In the following we describe the procedure to obtain a smoothed operator for the gradient and address the interested reader to (Colagrossi et al., 2009; Colagrossi et al., 2011; Español and Revenga, 2006; Quinlan et al. 2006; Amicarelli et al., 2011; Violeau, 2012) for more details on the theoretical foundations of SPH.

Applying the filtering formula (3) to the gradient of the function f, we obtain:

$$\langle \nabla f \rangle (\boldsymbol{r}) = \int_{\Omega} \nabla' f(\boldsymbol{r}') \, W(\boldsymbol{r}' - \boldsymbol{r}) \, dV' \tag{5}$$

where the prime on ∇ means that the derivatives are computed on the $\boldsymbol{r}'$ variable. Using Green's theorem, the above integral can be arranged as follows:

$$\langle \nabla f \rangle (\boldsymbol{r}) = \int_{\Omega} f(\boldsymbol{r}') \, \nabla W(\boldsymbol{r}' - \boldsymbol{r}) \, dV' + \int_{\partial \Omega} f(\boldsymbol{r}') \, W(\boldsymbol{r}' - \boldsymbol{r}) \, \boldsymbol{n}' \, dS' \tag{6}$$

In this expression, ∇ indicates the derivatives with respect to the variable $\boldsymbol{r}$ and $\boldsymbol{n}'$ is a unitary normal vector to $\partial \Omega$, pointing outwards Ω. Note that the antisymmetry property of the kernel gradient, namely $\nabla' W(\boldsymbol{r} - \boldsymbol{r}') = -\nabla W(\boldsymbol{r} - \boldsymbol{r}')$, has been used. Thanks to this reformulation of Equation (5), the gradient of any generic function is accessible from the knowledge of the function itself; this represents the key point of the SPH method. The modelling of the surface term in Equation (6) differs according to the single spatial operator under consideration (e.g., pressure gradient, velocity divergence, etc.) and is motivated by the requirement of specific conservation properties of the scheme or by stability reasons. For further details, the reader can refer to the references (Colagrossi et al., 2009; Colagrossi et al., 2011) where an in-depth analysis of the smoothed differential operators and surface integrals is provided.

If the smoothing procedure is applied to the differential operators of the governing Equations (1), shortening the notation $\langle f \rangle(\boldsymbol{r})$ by $\langle f \rangle$, the SPH continuous formulation of the Euler equations in its weakly compressible form is obtained:

$$
\begin{cases}
\dfrac{d\rho}{dt} = -\rho \left\langle \nabla \cdot \boldsymbol{u} \right\rangle, \\[2ex]
\dfrac{d\boldsymbol{u}}{dt} = -\dfrac{\langle \nabla p \rangle}{\rho} + \boldsymbol{g}, \\[2ex]
\dfrac{d\boldsymbol{r}}{dt} = \boldsymbol{u}, \qquad p = F(\rho).
\end{cases}
\tag{7}
$$

The consistency of (7) for the modeling of the Euler equations in the presence of solid boundaries and free surfaces has been thoroughly investigated in Colagrossi et al., 2009.

In the following section we describe the discrete model that stems from system (7). In particular, we introduce the δ-SPH scheme which is a diffusive variant of the standard SPH model defined in Monaghan, 1994; Monaghan, 2005.

3   The δ-SPH scheme

The differential operators adopted to discretize the system (7) are those typical of the so-called standard SPH model. Their forms are not only motivated by a direct discretization of the mollified operators but also by specific requirements for conservation properties of the final scheme (see Monaghan, 2005; Colagrossi et al., 2009). In the SPH literature, several variants of these operators have been proposed (see, for example, Monaghan, 2005) but the differences from one formulation to the other generally have negligible effects on the final model. In the present case, we adopt the following forms for the divergence of the velocity field and the pressure gradient:

$$
\langle \nabla \cdot \boldsymbol{u} \rangle_i = \sum_j (\boldsymbol{u}_j - \boldsymbol{u}_i) \cdot \nabla_i W_{ij} \, V_j,
\tag{8}
$$

$$
\langle \nabla p \rangle_i = \sum_j (p_j + p_i) \nabla_i W_{ij} \, V_j.
\tag{9}
$$

The subscript i indicates the quantities transported by i-th particle while the subscript j those transported by the neighbouring particles, that is, those particles inside the compact support of the kernel function centred at the i-th particle position. The latter is indicated through $\boldsymbol{r}_i$ while $W_{ij} = W(\boldsymbol{r}_i - \boldsymbol{r}_j)$ and ∇_i is the gradient with respect to the position $\boldsymbol{r}_i$. In the present case a C2-Wendland function is used (see Wendland, 2012) and about 50 neighbouring particles (in a 2D framework) are considered in the kernel support, which corresponds to $h = 2\Delta x$ where Δx is the initial particle distance.

In Equations (8) and (9) the symbol V_i denotes the elementary volume, $\boldsymbol{u}_i$ the fluid velocity and p_i the pressure field. Since the fluid is assumed to be weakly-compressible, the state equation can be linearized as follows:

$$
p_i = c_0^2 \left(\rho_i - \rho_0 \right),
\tag{10}
$$

where ρ_0 is a reference density value (usually, for free-surface flows this is the density at the interface) and c_0 is a reference sound velocity. To ensure that the density variations are kept small during the flow evolution, the inequality (2) has to be fulfilled. Since the time step of the SPH is essentially computed basing on the sound velocity (as explained later in the present section), the above requirement is relaxed as follows:

$$c_0 = 10 \max \left(U_{max}, \sqrt{\frac{\Delta p_{max}}{\rho_0}} \right), \qquad (11)$$

where U_{max} and Δp_{max} are the maximum expected velocity and pressure variation. The above choice guarantees that the density variations remain below 1% during the flow motion and, moreover, that the time-step of the numerical scheme is not too small. We note that in some particular impact situations, such as those in which water hammer pressure develops, the above estimation is not sufficient to guarantee negligible density variations. The interested reader can find more details about this topic in Marrone et al., 2015.

As explained in the Introduction, the pressure and density fields of the standard SPH model are generally affected by a spurious high-frequency noise that is caused by the weak-compressibility assumption. A reliable strategy to overcome such an issue is based on the addition of a suitable numerical diffusive term, hereinafter $\mathcal{D}$, in the continuity equation. Over the years, different diffusive variants of the standard SPH scheme have been introduced (see, for example, Ferrari et al., 2009; Molteni and Colagrossi, 2009). The δ-SPH model, which has been firstly defined in Antuono et al., 2010 and further inspected in Antuono and Colagrossi, 2012; Antuono et al., 2015, belongs to this family of diffusive SPH schemes. Similar to standard SPH, it preserves the mass and the linear/angular momenta and, differently from the diffusive variants proposed in Ferrari et al., 2009; Molteni and Colagrossi, 2009, it is compatible with the presence of a free surface. The δ-SPH scheme has been successfully applied to several hydrodynamics problems, proving to be accurate and robust and, over the last few years, it has become a popular variant of SPH, being used by several researchers and engineers in different scientific areas (see, e.g., Chowdhury and Sannasiraj, 2014; Valizadehand Monaghan, 2012; Marrone et al., 2013; Meringolo et al., 2015). For these reasons, in the following we just focus on this SPH variant, showing its main applications in the field of hydrodynamics.

When the diffusive term $\mathcal{D}$ is added in the continuity equation, the discrete system for the δ-SPH scheme reads:

$$\begin{cases} \dfrac{d\rho_i}{dt} = -\rho_i \sum_j (\boldsymbol{u}_j - \boldsymbol{u}_i) \cdot \nabla_i W_{ij} V_j + \delta\, h\, c_0\, \mathcal{D}_i, \\[3mm] \dfrac{d\boldsymbol{u}_i}{dt} = -\dfrac{1}{\rho_i} \sum_j (p_j + p_i)\, \nabla_i W_{ij}\, V_j + \boldsymbol{g} + \alpha\, h\, c_0\, \dfrac{\rho_0}{\rho_i} \sum_j \pi_{ij}\, \nabla_i W_{ij}\, V_j, \\[3mm] \dfrac{d\boldsymbol{r}_i}{dt} = \boldsymbol{u}_i, \qquad p_i = c_0^2\,(\rho_i - \rho_0), \qquad V_i = m_i/\rho_i. \end{cases} \qquad (12)$$

Here m_i indicates the mass of the i-th particle. In particular, this remains constant during the flow evolution, implying intrinsic conservation of the global mass, while the particle volume is computed through the ratio m_i/ρ_i. The additional term in the momentum equation represents an artificial viscosity term and is generally adopted in weakly-compressible models to stabilize the scheme (see, for example, Monaghan,2005). Accordingly, the use of such a term does not correspond to an actual modelling of the boundary layer close to solid profiles or along the free-surface. In fact, for many applications in the context of coastal and marine engineering this may be neglected, because it plays a secondary role on the flow dynamics. The argument of such a viscous term is:

$$\pi_{ij} = \frac{(\boldsymbol{u}_j - \boldsymbol{u}_i) \cdot (\boldsymbol{r}_j - \boldsymbol{r}_i)}{\|\boldsymbol{r}_j - \boldsymbol{r}_i\|^2}, \qquad (13)$$

while the dimensionless parameter α is used to set the order of magnitude of the numerical viscosity. Generally, the range of variability of α is quite limited (usually, from 0.01 to 0.1). In particular, for the simulations shown in the present Chapter, $\alpha = 0.01$ has been chosen.

The diffusive term of the δ-SPH is given by:

$$\mathcal{D}_i = 2 \sum_j \psi_{ji} \frac{(\boldsymbol{r}_j - \boldsymbol{r}_i) \cdot \nabla_i W_{ij}}{\|\boldsymbol{r}_j - \boldsymbol{r}_i\|^2} V_j, \tag{14}$$

where

$$\psi_{ji} = \left\{ (\rho_j - \rho_i) - \frac{1}{2} \left(\langle \nabla \rho \rangle_j^L + \langle \nabla \rho \rangle_i^L \right) \cdot (\boldsymbol{r}_j - \boldsymbol{r}_i) \right\}. \tag{15}$$

and the symbol $\langle \nabla \rho \rangle_i^L$ indicates the renormalized density gradient (see Randles and Libersky, 1996). As shown in (Antuono and Colagrossi, 2012), the diffusive term defined by Equations (14) and (15) represents a fourth-order spatial differential operator inside the bulk fluid and reduces to a third-order operator close to the free-surface. Its intensity is controlled thought the dimensionless parameter δ (hence the name of this particular SPH variant) which, in any case, is not problem-dependent. In particular, Antuono et al., 2012 showed that its range of variation is quite narrow and only depends on $h/\Delta x$ (where Δx is the mean particle distance). In practise, for a given value of $h/\Delta x$, there is no need for heuristic tuning. For example, in all the simulations proposed in the present chapter δ has been set equal to 0.1. Before proceeding with the description of the model, we highlight that both the viscous and the diffusive terms are multiplied by h (see the system in Equation (12)). This ensures that, for increasing resolution (that is, for decreasing values of h), consistency with the Euler equations is recovered.

The system (12) is usually integrated in time by using a fourth-order Runge-Kutta scheme with a frozen diffusion approach as described in (Antuono and Colagrossi, 2012). The latter technique allows for a substantial reduction of the CPU-time costs and increases the robustness of the scheme. The time step is obtained as the minimum over the following bounds:

$$\Delta t \leq 0.44 \frac{h}{\delta c_0}, \quad \Delta t \leq 0.25 \min_i \sqrt{\frac{h}{\|\boldsymbol{a}_i\|}}, \quad \Delta t \leq CFL \min_i \left(\frac{h}{c_0 + \|\boldsymbol{u}_i\| + h \max_j |\pi_{ij}|} \right).$$

where $\|\boldsymbol{a}_i\|$ is the i-th particle acceleration and $CFL = 2$ for a Wendland Kernel. Generally, the last inequality is the most restrictive.

4 Modelling solid bodies

The presence of solid profiles causes a cut of the kernel domains of particles that are close to the body. To avoid inconsistencies and to assign the correct boundary conditions, the "missing" volume in the incomplete kernel domain has to be properly modelled. Incidentally, we underline that such a cut also occurs for particles close to the free surface but, in that case, this does not induce any problem, since the weakly-compressible SPH implicitly satisfies the dynamic boundary condition along the interface.

In the SPH literature several techniques have been proposed to impose the correct boundary conditions along solid profiles. Early techniques relied on repulsive (Monaghan, 1989) or dynamic particles (Dalrymple and Knio, 2001). Later on ghost fluid approaches (Randles and Libersky, 1996; Colagrossi and Landrini, 2003) were proposed and adapted for general body shapes as in the case of the fixed ghost particles Marrone et al., 2011 or dummy particles Adami et al., 2012. More recently semi-analytical (Kulasegaramet al., 2004; Leroy et al., 2014) approaches and methods based on the evaluation of the flux normal to the wall (Marongiu et al., 2012; De Leffe et al., 2009) have been proposed. Due to its flexibility, robustness and accuracy, the fixed ghost particle method is one of the best suited technique for the applications considered in the present Chapter. For this reason, we just restrict the discussion to such a method and address the interested readers to the above mentioned works for alternative approaches.

Before proceeding, we highlight that free-slip boundary conditions are imposed along the solid profiles. In fact, for the applications described here, the presence of the boundary layer may be neglected. This also means that the viscous term introduced in system (12) only represents an artificial bulk viscosity.

4.1 The ghost-fluid method

In the present section we describe how to enforce the appropriate boundary conditions on the body surface by using the ghost-fluid technique. This basically comprises two steps: (i) the solid domain is modelled through a set of "imaginary particles" (hereinafter denoted as "ghost particles" and labelled with the subscript "s"), (ii) the fluid fields (that is, velocity, pressure) are extended on such fictitious particles through proper mirroring techniques. In particular, different mirroring techniques are adopted to enforce different boundary conditions (e.g., Dirichlet or Neumann conditions).

At step (i), the solid surface is discretized in equispaced body nodes and several layers of ghost particles are disposed inside the solid up to a radius of the kernel domain (see the sketch in the left plot of Figure 2). At step (ii), the velocity and pressure assigned to the fixed ghost particles, namely $(\boldsymbol{u}_s, p_s)$, are computed by using the values interpolated at specific nodes internal to the fluid and uniquely associated with the fixed ghost particles. Hereinafter, the interpolated values are indicated through $(\boldsymbol{u}^*, p^*)$. More details about the way to dispose the ghost particles and define the interpolation nodes can be found in (Marrone et al., 2011).

The pressure field p_s is mirrored on the fixed ghost particles to enforce the following Neumann condition (derived from the momentum equation):

$$\frac{\partial p}{\partial \boldsymbol{n}} = \rho \left[\boldsymbol{f} \cdot \boldsymbol{n} - \frac{\mathrm{d}\boldsymbol{u_b}}{\mathrm{d}t} \cdot \boldsymbol{n} \right] , \qquad (16)$$

where $\boldsymbol{f}$ is a generic body force and $\boldsymbol{u_b}$ is the velocity of the solid boundary (for details see Marrone et al., 2011). This leads to:

$$p_s = p^* + \frac{\partial p}{\partial \boldsymbol{n}} \cdot (\boldsymbol{r}^* - \boldsymbol{r}_s) \qquad (17)$$

where $\boldsymbol{r}^*$ indicates the position of the interpolation node.

Conversely, the velocity field is the object of a specific treatment. As sketched in the right plot of Figure 2, the ghost velocity $\boldsymbol{u}_s$ depends on both $\boldsymbol{u}^*$ and $\boldsymbol{u}_b$, the latter being the velocity of the nearest body node. De Leffe et al., 2017 found that different mirroring techniques have to be used to evaluate different differential operators (for example, the mirroring technique for the velocity divergence is different from that used for the Laplacian of the velocity). This is needed to avoid inconsistencies, loss of accuracy or, in the worst case, instabilities. The specific mirroring techniques depend on the components of $\boldsymbol{u}^*$ in the normal and tangential direction to the solid surface (right plot of Figure 2). In particular, De Leffe et al., 2011 proved that the velocity-divergence operator is convergent and consistent

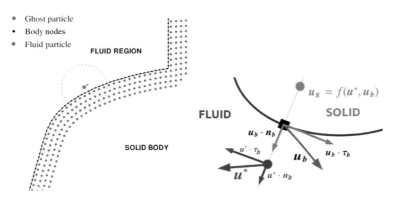

Figure 2: Sketch of the ghost-fluid approach. Left: Discretization of the ghost-fluid through ghost particles. Right: Mirroring of the velocity.

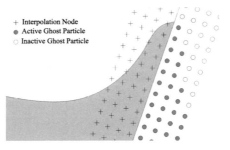

Figure 3: Sketch of the activated ghost particles at the intersection between the free surface and the body surface.

if the normal component of $\boldsymbol{u}^*$ is mirrored in the frame of reference of the solid profile (see Colagrossi and Landrini, 2003), leaving the tangential component unaltered:

$$\langle \nabla \cdot \boldsymbol{u} \rangle \qquad \begin{cases} \boldsymbol{u}_s \cdot \boldsymbol{n} = 2\,\boldsymbol{u}_b \cdot \boldsymbol{n} - \boldsymbol{u}^* \cdot \boldsymbol{n} \\ \boldsymbol{u}_s \cdot \boldsymbol{\tau} = \boldsymbol{u}^* \cdot \boldsymbol{\tau} \end{cases} \tag{18}$$

Regarding the viscous operator in the system (12), this approximates the Laplacian of the velocity field and, therefore, a different mirroring technique has to be implemented. Since this is actually an artificial viscous term (which is just used to regularize the velocity field), the adopted mirroring technique is designed to give a negligible contribution along the solid profiles, thus resembling free-slip conditions. In particular, we use:

$$\langle \nabla^2 \boldsymbol{u} \rangle \qquad \begin{cases} \boldsymbol{u}_s \cdot \boldsymbol{n} = \boldsymbol{u}^* \cdot \boldsymbol{n} \\ \boldsymbol{u}_s \cdot \boldsymbol{\tau} = \boldsymbol{u}^* \cdot \boldsymbol{\tau} \end{cases} \tag{19}$$

which simply represents a uniform extension of the velocity field in the solid domain.

Before proceeding to the evaluation of forces and torques through the ghost-fluid technique, we highlight that some care has to be given to the numerical treatment of the intersection between the free surface and the body. In fact, in this case some interpolations nodes may be outside of the fluid region and, therefore, may cause an inaccurate mirroring. To avoid this issue, these nodes (and the related ghost particles) are switched off, as shown in Figure 3. A reliable technique for the detection of these nodes is described, for example, in (Bouscasse et al., 2013).

4.2 Evaluation of Forces and Torques through the ghost-fluid method

The forces on solid bodies are generally calculated by using two main approaches: (*i*) the integration of the local stresses along the body surface (see, e.g., Fourey et al., 2017), (*ii*) the definition of a volume integral over the ghost particle domain Doring, 2005; Bouscasse et al., 2013. Consistent with the introduction of fixed ghost particles in the previous section, we only deal with the latter approach (see Bouscasse et al., 2013 for more details).

To find out the formulation for the global loads exerted by the fluid on solid structures, it is convenient to develop the analysis at the continuum level and, later, derive the discrete equations. In the following, the fluid and solid domains are denoted by Ω_f and Ω_s respectively while $\langle \mathbb{T} \rangle$ indicates the smoothed stress tensor. Then, the global force on the body is expressed by:

$$\boldsymbol{F}_{\text{fluid-solid}} = \int_{\partial \Omega_s} \langle \mathbb{T} \rangle \cdot \boldsymbol{n} \, dS \tag{20}$$

where $\boldsymbol{n}$ is the unit outward normal to the solid profile. Assuming the flow field to be mirrored on the solid body through a suitable ghost-fluid technique, the stress tensor can be decomposed in:

$$\langle \mathbb{T} \rangle(\boldsymbol{r}) = \int_{\Omega_f} \mathbb{T}' \, W(\boldsymbol{r}' - \boldsymbol{r}) \, dV' + \int_{\Omega_s} \mathbb{T}^* \, W(\boldsymbol{r}^* - \boldsymbol{r}) \, dV^* \tag{21}$$

where the starred variables indicate quantities mirrored over the solid domain Ω_s. Substituting (21) into (20) and using the divergence theorem and the symmetry properties of the kernel function, we obtain the following equality:

$$\boldsymbol{F}_{\text{fluid-solid}} = \int_{\Omega_f} dV \int_{\Omega_s} (\mathbb{T}^* + \mathbb{T}) \cdot \nabla W(\boldsymbol{r}^* - \boldsymbol{r}) \, dV^* + \mathcal{O}(h) \tag{22}$$

where ∇ denotes the differentiation with respect to the position $\boldsymbol{r}$. The terms of order $\mathcal{O}(h)$ indicate the contributions due to the presence of the free surface. These terms are small since, by definition, the tension along the free surface is zero. When Equation (22) is discretized, it reads:

$$\boldsymbol{F}_{\text{fluid-solid}} = \sum_{i \in fluid} \sum_{j \in solid} (\mathbb{T}_j + \mathbb{T}_i) \cdot \nabla_i W_{ij} \, V_i \, V_j \tag{23}$$

where i and j denote quantities associated with the fluid particles and the ghost particles respectively. One of the advantages of Equation (23) is that it directly uses the flow quantities mirrored inside the body while it does not require interpolation on body nodes (i.e., along the body profile). For this reason, it is simpler and faster to use in practical applications. Since the inner summation of (23) approximates the divergence of the stress tensor, in practical simulations it is sufficient to substitute the corresponding operator of the SPH scheme at hand. In the present case, this leads to:

$$\boldsymbol{F}_{\text{fluid-solid}} = \sum_{i \in fluid} \sum_{j \in solid} \left[-(p_j + p_i) + \alpha \, h \, c_0 \, \rho_0 \, \pi_{ij} \right] \nabla_i W_{ij} \, V_i \, V_j \, . \tag{24}$$

Note that, despite the fact that fluid is assumed to be inviscid, the artificial viscosity term has to be included in the formula above. This is required to ensure the correct energy balance between the solid and fluid phases (see the next section for more details). In any case, the mirroring technique defined in Equation (19) guarantees that the viscous contribution in formula (24) is generally small.

The evaluation of the torque $\boldsymbol{T}_{\text{fluid-solid}}$ acting on the solid body is derived following the same approach shown above. Let us consider a fixed point $\boldsymbol{r_0}$. Then, the torque with respect to it is:

$$\boldsymbol{T}_{\text{fluid-solid}} \quad = \quad \int_{\partial \Omega_s} (\boldsymbol{r} - \boldsymbol{r_0}) \times \langle \mathbb{T} \rangle \cdot \boldsymbol{n} \, dS \tag{25}$$

and, following the same procedure adopted for the evaluation of the force, it is possible to rearrange the above expression as follows:

$$\boldsymbol{T}_{\text{fluid-solid}} = \int_{\Omega_f} dV \int_{\Omega_s} \Big\{ \; (\boldsymbol{r}^* - \boldsymbol{r_0}) \times \Big[\mathbb{T} \cdot \nabla W(\boldsymbol{r}^* - \boldsymbol{r}) \Big] + \\ + \; (\boldsymbol{r} - \boldsymbol{r_0}) \times \Big[\mathbb{T}^* \cdot \nabla W(\boldsymbol{r}^* - \boldsymbol{r}) \Big] \Big\} \, dV^* + \mathcal{O}(h) \, . \tag{26}$$

By analogy with formula (24), the above equation is discretized as follows:

$$\boldsymbol{T}_{\text{fluid-solid}} = \sum_{i \in fluid} \sum_{j \in solid} \Big\{ \; (\boldsymbol{r}_j - \boldsymbol{r_0}) \times \Big[(-p_i + \alpha \, h \, c_0 \, \rho_0 \, \pi_{ij}/2) \, \nabla_i W_{ij} \Big] + \\ + \; (\boldsymbol{r}_i - \boldsymbol{r_0}) \times \Big[(-p_j + \alpha \, h \, c_0 \, \rho_0 \, \pi_{ij}/2) \, \nabla_i W_{ij} \Big] \Big\} V_i \, V_j \, . \tag{27}$$

The above expressions for the force and torque are reminiscent of the technique proposed in (Monaghan et al., 2003) and (Kajtar and Monaghan, 2008). In any case, we underline that, apart from the different enforcement of solid boundary conditions, in those cases the formulation is directly obtained from a momentum balance between the fluid and the repulsive body particles. Conversely, in the present case, the global loads are derived from the evaluation of the stress tensor on the body surface by means of a proper ghost fluid extension in the solid region.

4.3   Algorithm for fluid-body coupling

In the present section, we briefly describe an algorithm for the coupling of the fluid-body dynamics. The solid dynamics are modelled by Newton's law of motion which, for the sake of simplicity, is introduced here in a 2D framework:

$$\begin{cases} M \dfrac{d\boldsymbol{V}_G}{dt} = M\boldsymbol{g} + \boldsymbol{F}_{\text{fluid-solid}} \\[4mm] I_G \dfrac{d\Omega_G}{dt} = T_{\text{fluid-solid}} \end{cases} \tag{28}$$

where $\boldsymbol{V}_G$ and Ω_G are the velocity of the centre of gravity and the angular velocity of the body, M and I_G are the mass and the moment of inertia around the centre of gravity and, finally, $\boldsymbol{F}_{fluid-solid}$ is the hydrodynamic force acting on the body. Here, $T_{fluid-solid}$ is the projection of the hydrodynamic torque along the unit vector $\boldsymbol{k}$ normal to the 2D plane, that is $T_{fluid-solid} = \boldsymbol{T}_{\text{fluid-solid}} \cdot \boldsymbol{k}$.

The dynamical state of the system made of fluid particles, ghost particles and body nodes can be expressed through the vectors y_f, y_s and y_b respectively:

$$\begin{aligned} y_f &= (\dots, \rho_i, \boldsymbol{u}_i, \boldsymbol{r}_i, \dots) & i &\in \text{Fluid} \\ y_s &= (\dots, \rho_j, \boldsymbol{u}_j, \boldsymbol{r}_j, \dots) & j &\in \text{Solid} \\ y_b &= (\dots, \boldsymbol{r}_k, \boldsymbol{u}_k, \boldsymbol{a}_k, \boldsymbol{n}_k, \dots) & k &\in \text{Body Surface} \end{aligned} \tag{29}$$

while the dynamical state of the rigid body is expressed by:

$$y_g = (\boldsymbol{r}_G, \boldsymbol{V}_G, \theta_G, \Omega_G) \tag{30}$$

where $\boldsymbol{r}_G$ and θ_G are respectively the position vector of the centre of gravity and the related angle of rotation. The coupling between the two systems from Equations (12) and (28) is given by:

$$
\begin{cases}
\dot{y}_f &= F_f(y_f, y_s, t) & y_f(t_0) = y_{f0} \\
\dot{y}_g &= F_g(y_g, y_f, y_s, t) & y_g(t_0) = y_{g0} \\
y_s &= F_s(y_b, y_f, t) & \\
y_b &= F_b(y_g, \dot{y}_g, y_{b0}) & y_{b0} = y_b(t_0)
\end{cases} \tag{31}
$$

where the last two equations represent, respectively, the dependence of the ghost state [see Equations (17), (18) and (19)] and of the body nodes state from the rigid motion equations.

Since the δ-SPH equations march in time using an explicit scheme (a fourth-order Runge-Kutta scheme in the present case), the same integration scheme is adopted for the whole system (31). Note that the acceleration of the body appears on both sides of the system (31). This is common in explicit schemes for fluid-body coupling. In particular, in potential flow solvers the body acceleration is generally taken into account through the added mass term (see, e.g., Vinje and Brevig, 1981), enabling one to move such an acceleration term from the right side to the left side. The added mass approach is also applied in other numerical solvers as an under-relaxation correction in the body motion (see, e.g., Adami et al., 2012). In the present scheme the acceleration of the body on the right side is taken from the previous Runge-Kutta substep. This procedure is justified by the use of a very small time-step that is required by the weakly-compressible assumption, as briefly recalled below.

At the generic time instant t^n, the state vector y_b is determined through $\dot{y}_g$ (predicted at the previous Runge-Kutta substep) and y_g. Then, the ghost fluid state y_s is obtained through the interpolation on the fluid particles, y_f, and the mirroring procedure (which requires y_b, as discussed in Section 4.1). The global loads $(\boldsymbol{F}, T)_{\text{fluid-solid}}$ are evaluated through Equations (23) and (27) and $\dot{y}_g$ is obtained by (28). Finally, the interaction between fluid and ghost particles, namely $\dot{y}_f$, is computed through the equations of system (12). The iteration substep ends with the integration of $\dot{y}_g$ and $\dot{y}_f$ to obtain respectively y_g and y_f at the time instant t^{n+1}.

As far as the time integration of system (31) is concerned, the time step (see the last equation in Section 3) has to account for the maximum acceleration over both body nodes and fluid particles, namely $|\boldsymbol{a}| = \max(\|\boldsymbol{a}_f\|_\infty, \|\boldsymbol{a}_b\|_\infty)$.

5   Energy balance

A strong point of the δ-SPH method is that its energy equation can be cast in a conservative format, thus implying an accurate account for the energy exchanges between the fluid and solid phase.

The global mechanical energy equation is obtained by multiplying the momentum equation for $\boldsymbol{u}_i$ scalarly and by summing all over the fluid particles. Using the symmetry properties of the arguments of the summations and assuming that the body force is conservative, i.e., $\boldsymbol{f} = \nabla\phi$ with $\phi = \phi(\boldsymbol{r})$, we obtain:

$$
\frac{\mathrm{d}\mathcal{E}_M}{\mathrm{d}t} = -\frac{1}{2}\sum_i^* V_i \sum_j (p_j + p_i)(\boldsymbol{u}_i - \boldsymbol{u}_j)\cdot\nabla_i W_{ij}\, V_j +
$$

$$
+ \frac{\alpha h c_0 \rho_0}{2}\sum_i^* V_i \sum_j \pi_{ij}(\boldsymbol{u}_i - \boldsymbol{u}_j)\cdot\nabla_i W_{ij}\, V_j, \tag{32}
$$

where the starred summations indicate the summations over the fluid particles and the global mechanical energy is given by:

$$\mathcal{E}_M = \mathcal{E}_k + \mathcal{E}_p = \sum_i^* m_i \frac{\|\boldsymbol{u}_i\|^2}{2} - \sum_i^* m_i \, \phi_i \,. \tag{33}$$

Here, $\mathcal{E}_k$ is the global kinetic energy and $\mathcal{E}_p$ is the global potential energy. Note that $\mathcal{E}_M$, $\mathcal{E}_k$ and $\mathcal{E}_p$ only refer to the fluid phase. To guarantee the conservation of the total energy of the fluid-solid system, we introduce the global internal energy $\mathcal{E}_i$ and require:

$$\frac{\mathrm{d}\mathcal{E}_{tot}}{\mathrm{d}t} = \frac{\mathrm{d}}{\mathrm{d}t}\left(\mathcal{E}_M + \mathcal{E}_i\right) = 0 \,. \tag{34}$$

By definition, the total energy of the fluid-solid system remains constant during the evolution while energy exchanges may occur between mechanical and internal energy. As a consequence of (34), the equation for the internal energy reads:

$$\frac{\mathrm{d}\mathcal{E}_i}{\mathrm{d}t} = \frac{1}{2}\sum_i^* V_i \sum_j \left(p_j + p_i\right)\left(\boldsymbol{u}_i - \boldsymbol{u}_j\right) \cdot \nabla_i W_{ij} \, V_j +$$

$$- \frac{\alpha h c_0 \rho_0}{2}\sum_i^* V_i \sum_j \pi_{ij}\left(\boldsymbol{u}_i - \boldsymbol{u}_j\right) \cdot \nabla_i W_{ij} \, V_j \,. \tag{35}$$

The expression on the right hand-side may be regarded as a source/sink term dragging energy from/to the mechanical energy to/from the internal energy. To show this, it is convenient to focus on the right-hand side of Equation (32). Using the continuity equation and the symmetry properties of arguments in the double summations, it is possible to rearrange it in the following form (see Antuono et al., 2015 for details):

$$\frac{\mathrm{d}\mathcal{E}_M}{\mathrm{d}t} = -\frac{\mathrm{d}\mathcal{E}_C}{\mathrm{d}t} + \mathcal{P}_\delta + \mathcal{P}_\alpha + \mathcal{P}_s \,, \tag{36}$$

where

$$\mathcal{P}_\delta = \delta \, h \, c_0 \sum_i^* \frac{p_i}{\rho_i} \mathcal{D}_i V_i \tag{37}$$

$$\mathcal{P}_\alpha = \frac{\alpha h c_0 \rho_0}{2}\sum_i^* V_i \sum_j^* \pi_{ij}\left(\boldsymbol{u}_i - \boldsymbol{u}_j\right) \cdot \nabla_i W_{ij} \, V_j \,. \tag{38}$$

$$\mathcal{P}_s = -\sum_i^* V_i \sum_j \overline{\left(p_j + p_i\right)} \, \boldsymbol{u}_i \cdot \nabla_i W_{ij} \, V_j + \alpha h c_0 \rho_0 \sum_i^* V_i \sum_j \overline{\pi_{ij}} \, \boldsymbol{u}_i \cdot \nabla_i W_{ij} \, V_j +$$

$$- \sum_i^* V_i \, p_i \sum_j \overline{\left(\boldsymbol{u}_j - \boldsymbol{u}_i\right)} \cdot \nabla_i W_{ij} \, V_j \,, \tag{39}$$

and the barred summations denote the summation over the ghost particles in the solid domain. Here, $\mathcal{P}_\delta$ represents the power associated with the diffusive term, $\mathcal{P}_\alpha$ is the power dissipated by the artificial viscosity and $\mathcal{P}_s$ indicates the power due to the interaction between the fluid with the solid. Finally, the symbol $\mathcal{E}_C$ indicates the energy due to compressibility, namely

$$\mathcal{E}_C = \sum_i^* m_i \int_{\rho_0}^{\rho_i} \frac{p(s)}{s^2} \, ds \,. \tag{40}$$

For a linear state equation like that adopted in (12), we obtain:

$$\mathcal{E}_C = \sum_i^* m_i c_0^2 \left[\log\left(\frac{\rho_i}{\rho_0}\right) + \frac{\rho_0}{\rho_i} \right] \tag{41}$$

Then, in comparison to the standard SPH scheme, the δ-variants predicts a further term in the global equations of mechanical and internal energy. Specifically, the global mechanical energy equation can be rearranged in the following way:

$$\frac{\mathrm{d}}{\mathrm{d}t}\left[\mathcal{E}_M + \mathcal{E}_C\right] = \mathcal{P}_\delta + \mathcal{P}_\alpha + \mathcal{P}_s . \tag{42}$$

The power term $\mathcal{P}_\alpha$ is always negative from the second principle of thermodynamics while $\mathcal{P}_\delta$ is always negative by construction (see Antuono et al., 2015 for details). As a consequence, these contributions represent purely dissipative terms. Conversely, $\mathcal{P}_s$ has not a defined sign, since it takes into account the energy exchange between the solid and fluid phase. As a consequence of Equation (34), we can also write:

$$\frac{\mathrm{d}}{\mathrm{d}t}\left[\mathcal{E}_i - \mathcal{E}_C\right] = -\mathcal{P}_\delta - \mathcal{P}_\alpha - \mathcal{P}_s , \tag{43}$$

which represents a rearrangement of Equation (35). Regarding the energy exchanges between the solid and fluid phase, a detailed study on the consistency of the mirroring techniques described in Equations (17), (18) and (19) has been tackled, for example, in (Cercos-Pita et al., 2017).

6   Applications

In this section we show some applications of the δ-SPH scheme to hydrodynamic problems of interest in the coastal and marine engineering fields. The first example provides simulations of violent water impacts against solid walls in both two and three dimensions. These problems highlight the ability of the δ-SPH in the prediction of complex free-surface dynamics, as well as in the evaluation of local loads on solid structures. Then, we consider an extreme water-impact event against a flap-type wave energy converter. This application encloses all the main topics/features discussed in the previous test cases, i.e., wave propagation, body motions and violent water impacts, in a real engineering application.

6.1   Prediction of water impacts

Here the robustness of this scheme is tested by simulating violent free-surface flows, characterized by water impact events where the air-water interface is subjected to rapid dynamics inducing high pressure peaks.

The first problem consists in a dam-break flow impacting against a vertical wall. This benchmark test cases has been studied through the δ-SPH model in the paper (Marrone et al., 2011). Figure 4 shows the dam-break flow generated by the gravity collapse of a water column of height H and width $2H$. The fluid is confined in a rectangular tank and, after the dam break, it evolves rightwards, impacts against the tank wall and generates a plunging breaking wave.

In (Marrone et al., 2011) different resolutions were adopted with a maximum of $H/\Delta x = 320$ (corresponding to about 200,000 particles), showing a good match with the experimental and reference data. In fact, the pressure signals recorded at the probes P_1 and P_4 (see Figure 4) did not show sensible variations for spatial resolutions finer than $H/\Delta x = 320$. In

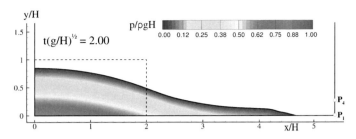

Figure 4: Dam-break flow of a water column of height H and width $2H$. The colors are representative of the pressure field (made dimensionless with respect to the initial hydrostatic value). During the simulation the pressure time histories at the probes P_1 ($y = 0.01H$) and P_4 ($y = 0.267H$) are recorded.

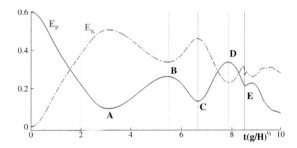

Figure 5: Dam-break flow of a water column of height H and width $2H$. Time histories of the kinetic and potential energy of the flow (made dimensionless with respect to the reference energy $\rho g H^3$).

the present section the ratio is increased up to $H/\Delta x = 1600$ (about 5.2 million particles) in order to get results with an higher accuracy in the whole domain.

Figure 5 depicts the time histories of kinetic and potential energy. During the early stages of the evolution, namely for $t(g/H)^{1/2} \in [0, 3]$, the potential energy is converted in kinetic energy. The left plot of Figure 6 displays the flow configuration at $t(g/H)^{1/2} = 3$, when the maximum of the kinetic energy is reached. At this time instant the water is generating a run-up along the right wall of the tank while a high pressure region develops at the bottom with a maximum pressure value of about 1.3 $\rho g H$.

Figure 7 shows the time histories recorded at the probe P_1. The maximum pressure along the vertical wall occurs at time $t(g/H)^{1/2} = 2.385$ and corresponds to approximately $\sim 2.9 \rho g H$. This is in good agreement with the prediction obtained by the potential theory, namely 2.8 $\rho g H$ (see for details Dobrovol'Skaya, 1969 and Marrone et al., 2011). In the same figure, the numerical output is also compared to the experimental data by Lobovský et al. (2013). The maximum pressure peak of the experiments is close to the δ−SPH prediction even if the whole pressure peak appears more localized in time. This is likely due to the fact that the numerical simulation is two-dimensional while in the experiments 3D effects were not completely negligible. Further, the advance in time of the δ-SPH signal with respect to the measurements is due to the fact that friction along the bottom is not modelled (e.g., a free slip condition is imposed along the bottom).

After $t(g/H)^{1/2} = 3$ the flow starts to decelerate until the time $t(g/H)^{1/2} = 5.5$ when the potential energy reaches a local maximum. This instant corresponds to the formation of a plunging jet (see the right plot of Figure 6). As shown in Figure 8 the latter hits the underlying layer of water at about $t(g/H)^{1/2} = 6.10$ and generates acoustic pressure

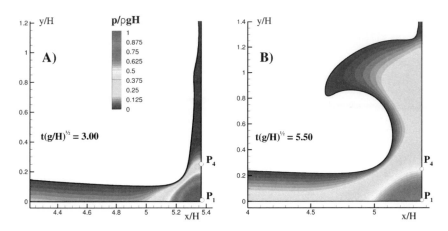

Figure 6: Dam-break flow of a water column of height H and width $2H$ impacting on a vertical wall. Enlarged views of the pressure field during two time instants: left) the water evolves upward along the wall, right) the water reaches its maximum vertical height and collapses under the gravity action.

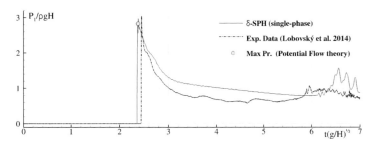

Figure 7: Dam-break flow of a water column of height H and width $2H$. Time histories of the pressure recorded on the probe P_1.

waves. This is a consequence of the weakly-compressible model adopted in the δ-SPH scheme which, during impacts, converts mechanical energy in acoustic wave energy (see, for example, Marrone et al., 2015).

On the top-left panel of Figure 9 the pressure field at time $t(g/H)^{1/2} = 6.65$ is shown. This time instant corresponds to the second local maximum of the kinetic energy (see Figure 5) and is related to the splash-up stage caused by the ricochet of the plunging jet. Since the simulation is performed with a single-phase model (and, consequently, no air-cushioning effects are accounted for), the cavity entrapped by the plunging jet is subjected to large volume variations. In turn, these lead to large changes in the pressure field which, for example, rises up to 1.3 $\rho g H$ at about $t(g/H)^{1/2} = 6.65$ and almost decreases to zero at $t(g/H)^{1/2} = 7.85$ (see the top-right plot of Figure 9). Figure 7 shows that the pressure variations measured in the experiments are generally smaller during this stage. This is due to the presence of the air in the cavity (i.e., air-cushioning effects) and to three-dimensional effects.

Figure 10 shows the pressure time histories recorded at the probe P_4, placed at $y = 0.267H$. As stressed in (Lobovský et al., 2014), it was not possible to obtain a good repeatability for the measures at P_4, since the pressure signals showed large fluctuations

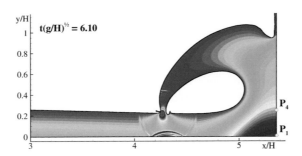

Figure 8: Dam-break flow of a water column of height H and width $2H$ impacting on a vertical wall. Enlarged view of the pressure field during the vertical drop of the water.

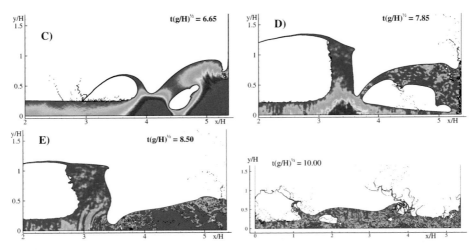

Figure 9: Dam-break flow of a water column of height H and width $2H$ impacting on a vertical wall. Enlarged views of the pressure field during four time instants.

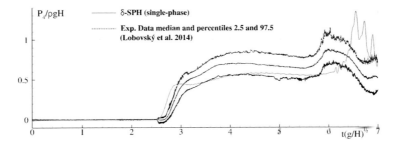

Figure 10: Dam-break flow of a water column of height H and width $2H$. Time histories of the pressure recorded on the probe P_4.

between the different runs. For this reason, in Figure 10 we also reported the curves of the averaged value and of the percentiles 2.5 and 97.5 from Lobovský et al., 2014. The numerical solution predicted by the δ-SPH scheme lays between these curves. In the past, repeatability issues with the experiments created several problems for the validation of the SPH model on this benchmark test-case. For example, in (Buchner et al., 2011) the same experiments were performed by using pressure probes characterized by a large diameter, namely 9 cm.

As a consequence, those probes in such a critical region (characterized by very local loads and large oscillations) provided measures that were difficult to use for a reliable validation procedure (see, e.g., Colicchio et al., 2002; Marrone et al., 2011).

At $t(g/H)^{1/2} = 7.85$ the potential energy reaches its second peak which corresponds to the high water column fed during the splash-up stage (see the top-right plot of Figure 9). At this time instant the cavity reaches is maximum extension and the pressure field becomes noisy, because of the impacts of falling water drops and because of the closure of many small cavities generated during the run-down of the flow on the vertical wall.

The bottom-left plot of Figure 9 displays the collapse of the cavity at time $t(g/H)^{1/2} = 8.50$. This causes a large pressure peak of intensity $\sim 2.5\rho g H$ (see, for example, Figure 13). The cavity closes with a velocity of about $v = 0.25\sqrt{gH}$, and the collapse induces a hammer pressure close to $\rho v c_0$.

After this stage the flow evolves in a complex way, moving from right to left, with the fragmentation of the free-surface and the generation of several small jets and drops (see the bottom-right plot of Figure 9).

Finally, Figure 11 sketches the time histories of the forces evaluated on the left and right vertical wall and along the bottom. The forces are made dimensionless through the water weight $W = \rho g 2 H^2$ acting on the bottom before the dam release. Note that the water weight vector, namely $-W\hat{j}$, has been subtracted to the vertical force F_y, and, consequently, the signal for F_y oscillates around the zero level. Figure 11 shows that the force along the bottom is subjected to the largest variations even in comparison to the force along the right wall (where the first water impact occurs). In particular, the cavity collapse induces an overload of $\sim 1.8W$ on the bottom while its effect on the right wall is more modest.

In references Colagrossi and Landrini, 2003 and Marrone et al., 2016 the same simulation has been performed including the presence of air. In fact, the latter substantially modifies the dynamics of the flow after the cavity closure, i.e., for $t(g/H)^{1/2} > 6$. In (Colagrossi and Landrini, 2003) different values of the reference pressure of the air were considered and their influence on the cavity evolution and on the pressure recorded on the right wall was discussed. Conversely, in (Marrone et al., 2016) the reference pressure was set to $P_{0air} = 17.2\,\rho g H$, that corresponds to $P_{0air} = 1\,atm$ in the length scale experiment of Buchner et al., 2011. In this latter case, the dynamic of the post-breaking stage is reported in Figure 12. Compared to the single-phase simulation, the presence of the air prevents the reduction of the cavity (top-right plot of Figure 12) which at $t(g/H)^{1/2} = 8.43$ split in two parts (bottom-left plot of Figure 12).

Figure 13 depicts the time histories of the pressure recorded at the probe P_3 positioned at $y = 0.17H$ for both the single- and two-phase model. The pressure signal provided by the single-phase simulation has been filtered in order to remove the acoustic components (see, for example, Meringolo et al., 2017). No filtering procedure has been applied to the signal

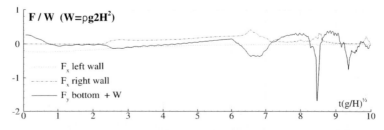

Figure 11: Dam-break flow of a water column of height H and width $2H$. Time histories of the forces predicted by the $\delta-$SPH acting on the left and right vertical walls and along the bottom.

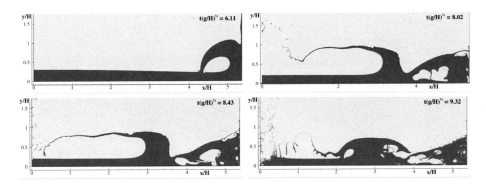

Figure 12: Dam-break flow of a water column of height H and width $2H$ impacting on a vertical wall. Snapshots of the evolution using an air-water δ−SPH model.

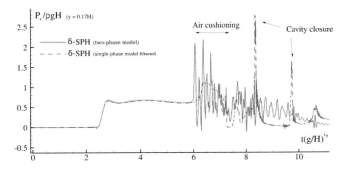

Figure 13: Dam-break flow of a water column of height H and width $2H$. Time histories of the pressure recorded at the probe P_3 ($y = 0.17H$) evaluated through a single- and a two-phase δ−SPH model.

of the two-phase model whose oscillations are mainly due to the air cushion effect caused by the gas entrapped in the cavity. Incidentally, we observe that this behaviour is not visible in the experimental pressure records of (Lobovský et al., 2014). A possible reason is that 3D effects lead to a complex cavity shape which is fragmented much faster than the cavity predicted by the 2D models.

Finally to close this section a more realistic case is considered moving to a three-dimensional framework. As in (Marrone et al., 2011) the dam-break flow against a prismatic column is here discussed. Figure 14 shows some snapshots of the free-surface evolution after the water impact against the column. At $t = 0.36\,s$ (top-left plot of Figure 14) part of the water hits the column and generates a run-up while the the remaining bulk of fluid moves forward. At $t = 0.51\,s$ (top-right plot) the run-up stage ends and the water along the column starts its descent, while the remaining part of the flow hits against the rear vertical wall. At $t = 0.83\,s$ (bottom-left plot) a plunging jet is formed just in front of the column with a dynamic similarly to the one discussed for the 2D case. The water impact on the vertical wall is more complicated because of the large 3D effects and the free-surface is characterized by a complex shape because of the fragmentation. The bottom-right plot, at $t = 1.05$, shows the formation of a splash-up in front of the column and a violent flow generated behind it by the flow reflected at the rear vertical wall.

Here, for the sake of brevity, the description is just limited to a qualitative point of view, while in (Marrone et al., 2011) a validation against experimental data is also provided. In (Marrone et al., 2011) about 1.1 millions particles were used (i.e., $H/\Delta x = 75$ where H is the

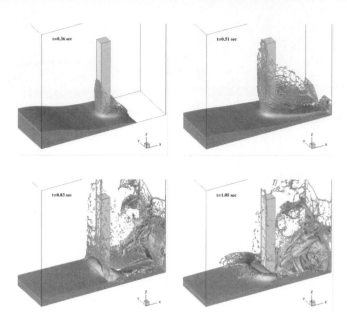

Figure 14: Dam-break flow against a vertical column. Snapshots of the free-surface evolution.

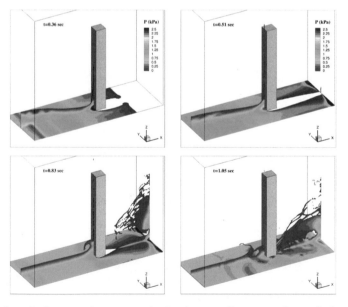

Figure 15: Dam-break flow against a vertical column. Contour plots of the pressure field on a vertical and a horizontal plane.

water column height at the beginning of the simulation). This resolution was fine enough to obtain a good match with the experimental data in terms of the forces acting on the column and local measures of the velocity in front of the column. Here, the simulation is repeated increasing the number of particles to 40 millions, i.e., $H/\Delta x = 245$. The time steps has been set equal to 0.2 milliseconds and the code marched for 10,000 steps. The simulation ran on 300 cores for 24 hours. Figure 15 reports a vertical and a horizontal slice of the flow field containing pressure contour plots.

6.2 Extreme loads on a Wave Energy Converter (WEC)

As a final application the 3D simulation of an extreme wave impacting a flap-type wave energy converter is considered. Specifically, the survivability of a bottom-mounted pitching device, consisting of a partly submerged flap placed in front of a breakwater, is studied numerically. To this aim, a realistic site is considered: a flap-type WEC positioned ahead the north dike in the city of Bayonne, South west Atlantic coast of France (further details can be found in Baudry et al., 2015). Wave measurements from the closest buoy are considered in order to have realistic characteristics of the extreme events occurring in the site. For the generation of the wave, the adopted numerical set-up is composed by a piston wave-maker inside a tank limited by a sloping dike (see the sketch in the top plot of Figure 16). In order to produce the desired wave characteristics in terms of period and height, a sinusoidal motion law has been prescribed to the piston following the formulae described in (Dean and Dalrymple, 1991). In the bottom plot of Figure 16 the adopted geometrical configuration is shown. For the sake of completeness, a view of the full numerical domain adopted for the simulation is provided in Figure 17. The tank width has been set equal to 60 m in order to avoid possible reflections from the lateral walls.

In Figures 18 and 19 the free-surface evolution is depicted from two different perspectives before and after the impact. The wave breaking starts just before the impact and, then, the

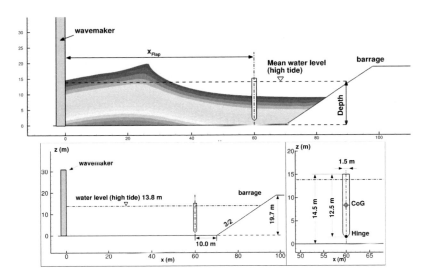

Figure 16: Top: Sketch of the numerical domain and definition of the main parameters. Bottom: Geometrical configuration adopted for simulation.

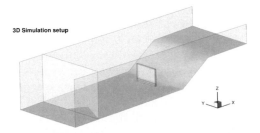

Figure 17: Numerical set-up for the 3D simulations.

Figure 18: Free-surface evolution of the wave impacting a bottom-hinged wave energy converter from two different views, just before the impact.

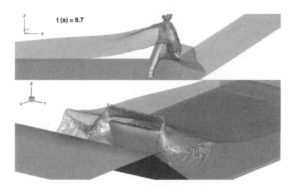

Figure 19: Free-surface evolution of the wave impacting a bottom-hinged wave energy converter from two different views, just after the impact.

wave impacts the flap with a front almost parallel to the body surface, causing a complex 3D splashing process with several jet ejections and fragmentations. During the impact stage the flap rotation remains quite small. As a consequence, the loading process evolves as a sort of flip-through phenomenon Lugni et al., 2006, similar to what documented by Peregrine, 2003 for waves impacting vertical walls. For this kind of dynamics, the impact is characterized by very small time and space scales.

This behaviour can be better observed performing the same simulation within a 2D framework. The wave steepens as it approaches the flap and its profile becomes almost parallel to that of the flap (see Figure 20). As a consequence, the contact point between liquid and solid settles very rapidly along the body and the momentum transfer of the wave focuses in a very small region. This behaviour can be better observed in Figure 21 where the pressure measured along the flap is plotted on the time-space plane. A very narrow pressure peak is observed around $t = 9.25$ s. This pressure increase expires in few tenths of seconds and involves a space scale of about 0.1 m. Apart from the large force and torque experienced, this kind of loading is potentially harmful for the device, as it can imply local permanent deformations of the structure.

The time histories of the horizontal force and torque around the hinge are reported in Figure 22 for both 3D and 2D simulations (the latter multiplied by the flap width). In the 2D model the torque peak has a narrower shape and is overestimated by a factor close to 1.5. This can be justified by the fact that the flap shape is close to a square and, therefore, 3D effects can be non-negligible.

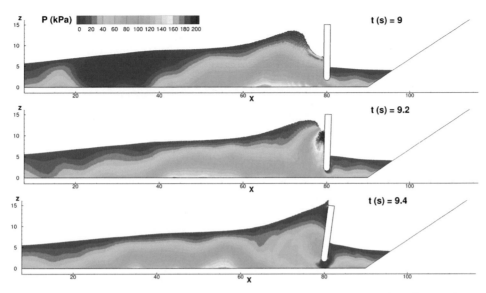

Figure 20: Pressure field evolution before and after the impact, in the 2D simulation.

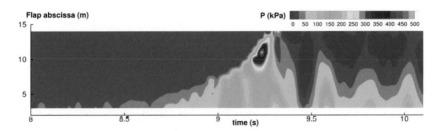

Figure 21: Evolution of the measured pressure on a section of the flap wall in the time-space plane.

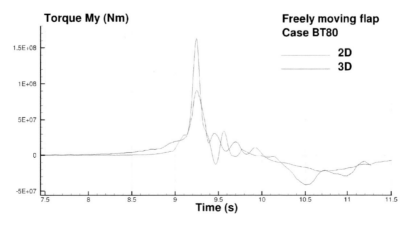

Figure 22: Comparison of the torque around the hinge evaluated by the 2D and 3D models (2D values are multiplied by the flap width for consistency with the 3D model).

7 Conclusions

In many practical applications of coastal and marine engineering an accurate description of complex free-surface flows and violent impacts against structures is of fundamental importance for the assessment of local loads. In such a framework the δ-SPH provides a powerful tool for the modelling of both the free-surface dynamics (including wave breaking and interface fragmentation) and the fluid-body interaction. An overall description of the method and of its main features has been provided, along with several applications to hydrodynamics. These show the accuracy and the robustness of the δ-SPH scheme and its versatility in a wide range of practical problems.

Incidentally, it has been recently shown that diffusive variants of the SPH scheme can be used to model turbulent flows by recasting the SPH model in a Large-Eddy-Simulation framework (see, for example, Di Mascio et al., 2017). In fact, in the SPH community there is a continuous research on enhanced SPH schemes which generally lead to the introduction of a sort of diffusion in the numerical model (see, e.g., Oger et al., 2016; Lind et al., 2012; Adami et al., 2013; Monaghan, 2002).

References

Adami, S., Hu, X. and Adams, N. (2012). A generalized wall boundary condition for smoothed particle hydrodynamics. Journal of Computational Physics 231(21): 7057–7075.

Adami, S., Hu, X. and Adams, N. (2013). A transport-velocity formulation for smoothed particle hydrodynamics. Journal of Computational Physics 241: 292–307.

Altomare, C., Domínguez, J., Crespo, A., González-Cao, J., Suzuki, T., Gómez-Gesteira, M. and Troch, P. (2017). Long-crested wave generation and absorption for sph-based duals physics model. Coastal Engineering 127: 37–54.

Amicarelli, A., Marongiu, J. -C., Leboeuf, F., Leduc, J. and Caro, J. (2011). SPH truncation error in estimating a 3D function. Computers & Fluids 44(1): 279–296.

Antuono, M., Colagrossi, A., Marrone, S. and Molteni, D. (2010). Free-surface flows solved by means of SPH schemes with numerical diusive terms. Computer Physics Communications 181(3): 532–549.

Antuono, M., Colagrossi, A., Marrone, S. and Lugni, C. (2011). Propagation of gravity waves through an SPH scheme with numerical diffusive terms. Computer Physics Communications 182(4): 866–877.

Antuono, M. and Colagrossi, A. (2012). The damping of viscous gravity waves. Wave Motion 50(0): 197–209.

Antuono, M., Colagrossi, A. and Marrone, S. (2012). Numerical diffusive terms in weakly-compressible SPH schemes. Computer Physics Communications 183(12): 2570–2580.

Antuono, M., Marrone, S., Colagrossi, A. and Bouscasse, B. (2015). Energy balance in the δ-SPH scheme. Computer Methods in Applied Mechanics and Engineering 289: 209–226.

Baudry, V., Marrone, S., Babarit, A., Le Touzé, D. and Clément, A. H. (2015). Power matrix assessment and extreme loads estimation on a flap type wave energy converter in front of a dike. In Proc. 11th Eur. Wave Tidal Energy Conf, pp. 1–10.

Ben Moussa, B. and Vila, J. (2000). Convergence of SPH method for scalar nonlinear conservation laws. SIAM Journal on Numerical Analysis 37(3): 863–887.

Bouscasse, B., Antuono, M., Colagrossi, A. and Lugni, C. (2013). Numerical and experimental investigation of nonlinear shallow water sloshing. International Journal of Nonlinear Sciences and Numerical Simulation 14(2): 123–138.

Bouscasse, B., Colagrossi, A., Marrone, S. and Antuono, M. (2013). Nonlinear water wave interaction with floating bodies in SPH. Journal of Fluids and Structures 42: 112–129.

Buchner, B., van Dijk, A. and deWilde, J. (2011). Numerical Multiple-Body Simulations of Side-by-Side Mooring to a FPSO. In Proceedings of the 21st ISOPE Conference, (495 North Whisman Road, Suite 300 Mountain View, CA 94043-5711, USA), pp. 343–353, ISOPE.

Canelas, R. B., Crespo, A. J., Domínguez, J. M., Ferreira, R. M. and Gómez-Gesteira, M. (2016). SPH-DCDEM model for arbitrary geometries in free surface solid-fluid flows. Computer Physics Communications 202: 131–140.

Carrica, P. M., Wilson, R. V., Noack, R. W. and Stern, F. (2007). Ship motions using single-phase level set with dynamic overset grids. Computers & Fluids 36(9): 1415–1433.

Cercos-Pita, J., Antuono, M., Colagrossi, A. and Souto-Iglesias, A. (2017). {SPH} energy conservation for fluid-solid interactions. Computer Methods in Applied Mechanics and Engineering 317: 771–791.

Chowdhury, S. D. and Sannasiraj, S. (2014). Numerical simulation of 2D sloshing waves using SPH with diffusive terms. Applied Ocean Research 47: 219–240.

Clausen, J. R. (2013). Entropically damped form of artificial compressibility for explicit simulation of incompressible flow. Physical Review E 87(1): 013309.

Colagrossi, A. and Landrini, M. (2003). Numerical simulation of interfacial flows by smoothed particle hydrodynamics. Journal of Computational Physics 191: 448–475.

Colagrossi, A., Antuono, M. and Le Touzé, D. (2009). Theoretical considerations on the free-surface role in the Smoothed-particle-hydrodynamics model. Physical Review E 79(5): 056701.

Colagrossi, A., Antuono, M., Souto-Iglesias, A. and Le Touzé, D. (2011). Theoretical analysis and numerical verification of the consistency of viscous smoothed-particle-hydrodynamics formulations in simulating free-surface flows. Physical Review E 84: 026705.

Colicchio, G., Colagrossi, A., Greco, M. and Landrini, M. (2002). Free-surface flow after a dam break: a comparative study. Shistechnik (Ship Technology Research) 49(3): 95–104.

Crespo, A. C., Dominguez, J. M., Barreiro, A., Gómez-Gesteira, M. and Rogers, B. D. (2011). GPUs, a new tool of acceleration in CFD: Eciency and reliability on smoothed particle hydrodynamics methods. PLoS ONE 6(6).

Crespo, A., Altomare, C., Domínguez, J., González-Cao, J. and Gómez-Gesteira, M. (2017). Towards simulating floating offshore oscillating water column converters with Smoothed Particle Hydrodynamics. Coastal Engineering 126: 11–26.

Dalrymple, R. A. and Knio, O. (2001). SPH modelling of water waves. In Coastal dynamics' 01: 779–787.

De Lee, M., Le Touzé, D. and Alessandrini, B. (May 2009). Normal flux method at the boundary for SPH. In 4th SPHERIC, pp. 149–156.

De Lee, M., Le Touzé, D. and Alessandrini, B. (2011). A modified no-slip condition in weakly-compressible SPH. In 6th ERCOFTAC SPHERIC Workshop on SPH Applications, pp. 291–297.

Dean, R. and Dalrymple, R. (1991). Water wave mechanics for engineers and scientists. Advanced Series on Ocean Engineering, World Scientific.

Di Mascio, A., Antuono, M. Colagrossi, A. and Marrone, S. (2017). Smoothed particle hydrodynamics method from a large eddy simulation perspective. Physics of Fluids 29(3): 035102.

Dobrovol'Skaya, Z. (1969). On some problems of similarity flow of fluid with a free surface. Journal of Fluid Mechanics 36(4): 805–829.

Doring, M. (2005). Développement d'une méthode SPH pour les applications à surface libre en hydrodynamique. Ph.D. thesis, Ecole Centrale de Nantes.

Español, P. and Revenga, M. (2003). Smoothed dissipative particle dynamics. Physical Review E 67(2): 026705.

Ferrari, A., Dumbser, M., Toro, E. F. and Armanini, A. (2009). A new 3D parallel SPH scheme for free surface flows. Computers & Fluids 38(6): 1203–1217.

Fourey, G., Hermange, C., Touzé, D. L. and Oger, G. (2017). An efficient FSI coupling strategy between Smoothed Particle Hydrodynamics and Finite Element methods. Computer Physics Communications 217: 66–81.

Gingold, R. and Monaghan, J. (1977). Smoothed Particle Hydrodynamics: theory and application to non-spherical stars. Mon. Not. Roy. Astron. Soc. (MNRAS) 181: 375–389.

Hadžić, I., Hennig, J., Perić, M. and Xing-Kaeding, Y. (2005). Computation of flow-induced motion of floating bodies. Applied Mathematical Modelling 29(12): 1196–1210.

Kajtar, J. and Monaghan, J. (2008). SPH simulations of swimming linked bodies. Journal of Computational Physics 227(19): 8568–8587.

Koukouvinis, P., Anagnostopoulos, J. and Papantonis, D. (2013). An improved MUSCL treatment for the SPH-ALE method: comparison with the standard SPH method for the jet impingement case. International Journal for Numerical Methods in Fluids 71: 1152–1177.

Kulasegaram, S., Bonet, J., Lewis, R. and Profit, M. (2004). A variational formulation based contact algorithm for rigid boundaries in two-dimensional SPH applications. Computational Mechanics 33(4): 316–325.

Lattanzio, J., Monaghan, J., Pongracic, H. and Schwarz, M. (1989). Interstellar cloud collisions. Mon. Not. R. Astron. Soc. 215: 125.

Leroy, A., Violeau, D., Ferrand, M. and Kassiotis, C. (2014). Unified semi-analytical wall boundary conditions applied to 2-D incompressible SPH. Journal of Computational Physics 261: 106–129.

Lind, S., Xu, R., Stansby, P. and Rogers, B. (2012). Incompressible smoothed particle hydrodynamics for free-surface flows: A generalised diffusion-based algorithm for stability and validations for impulsive flows and propagating waves. Journal of Computational Physics 231(4): 1499–1523.

Lobovský, L., Botia-Vera, E., Castellana, F., Mas-Soler, J. and Souto-Iglesias, A. (2014). Experimental investigation of dynamic pressure loads during dam break. Journal of Fluids and Structures 48: 407–434.

Löhner, R., Cebral, J. R., Camelli, F. E., Appanaboyina, S., Baum, J. D., Mestreau, E. L. and Soto, O. A. (2008). Adaptive embedded and immersed unstructured grid techniques. Computer Methods in Applied Mechanics and Engineering 197(25-28): 2173–2197.

Lugni, C., Brocchini, M. and Faltinsen, O. M. (2006). Wave impact loads: The role of the flip-through. Physics of Fluids 18(12): 101–122.

Marongiu, J. C., Leboeuf, F. and Parkinson, E. (2007). A new treatment of solid boundaries for the SPH method. In SPHERIC—Smoothed Particle Hydrodynamics European Research Interest Community, pp. 165+.

Marongiu, J., Leboeuf, F., Caro, J. and Parkinson, E. (2010). Free surface flows simulations in Pelton turbines using an hybrid SPH-ALE method. Journal of Hydraulic Research 48(S1): 40–49.

Marrone, S., Antuono, M., Colagrossi, A., Colicchio, G., Le Touzé, D. and Graziani, G. (2011). Delta-SPH model for simulating violent impact flows. Computer Methods in Applied Mechanics and Engineering 200(13-16): 1526–1542.

Marrone, S., Colagrossi, A., Antuono, M., Colicchio, G. and Graziani, G. (2013). An accurate SPH modeling of viscous flows around bodies at low and moderate Reynolds numbers. Journal of Computational Physics 245: 456–475.

Marrone, S., Colagrossi, A., Di Mascio, A. and Le Touzé, D. (2015). Prediction of energy losses in water impacts using incompressible and weakly compressible models. Journal of Fluids and Structures 54: 802–822.

Marrone, S., Mascio, A. D. and Le Touzé, D. (2016). Coupling of Smoothed Particle Hydrodynamics with Finite Volume method for free-surface flows. Journal of Computational Physics Volume 310, 1 April 2016, pp. 161–180.

Meringolo, D. D., Aristodemo, F. and Veltri, P. (2015). SPH numerical modeling of wave-perforated breakwater interaction. Coastal Engineering 101: 48–68.

Meringolo, D., Colagrossi, A., Marrone, S. and Aristodemo, F. (2017). On the filtering of acoustic components in weakly-compressible SPH simulations. Journal of Fluids and Structures 70: 1–23.

Mittal, R. and Iaccarino, G. (2005). Immersed boundary methods. Annu. Rev. Fluid Mech. 37: 239–261.

Molteni, D. and Colagrossi, A. (2009). A simple procedure to improve the pressure evaluation in hydrodynamic context using the SPH. Computer Physics Communications 180: 861–872.

Monaghan, J. (1994). Simulating free surface flows with SPH. Journal of Computational Physics 110(2): 39–406.

Monaghan, J. (1996). Gravity currents and solitary waves. Physica D. Nonlinear Phenomena 98(2-4): 523–533.

Monaghan, J. (2002). SPH compressible turbulence. Mon. Not. R. Astrom. Soc. 335: 843–852.

Monaghan, J., Kos, A. and Issa, N. (2003). Fluid motion generated by impact. Journal of Waterway, Port, Coastal, and Ocean Engineering 129(6): 250–259.

Monaghan, J. J. (2005). Smoothed particle hydrodynamics. Reports on Progress in Physics 68: 1703–1759.

Muscari, R., Felli, M. and Di Mascio, A. (2011). Analysis of the flow past a fully appended hull with propellers by computational and experimental fluid dynamics. Journal of Fluids Engineering 133(6): 061104.

Oger, G., Le Touzé, D., Guibert, D., de Lee, M., Biddiscombe, J., Soumagne, J. and Piccinali, J. (2016). On distributed memory MPI-based parallelization of SPH codes in massive HPC context. Computer Physics Communications 200: 1–14.

Oger, G., Marrone, S., Le Touzé, D. and de Lee, M. (2016). SPH accuracy improvement through the combination of a quasi-Lagrangian shifting transport velocity and consistent ALE formalisms. Journal of Computational Physics 313: 76–98.

Peregrine, D. (2003). Water-wave impact on walls. Annu. Rev. Fluid Mech. 35: 23–43.

Peskin, C. S. (1972). Flow patterns around heart valves: a numerical method. Journal of Computational Physics 10(2): 252–271.

Quinlan, N., Lastiwka, M. and Basa, M. (2006). Truncation error in mesh-free particle methods. International Journal for Numerical Methods in Engineering 66(13): 2064–2085.

Randles, P. and Libersky, L. (1996). Smoothed Particle Hydrodynamics: some recent improvements and applications. Computer Methods in Applied Mechanics and Engineering 39: 375–408.

Sibilla, S. (2007). SPH simulation of local scour processes. In Proc. SPHERIC, 2nd International Workshop, Universidad Politécnica de Madrid, Spain.

Valizadeh, A. and Monaghan, J. J. (2012). Smoothed particle hydrodynamics simulations of turbulence in fixed and rotating boxes in two dimensions with no-slip boundaries. Physics of Fluids 24(3): 035107.

Vila, J. (1999). On particle weighted methods and smooth particle hydrodynamics. Mathematical Models & Methods in Applied Sciences 9(2): 161–209.

Vinje, T. and Brevig, P. (1981). Nonlinear ship motions. In Proc. 3rd Int. Symp. Num. Ship Hydrodyn., Paris.

Violeau, D. (2012). Fluid Mechanics and the SPH Method: Theory and Applications. Oxford University Press.

Wendland, H. (1995). Piecewise polynomial, positive definite and compactly supported radial functions of minimal degree. Adv. Comput. Math. 4(4): 389–396.

Chapter 6

Wave and Structure Interaction
Porous Coastal Structures

Pablo Higuera

1   Introduction

Most coastal structures rely on an energy-dissipation function to fulfil their defence role, therefore it is not surprising that they are tightly linked with porous materials. Energy dissipation, which is enhanced by the porous materials, aids in several important tasks such as reducing the reflection coefficient, reducing runup and overtopping, and protecting the structure from direct wave impacts or scour. Furthermore, the porous layers or aprons may constitute a dynamic part of the structure, which depending on the typology, can help to identify structural failure at early stages and allow to repair it in time.

Although a wide range of coastal structure typologies comprise porous media elements but in this introduction only the most relevant breakwater types will be discussed. The reader is referred to Oumeraci and Kortenhaus (1997) for a more thorough classification of traditional coastal defence structures. Other types of structures that will not be covered in this chapter but are relevant are, for example fish cages or mussel rafts, which are being widely applied in aquaculture and can be assimilated to highly porous structures.

In rubble mound breakwaters (emerged or submerged), layers of different rock materials and/or pre-cast concrete units are arranged forming slopes. In some cases a crown wall is placed on top of the porous mantles to offer additional protection against overtopping. In vertical and composite breakwaters, a concrete caisson is supported by a foundation made of porous materials which may also be protected with external porous mantles. Furthermore, aprons made of porous materials can be placed around the lower part of other types of structures (e.g., pile foundations, seawalls), to protect them against scour. In all these cases, porous media need to be accounted for in order to, for example, obtain accurate flow patterns in the vicinity of the structure (for functionality calculations) and accurate pressure distributions around the monolithic elements (for stability calculations).

In the last decades Reynolds-Averaged Navier-Stokes-based (RANS) modelling, also called Computational Fluid Dynamics (CFD), has become widespread boosted, in some sense, by the release of open source models validated by the coastal community (Jacobsen et al., 2012; Higuera et al., 2013a) and the increase in computational capacity. In view of the large number of structural typologies that include porous media and the significant role that the porous materials play in the structural function, it can be concluded that it is of utmost importance to have the capability to simulate porous media numerically. As a matter of fact,

Department of Civil and Environmental Engineering, Faculty of Engineering, University of Auckland, Auckland, 1010, New Zealand. Email: pablo.higuera@auckland.ac.nz there are numerous references in the literature that deal with wave interaction with porous structures using different approaches, from which the most relevant will be reviewed in the next chapter. Furthermore, the importance of numerical modelling in the design process of coastal structures is also increasing. For example, CFD is already a standard design tool, complementary to physical modelling, in the Spanish design guidelines for coastal structures (ROM[1]).

The numerical simulation of flow through porous media is complex and presents several challenges. The most obvious complexity is how to represent the porous materials, which often present an intricate (random) internal geometry. All the factors required need to be correctly defined, in a unique and physical way, as they will affect the flow rate that can go through the materials. Another challenging task is two-phase flow advection, because as it will be discussed later in this chapter, not only the basic flow motion equations need to be modified, but also the surface tracking techniques to ensure mass conservation when the flow interface crosses these materials.

This chapter is organized as follows. A brief literature review is included in the next section. Afterwards, the mathematical formulation for multiphase flow through porous media is introduced. Next, a CFD numerical model is described and two applications are presented. The first application demonstrates the advantages of using numerical modelling to pre-design coastal structure modifications, while the second shows the capability of simulating sediment by coupling a Discrete Element Method model with CFD via a set of equations with time-varying porosity. Finally, the concluding remarks are enunciated.

2 Literature review

In this section we will review the literature on numerical simulation of porous materials with a special focus on coastal structures. Due to space limitations, only the most relevant references will be covered. For a more extensive review the reader is referred to Higuera (2015) and Losada et al. (2016).

There are two main approaches to simulate porous structures numerically. First, in the direct approach, all the individual elements that form the porous materials are represented in the numerical domain, either meshed around (Dentale et al., 2014, 2018), or placed as solids in meshless methods (Altomare et al., 2014). Although this approach can offer valuable insights and detailed flow patterns around individual elements, it is limited to the largest elements only (e.g., pre-cast concrete units). The smaller materials in the inner layers will often present random sizes and arrangements, making it impossible to represent such geometries numerically.

The second method is the averaged approach. In it, some simplifications are made (e.g., homogeneous and isotropic materials) and the Navier-Stokes equations are modified accordingly, via averaging. Two main averaging procedures can be found in literature, namely, time-averaged volume-averaged Navier-Stokes equations (de Lemos and Pedras, 2001) and volume-averaged RANS (VARANS) equations (Liu et al., 1999). In this section will be focussing on the second approach, as it is more widely used in our field.

Flow through porous media has been studied in multiple science and engineering branches. The initial advances in volume-averaged RANS equations, which will be introduced in the next section, were developed in the late 60s for the petroleum industry Whitaker (1967); Slattery (1967). This link is natural, as petroleum extraction involves extremely complex multiphase flow through porous media. Other fields, as the chemical and manufacturing industries, have also made significant advances during the years, as porous media is widely

---

[1]http://www.puertos.es/es-es/ROM, in Spanish. Some documents are available in English.

used in, e.g., catalysts. Therefore, significant developments can also be found in heat transfer (Kaviani, 1995).

The application of porous media flow in coastal engineering problems started as early as Madsen (1974), based on eigenfunction expansions. This approach has been applied for a long time (Dalrymple et al., 1991; Losada et al., 1993). Other solutions based on the mild-slope equation and depth-integrated models were also developed. The reader is referred to Losada et al. (2016) for a complete discussion on these methods.

Although the first truly numerical modelling application to coastal structures was implemented by Sakakiyama and Kajima (1992), it was not until Lin and Liu (1998); Liu et al. (1999), applied a set of Navier-Stokes equations, similar to the ones used today, and the Volume Of Fluid (VOF) technique (Hirt and Nichols, 1981), with the introduction of CO-BRAS (Cornell breaking waves). Prior to this breakthrough work, similar models as VOF-break (Troch and De Rouck, 1998) did not include the porosity inside the equations, merely porous media-induced drag forces. However, Liu et al. (1999) presented a main limitation: porosity had not been accounted for in the differential operators during the development, thus, spatial variations in it were not taken into account. Improvements such as including a volume-averaged $k - \epsilon$ turbulence model were introduced in Hsu et al. (2002), based on the closure developed by Nakayama and Kuwahara (1999) for heat transfer applications. Nevertheless, the differential operators were kept independent of porosity. The model IH2VOF, derived from COBRAS, was later applied to simulate all types of coastal structures (Lara et al., 2006, 2008), also with porosity outside of the differential operators. Finally, Hur et al. (2008) presented a 3D model in which the RANS equations were not volume-averaged, but included drag terms and face and volume fractions in the porous cells to account for the effects of the porous materials.

A boom in numerical modelling started from 2010, boosted by more accessible and efficient computations. The new generation of models solved the previous issues with the porosity gradients, extended the range of applicability to three-dimensions and shifted in most cases to a finite volume discretization. Examples of the newly developed models include COMFLOW (Wellens et al., 2010), IH3VOF (Lara et al., 2012) and OpenFOAM® (Higuera et al., 2014a; Jensen et al., 2014). Other models with different discretizations are Flow3D (finite differences) (Vanneste, 2012; Vanneste and Troch, 2012) and the tool presented in Larese et al. (2015) (finite elements).

The development of these models was accompanied by a thorough study of the volume-averaged equations. In that sense del Jesus (2011) performed an extensive work and derived a set of VARANS equations, including a new volume-averaged $k - \omega$ SST turbulence closure model. His work was applied in Lara et al. (2012) and del Jesus et al. (2012) and later implemented in OpenFOAM® (Higuera et al., 2014a,b). Jensen et al. (2014) re-developed the equations in a different way, comparing the results with Lara et al. (2012) and del Jesus et al. (2012), concluding that they are almost identical. The set of VARANS equations presented in Higuera (2015), currently implemented in the *olaFlow* model (Higuera, 2017), follow the same approach in Jensen et al. (2014), with the difference that the derivation includes the time variation of porosity, an important feature in some cases that will be applied in Section 6.

Although this review has so far been centred in Eulerian Navier-Stokes type of modelling, it must be noted that other methods exist in literature too. For example, porous media flow has been modelled following the averaged approach with the Smooth Particle Hydrodynamics (SPH) method (Shao, 2010; Ren et al., 2016) and with the explicit Moving Particle Semi-implicit (MPS) method (Sun et al., 2018). The Lattice Boltzmann method has also been applied (Espinoza-Andaluz et al., 2017), but in this case for only direct approach simulations.

3 Mathematical formulation

In this section we will present the mathematical formulation to represent multiphase flows through porous media in one of the most general ways available (i.e., with the least number of assumptions). As mentioned before, we will be volume-averaging the Navier-Stokes equations to obtain the so-called VARANS equations. For the sake of brevity, not all the steps of the mathematical derivation will be shown here. The reader is referred to Higuera (2015) for the complete details.

3.1 Definitions

Before presenting the governing equations, we need to understand the physical processes involved and define the mathematical operations to average the RANS equations.

When flow passes through a porous medium it loses momentum. Energy is dissipated via viscous effects (boundary layer around all the individual components that comprise the medium) and wakes (laminar or turbulent). The size, shape and arrangement of the solids inside the medium also play an important role. For example, a uniformly distributed and regular pattern of solids may present preferential flow directions, while a random distribution of the same elements can yield higher tortuosity, thus enhancing the dissipative effects. All these factors need to be considered when representing the effects that the porous media exert on the flow.

In the volume-averaging technique, a volume of fixed size (V) acts as an averaging filter for the porous media while swept all over the domain of interest. There are three requirements for the volume-averaging volume. First, V needs to be large enough to capture the internal structure of the porous media. At the same time, the volume needs to be small enough so that physical properties can be assumed constant within it, as defined in Gray (1975) and Whitaker (1996). Finally, V also needs to be small enough to capture the scale of the relevant flow features.

The volume averaging operator ($\langle\,\rangle$) for a given field f inside V is called extended or superficial averaging, and is defined as follows:

$$\langle f \rangle = \frac{1}{V} \int_{V_v} f \, \mathrm{d}V \tag{1}$$

in which V_v is the volume of voids, i.e., the volume outside the solid phases (capable of containing fluid), located within the averaging volume, as sketched in Figure 1.

The volume averaging operator defined in Equation 1 includes the variation of V_v in time and space. While V is fixed, it can be placed at different locations in which the solid geometry enclosed may vary. This means that the extended volume-average of a constant

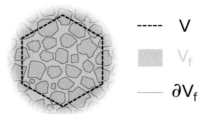

Figure 1: Sketch of the VARANS domains.

field may present space or time gradients due to the variations of V_v itself. To correct this behaviour, the intrinsic volume average ($\langle\ \rangle^f$) is defined:

$$\langle f\rangle^f = \frac{1}{V_v} \int_{V_v} f\, \mathrm{dV} \tag{2}$$

In order to link the extended and intrinsic averages we need to introduce the variable porosity (ϕ), which is defined as the quotient between the volume of voids and the control volume: $\phi = \frac{V_v}{V}$. Therefore, the two average types are linked in this way: $\langle a\rangle = \phi\,\langle a\rangle^f$.

The following variable decomposition will be adopted:

$$\bar{f} = \langle\bar{f}\rangle^f + f'' \tag{3}$$

in which f is the real value (obtained by measurement or via Direct Numerical Simulation). The over-bar denotes that f is a Reynolds-averaged variable. $\langle\bar{f}\rangle^f$ is the intrinsic volume-averaged value and f'' is the spatial fluctuation. The spatial fluctuation term represents the small-scale features of the flow, lost when applying volume-averaging. Moreover, f'' has a spatial mean equal to zero by definition, similar to the turbulent fluctuation in the Reynolds decomposition, which has a zero time mean, namely $\langle f''\rangle^f = 0$. However, for two arbitrary fields, generally $\langle f''g''\rangle^f \neq 0$, thus, the spatial fluctuation components will originate terms that cannot be solved, and need to be modelled to close the equations.

For convenience the over-bars will be dropped from now on, understanding that volume-averaging is applied on the Reynolds-averaged variables.

There are a number of techniques and theorems that are an important part of the derivations. The reader is referred to Whitaker (1967) and Slattery (1967) for the theorems for the local volume average of a gradient and the theorem for the local volume average of a time derivative, and to the appendices in Higuera (2015) for additional materials.

A single porosity (ϕ) only suffices to represent a system with a unique fluid phase and an obstacle. However, for the sake of completeness, we will propose a general framework for multiphase systems and different porous materials. As shown in Figure 2, the fluid phases (immiscible fluids) are denoted by β_i. Solids that have a rigid matrix which cannot deform or move are denoted by γ_j, whereas moving solids are denoted by δ_k.

A new variable is required to link all the materials together: volume fraction (ϵ). This variable is also known as saturation if the material is a fluid. By definition, volume fraction is the unit volume of a given material within a control volume, therefore:

$$\sum_i \epsilon_{\beta_i} + \sum_j \epsilon_{\gamma_j} + \sum_k \epsilon_{\delta_k} = 1 \tag{4}$$

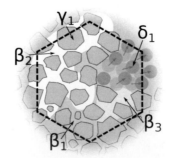

Figure 2: Sketch of the VARANS phases.

Given this definition, porosity can now be expressed as:

$$\phi = \sum_i \epsilon_{\beta_i} = 1 - \sum_j \epsilon_{\gamma_j} - \sum_k \epsilon_{\delta_k} \tag{5}$$

Furthermore, it is convenient to separate the materials that will not deform or move from those that will, and derive an independent porosity term for each of them. This will allow us, for example, to distinguish between the effects of the materials forming the armour layers of breakwaters and solids that may go into suspension, as sediment grains. We will call these static porosity and dynamic porosity, respectively, and define them as follows:

$$\phi_{ST} = 1 - \sum_j \epsilon_{\gamma_j} \quad \text{and} \quad \phi_{DY} = 1 - \sum_k \epsilon_{\delta_k}. \tag{6}$$

Finally, the total porosity can also be posed in terms of static and dynamic porosities, namely:

$$\phi = \phi_{ST} + \phi_{DY} - 1 \tag{7}$$

3.2   RANS equations

The starting point of the derivation are the classical Reynolds-Averaged Navier-Stokes equations, the mass conservation and the momentum conservation equations.

Conservation of mass can be posed in two different forms, first as a single fluid approach:

$$\frac{\partial \rho \, u_i}{\partial x_i} = 0 \tag{8}$$

in which u is the Reynolds-averaged velocity vector and x_i denotes the *i-th* direction in space.

It is often the case that for free surface applications there is a need to simulate two fluids, therefore the alternative form of the equation, also called Volume Of Fluid (VOF) equation, is:

$$\frac{\partial \alpha}{\partial t} + \frac{\partial \alpha \, u_i}{\partial x_i} = 0 \tag{9}$$

where α is the so-called VOF or color function and t denotes time. In the VOF technique (Hirt and Nichols, 1981), α represents the amount of fluid per unit volume. By convention, for systems with air and water phases, α denotes the unit volume of water, while the unit volume of air would simply be $(1 - \alpha)$. In this way, the calculation of fluid properties, as density (ρ) or kinematic viscosity (ν), under the incompressible assumption can be simply achieved by means of a weighted average: $\rho = \rho_w \, \alpha + \rho_a \, (1 - \alpha)$, where the subscript w denotes water and a denotes air.

Finally, the momentum conservation equations are:

$$\frac{\partial \rho \, u_i}{\partial t} + u_j \frac{\partial \rho \, u_i}{\partial x_j} = -\frac{\partial p}{\partial x_i} + \rho \, g_i + \frac{\partial}{\partial x_j}\left[\mu \frac{\partial u_i}{\partial x_j}\right] - \frac{\partial}{\partial x_j}\left[\rho \, \overline{u_i' u_j'}\right] \tag{10}$$

in which the only new variables are p, pressure; g, the acceleration due to gravity; and μ, the dynamic viscosity. The last term, the Reynolds stresses, includes the effects of turbulence. Reynolds stresses need to be modelled because they cannot be simulated under the RANS assumptions.

3.3   Volume-Averaged RANS equations

By applying the volume-averaging procedure the RANS equations are transformed. Nevertheless, the new set of equations retains a high degree of similarity with respect to the original RANS equations. For example, the volume-averaged mass conservation equation in terms of extended variables is:

$$\frac{\partial \rho\, \phi_{\mathrm{DY}}}{\partial t} + \frac{\partial \rho\, \langle u_i \rangle}{\partial x_i} = 0. \tag{11}$$

A new term appears inside the time derivative. ϕ_{DY} accounts for the variation in dynamic porosity, therefore, it can change in time and must be kept inside the differential operator.

Before introducing the volume-averaged VOF equation, we need to recall that before introducing porosity into the equations α was defined as the amount of fluid per unit volume. However, some portion of the control volume may be blocked by solids. Therefore, the correct definition of α is the amount of fluid per unit of the volume that can be occupied by fluids. With the new definition, $0 \leq \alpha \leq 1$ still holds and $\langle \alpha \rangle^f = \alpha$, as demonstrated in (Higuera, 2015, Appendix A.2). Taking this into consideration, the volume-averaged VOF equation is:

$$\frac{\partial \phi\, \alpha}{\partial t} + \frac{\partial \alpha\, \langle u_i \rangle}{\partial x_i} = 0, \tag{12}$$

which resembles closely Equation 11, with *alpha* instead of ρ. However, only the dynamic component of porosity (ϕ_{DY}) appears inside the partial derivative with respect to time term in Equation 11, whereas ϕ appears in Equation 12. The reason behind this is that under the incompressible assumption, $\frac{\partial \rho}{\partial t} = 0$, while generally $\frac{\partial \alpha}{\partial t} \neq 0$, hence $\frac{\partial \phi \alpha}{\partial t} = \phi \frac{\partial \alpha}{\partial t} + \alpha \frac{\partial \phi_{\mathrm{DY}}}{\partial t}$.

Finally, the volume-averaged momentum equations involve an increasing complexity. New terms that cannot be simulated appear during the derivation, as a result of filtering out the unknown internal geometry of the porous media. Therefore, a closure model needs to be introduced to retain the contribution of such terms, as follows:

$$\frac{\partial \rho\, \langle u_i \rangle}{\partial t} + \frac{\partial}{\partial x_j} \left[\frac{1}{\phi} \rho\, \langle u_i \rangle \langle u_j \rangle \right] =$$
$$- \phi \frac{\partial \langle p \rangle^f}{\partial x_i} + \phi \rho g_i + \frac{\partial}{\partial x_j} \left[\mu \frac{\partial \langle u_i \rangle}{\partial x_j} \right] - \frac{\partial}{\partial x_j} \left[\rho \left\langle \overline{u_i' u_j'} \right\rangle \right]$$
$$+ [CT]_{\mathrm{ST}} + [CT]_{\mathrm{DY}} \tag{13}$$

The closure terms are divided in two, corresponding to those derived from static porosity ($[CT]_{\mathrm{ST}}$) and dynamic porosity ($[CT]_{\mathrm{DY}}$), and will be discussed in next subsection.

It is often the case that porosity does not change with time. This means that the porous structure is fixed and will not move due to the wave action. Under this condition, only static porosity is involved ($\phi = \phi_{\mathrm{ST}}$), and the conservation of mass and VOF equations are reduced to simpler expressions:

$$\frac{\partial \langle u_i \rangle}{\partial x_i} = 0 \tag{14}$$

$$\frac{\partial \alpha}{\partial t} + \frac{1}{\phi} \frac{\partial \alpha\, \langle u_i \rangle}{\partial x_i} = 0 \tag{15}$$

The only change in momentum conservation (Equations 13) is that the term $[CT]_{\mathrm{DY}} = 0$.

3.4 Closure

Similarly to what occurs with Reynolds stresses in RANS, the volume-averaging operation also yields a number of terms that cannot be simulated and need to be modelled instead. These terms represent the inertial (Polubarinova-Kochina, 1962) and drag forces (Forcheimer, 1901) due to the geometry of solids which have been filtered out. Different closure models can be found in literature, as previously referenced in Section 2. In this work the formulation developed by Engelund (1953), later applied in Burcharth and Andersen (1995), will be used. The inertial force, linear drag force and nonlinear drag force are as follows:

$$C\frac{\partial \rho \langle u_i \rangle}{\partial t} \tag{16a}$$

$$\alpha \frac{(1-\phi)^3}{\phi^2} \frac{\mu}{D_{50}^2} \langle u_i \rangle \tag{16b}$$

$$\beta \left(1 + \frac{7.5}{\text{KC}} \right) \frac{1-\phi}{\phi^2} \frac{\rho}{D_{50}} |\langle u \rangle| \langle u_i \rangle . \tag{16c}$$

Equation 16a is usually combined with the local derivative in the NS equations (i.e., first term in Equation 13). C is a calibration parameter, which is often taken as 0.34, given that it has been found to produce a negligible contribution when compared to the other two terms (del Jesus, 2011). Equation 16b is a drag force, which is linearly dependent on velocity. α is a calibration factor and D_{50} is the mean nominal diameter of the solid elements in the porous medium. Equation 16c is the nonlinear drag force, dependent on the β calibration factor. KC is the Keulegan-Carpenter number, defined as $\left(\frac{T_w}{D_{50}} \frac{u_M}{\phi} \right)$, where u_M is the maximum oscillatory velocity and T_w is the wave period. This element, first introduced in van Gent (1995), increases the friction under oscillatory conditions (i.e., in wave simulations).

The values of α and β are dependent on the case conditions, including the porous material and flow regime. Therefore, experimental data is usually required to obtain suitable values for both parameters. The process involves tuning the calibration factors until the numerical solution matches the experimental results. This procedure will be discussed in Section 3.6.

3.5 Turbulence modelling

As it occurs in the RANS approach, the volume-averaged Reynolds stresses cannot be simulated and need to be modelled instead. Therefore, turbulence modelling in VARANS requires volume-averaged turbulence equations that will yield a volume-averaged turbulent eddy viscosity. However, the volume-averaging process, similarly to what happened in the conservation of momentum equations, causes the appearance of a number of terms which cannot be simulated. Thus, an additional model is also needed to close the turbulence equations.

So far, the only volume-averaged turbulence model that has a published closure is $k - \epsilon$. The closure model was developed by Nakayama and Kuwahara (1999) for heat transfer studies and was later applied in Hsu et al. (2002) for wave interaction with a composite breakwater. Moreover, del Jesus et al. (2012) volume-averaged the $k - \omega$ SST turbulence model, although no closure model was developed. The reader is referred to Higuera (2015) for the expressions and the complete derivation of the volume-averaged $k - \epsilon$ and $k - \omega$ SST turbulence models.

3.6 Discussion

The VARANS equations just presented are a mathematical model that has been volume averaged. Consequently, it is important to understand the scope of applicability and the limitations that they pose.

First of all, from a physical perspective, porosity needs to be bounded between 0 and 1, both included. However, upon inspection of the equations it is clear that porosity cannot be zero, which means that this set of equations cannot represent an impermeable obstacle. For example, the convective term in Equation 15, will be infinity if $\phi = 0$. Moreover, practical experience indicates that the VARANS equations will also diverge even for small values of porosity, as the drag forces will grow unbounded even before ϕ approaches 0.

It must be remarked that the equations and the form of the closure terms presented in this chapter are not the only ones available in the literature, as other authors (Hsu et al., 2002; Hur et al., 2008; del Jesus et al., 2012; Nikora et al., 2013; Jensen et al., 2014) have followed different approaches and averaging techniques, obtaining accurate results in all cases. There is also a wide variety of friction factors to choose from in the literature. The reader is referred to Losada et al. (2016) for a catalogue of the most important ones used. A feasible explanation for the general good results in literature is the fact that the friction factors include variables that serve as tuning factors, and the authors are selecting the values that make the model results fit better with the experimental data. Unfortunately, currently there is no other way to estimate the parameters, as their physical significance is not completely understood and their calibration is flow-dependent. Moreover, as shown in Jensen et al. (2014) and Higuera (2015), there is often a problem selecting the friction factors, as there is no unique solution. Since both friction factors are linked together, an excess of the linear friction force can be balanced with a defect of the nonlinear friction force and vice versa.

Further research also needs to be pursued to shed light in the flow kinematics and dynamics through interfaces. It has been observed (Dimakopoulos, Allsop and Pullen, 2019, personal communication) that there are significant uncertainties in quantifying transitions of porosity both at the interface between the structure and the clear flow region (e.g., armour layers roughness) and between internal interfaces (e.g., transition between underlayer, core, filters, land fillings). For realistic applications, it is possible to calibrate the Darcy-Forchheimer coefficients against known empirical relations for run-up (e.g., Eurotop manual, van der Meer et al. (2018)) or internal pressure transmission, e.g., Troch et al. (2002). However, this calibration will reflect the uncertainties of the empirical relations and, in addition, it may not always be feasible and practical to apply, as flow kinematics at the interfaces are generally mesh dependent and there may be other elements involved (e.g., geotextiles) that produce additional challenges to model. Thus, for such studies, additional testing of sensitivity is recommended to explore different possible outputs. A better mathematical (and ultimately numerical) description of the flow dynamics at porous interfaces will certainly help quantify and address these uncertainties.

The real need for turbulent dissipation inside porous media (via a turbulence model) or even the turbulence enhancement produced by the closure terms in $k - \epsilon$ (which $k - \omega$ SST is lacking), is also an interesting open topic. Jensen et al. (2014) argues that "when the actual turbulence levels are of minor interest the effect of the turbulence can be included via the Darcy-Forchheimer equation". In fact, this is because slightly larger friction factors may account for the expected turbulent dissipation from a mathematical point of view due to their tuning factor behaviour, but from a physical point of view, it is clear that the turbulent terms should be included.

The final and most compromising factor is modelling prototype scale coastal structures with the VARANS equations. So far, detailed information about the flow inside structures in laboratory tests is widely available but the data inside real structures is very limited. Hence, modelling a prototype scale requires an extrapolation of the friction factors at laboratory scale. The fundamental difference between both scales, assuming Froude number similarity to make sure that accelerations (i.e., gravity forces) are scaled correctly, is that given a length scale factor λ, Reynolds number scales with a factor $\lambda^{3/2}$. Therefore, laminar or mildly turbulent cases at laboratory scale can often be fully turbulent at prototype scale, changing the physics driving the problem completely.

4 Numerical model

The computational fluid dynamics (CFD) model *olaFlow* (Higuera, 2017) is used in this work. *olaFlow* is an open source project conceived as a continuation of the work in Higuera (2015). The model is developed within the OpenFOAM® framework (Jasak, 1996; Weller et al., 1998) to enable the simulation of wave hydrodynamics and wave-structure interaction, including porous structures, in the coastal and offshore fields.

Wave generation is linked with active wave absorption at the boundaries (Higuera et al., 2013a), thus minimizing the computational cost with respect to relaxation zones and passive wave absorption approaches. The VARANS modelling was developed and validated in Higuera et al. (2014a), although the present implementation follows Higuera (2015). Further validation and applications of the model can be found in Higuera et al. (2013b, 2014b). Additional features, such as wave generation with moving boundaries to reproduce experimental wave tanks are also included and can be found in Higuera et al. (2015).

olaFlow solves the VARANS equations introduced in Section 3.3 for two incompressible phases (e.g., water and air) using a finite volume discretization. The implementation is as follows:

$$\frac{\partial \rho \langle u_i \rangle}{\partial x_i} = 0 \qquad (17)$$

$$\frac{1+C}{\phi} \frac{\partial \rho \langle u_i \rangle}{\partial t} + \frac{1}{\phi} \frac{\partial}{\partial x_j} \left[\frac{1}{\phi} \rho \langle u_i \rangle \langle u_j \rangle \right] =$$

$$-\frac{\partial \langle p^* \rangle^f}{\partial x_i} - g_j r_J \frac{\partial \rho}{\partial x_i} + \frac{1}{\phi} \frac{\partial}{\partial x_j} \left[\mu_{\text{eff}} \frac{\partial \langle u_i \rangle}{\partial x_j} \right] + \sigma \kappa \frac{\partial \alpha}{\partial x_i}$$

$$-\alpha \frac{(1-\phi)^3}{\phi^2} \frac{\mu}{D_{50}^2} \langle u_i \rangle - \beta \left(1 + \frac{7.5}{\text{KC}} \right) \frac{1-\phi}{\phi^2} \frac{\rho}{D_{50}} |\langle u \rangle| \langle u_i \rangle. \qquad (18)$$

The newly introduced variables are the dynamic pressure, defined as $p^* = p - \rho g_j r_j$, where p is the total pressure; g_j is the acceleration due to gravity and r_j is the position vector.

Reynolds stresses in Equation 13 are modelled using Boussinesq hypothesis, linking them with the velocity gradient through a turbulent eddy viscosity (ν_t) which follows the isotropy assumption and is given by the turbulent model selected. The viscous term in Equation 18 is computed with the effective dynamic viscosity (μ_{eff}), which comprises the molecular dynamic viscosity of the fluid (μ) and a turbulent dynamic viscosity ($\mu_t = \rho \nu_t$) components.

The last term in the second line of Equation 18 was introduced in Brackbill et al. (1992) to transform the surface tension force into a body force acting at the cells belonging to the interface between both fluids. Here σ is the surface tension coefficient, κ is the curvature of the free surface and α is the VOF indicator function introduced in Section 3.2.

The free surface between water and air is captured with the VOF technique (Hirt and Nichols, 1981). The implementation of VOF in OpenFOAM® is algebraic and only first order accurate. This method does not involve surface tracking or geometrical reconstruction, therefore, it is computationally efficient. The VOF advection equation is:

$$\frac{\partial \alpha}{\partial t} + \frac{1}{\phi} \frac{\partial \alpha \langle u_i \rangle}{\partial x_i} + \frac{1}{\phi} \frac{\partial \alpha (1-\alpha) \langle u_i^c \rangle}{\partial x_i} = 0 \qquad (19)$$

In order to obtain physical results, the scalar field α needs to be conservative, strictly bounded between 0 and 1 while maintaining a sharp interface. The first two conditions are achieved with a special solver called MULES (Multidimensional Universal Limiter with Explicit Solution). The reader is referred to Márquez (2013) for a complete description of the solver. The last term in the VOF equation does not appear in Equation 12. This additional term is a numerical artifact that produces compression forces at the interface ($0 < \alpha < 1$) in order to prevent the diffusive behaviour of Equation 19 (Rusche, 2002). The new variable $\langle u_i^c \rangle$ is a compression velocity normal to the interface, dependent on a user-defined compression enhancement constant and especially designed to avoid creating high artificial velocities.

Further details about *olaFlow* and the numerical implementation involved in OpenFOAM® can be found in Higuera (2015).

5 Applications: Solitary wave impacting into a rubble mound breakwater

In the first application case the numerical model will be applied as a design tool to evaluate different structural alternatives to protect a rubble mound breakwater against the action of a large solitary wave. Initially the original structure will be tested to establish the reference solicitations, including the instantaneous force on the caisson, overtopping and the reflection coefficient. The model will then be applied to simulate an array of cases, in which different parametrized alternatives are built on top of the original geometry to reduce the solicitations of the solitary wave. The goal is to determine the suitable alternatives in terms of reducing the forces, enhancing stability and limiting overtopping. The two selected alternatives will be compared with the original case in detail.

5.1 Numerical setup

The structure tested is similar to the one built for in the experiments by Guanche et al. (2009). The design of the caisson has been modified, because the original structure was designed to avoid any displacements during testing, thus it was heavily overdesigned. The new breakwater is shown in Figure 3.

The breakwater comprises a primary and secondary armour layers (1:2 slope, 12 cm and 10 cm thick, respectively) and a core. The physical properties of the porous media materials, measured in the laboratory, are gathered in Table 1. The new geometry of the

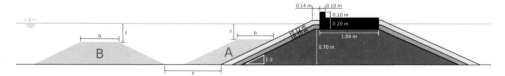

Figure 3: Sketch of the breakwater geometry including alternatives A and B.

Table 1: Physical properties and friction parameters of porous media materials.

Material	D_{50} (cm)	Porosity	α	β
Primary armour layer	12.0	0.50	50	0.6
Secondary armour layer	3.5	0.49	50	2.0
Core	1.0	0.49	50	1.2

concrete crown wall is L-shaped, which will reduce the high original safety factors (SF). The crown wall (1.04 m x 0.3 m) is located 25 m from the wave generation boundary and founded on top of the core at $z = 0.7$ m. The front vertical wall is 10 cm thick, and the bottom slab is 20 cm thick. The still water level is set at $h = 0.8$ m, therefore the breakwater has a positive freeboard of 20 cm.

The numerical wave flume is 30 m long (x-direction) and 1.3 m high (z-direction) and has a horizontal bottom. The mesh cell size in the vertical direction is constant and equal to 1 cm. The horizontal cell size varies in a geometric progression from the wave generation boundary (2 cm at $x = 0$ m) to $x = 20$ m, remaining constant and equal to 1 cm until the flume end ($x = 30$ m). These parameters yield a two-dimensional mesh of 350,000 cells.

A solitary wave of steepness $H/h = 0.5$ has been generated with the third order theory developed by Grimshaw (1971). The opposite end of the flume ($x = 30$ m) carries an active wave absorption boundary condition (Higuera et al., 2013a) to prevent the transmitted wave from reflecting back. The effective wavelength, given $k = \sqrt{(3H/4h^3)}$, is $L = 8.21$ m, therefore, the solitary wave propagates a distance of approximately 3 wavelengths before reaching the breakwater.

The friction parameters α and β used in the numerical simulations are included in Table 1. These values have been obtained by best-fit (Higuera, 2015, Chapter 7.2). The reader is referred to this reference for a complete validation study of the original case, including free surface elevation at different locations on the flume and pressure sensors mounted on the caisson, inside and outside the porous media.

Two different alternatives, shown in Figure 3, have been proposed and parametrized to enhance the initial design. The first alternative (A) consists of an additional armour layer on top of the primary layer. This alternative depends on 2 parameters: the freeboard (f) and the width of the berm (b). The second design (B) consists of a detached breakwater. This alternative also depends on f and b, plus an additional parameter s to control the separation between the toes of both structures. In either case the additional material is the same as in the primary armour layer, to avoid introducing further uncertainties in the case, and the new slopes are also 1:2.

The f, b and s parameters will be made dimensionless from here on in, dividing them by the water depth (h). The values considered are as follows: f range is [-0.8, -0.6, -0.4, -0.2, 0]; b range is [0.2, 0.4, 0.6, 0.8, 1]; and s range is [0, 1, 2, 3].

All the numerical experiments have been simulated in parallel using 6 cores of a Xeon (2.50 GHz) workstation. Turbulence modelling was connected and set to the volume-averaged $k - \omega$ SST model modified with the features presented in Larsen and Fuhrman (2018) and Devolder et al. (2017). The simulations are 15 seconds long, and each of the 125 cases computed takes less than 2 hours to compute.

5.2 Numerical results

The numerical simulations produce a large amount of data to analyse, considering the number of alternatives simulated and the fields available (VOF, velocity, pressure, turbulent variables...) Therefore, we have selected 5 variables to rationalize the analysis of the alternatives. F_x, F_z are the maximum wave-induced horizontal and vertical forces acting on the crown wall, measured in Newtons (N) and calculated by integrating the pressure field around the structure, filtering the time signal to eliminate impulsive loading. q_m is the mean overtopping rate over the event, measured in m^3/s; and $\sum Q$ is the total overtopping volume, measured in m^3. Consequently, the overtopping event duration can be calculated as $\sum Q/q_m$. Both overtopping variables have been measured at a gauge located on top of the crown wall. The values for all these variables are reported per unit width of the breakwater.

As indicated in Pedersen (1996), the four major reasons for crown wall failure are sliding, overturning, cracking and geotechnical failure. Therefore, the minimum safety factor against sliding (SF_S) throughout the simulation is also analysed. This parameter is defined as: $SF_S = f(F_W - F_U)/F_H$, where $f = 0.5$ is the friction coefficient between the concrete crown wall and the material of the core; F_W is the weight of the structure considering the emerged and submerged weights; F_U and F_H are the uplift and horizontal force components induced by wave loading (i.e., excess over the initial hydrostatic loading). SF_S is a dimensionless number that needs to be larger than 1 to ensure that the structure will not slide under the action of waves. Generally, a safety factor of 1.5 should be considered, thus, coastal structures are often designed to fulfil that $SF_S > 1.5$. The safety factor against sliding is studied instead of the safety factor against overturning (SC_O) because SF_S is significantly lower for this particular setup. The global minimum for SC_O obtained for all the simulations is 5.75, which is an extremely safe value. Nevertheless, the presence of impulsive loading could reduce the SC_O, particularly if impulsive loads were concentrated at the top of the crown wall. A dynamic response analysis has not been performed because although impulsive loads can cause structural failure, they have a high uncertainty associated with them. If preliminary design assessment shows that there is a risk of failure due to impulsive loads, this should be addressed primarily by physical modelling, using CFD modelling to gain further insights, as it is not yet a fully mature tool to address such complex processes.

5.2.1 Reference case

The base case is taken as a reference to analyse the performance of the new alternatives and the values are included in Table 2. The numerical experiments have been simulated at laboratory scale, therefore, some variables are scaled up to prototype scale for direct comparison with data available in literature. The model has been designed at a 1/10 length scale, which implies that the prototype structure is at 8 m of water depth. Froude number (defined as $Fr = u/\sqrt{g\,l}$, where u is velocity and l is a length scale) similarity scaling is applied to preserve the ratio between the inertial and gravitational forces. Therefore, assuming no scale effects, the laboratory-scale variables can be transformed into prototype-scale values by multiplying them by a factor. The factors for the different variables are as follows: 3.16 for time and velocity, 10 for length, 316.23 for flow rate and 1000 for volume and force.

The weight of the crown wall at laboratory scale is $F_W = 4112$ N, considering the emerged and submerged weights accordingly for the present still water level, and a dry concrete density of 2400 kg/m^3. The maximum horizontal force induced by the solitary wave is 28% of F_W, while the maximum uplift force totals 52% of F_W.

The overtopping results can be compared with the EurOtop manual (van der Meer et al., 2018). As referenced in Table 2, the equivalent prototype-scale mean discharge rate is $q_m = 60$ m^3/s per unit metre and the total overtopping volume is 250 m^3 per unit metre. These values are one to three orders of magnitude larger than the maximum values reported in the EurOtop manual. For example, a $q_m = 0.05$ m^3/s and $\sum Q$ between 5 and 50 m^3 would already pose a major threat, causing "significant damage or sinking of larger yachts". This

Table 2: Reference variables and values obtained for the base case. LS stands for laboratory scale. PS stands for prototype scale.

Variable	F_x (kN)	F_z (kN)	q_m (m^3/s)	$\sum Q$ (m^3)	SC_S
Value (LS)	1.149	2.123	0.19	0.25	0.87
Value (PS)	1149	2123	60	250	0.87

makes clear that the solitary wave conditions chosen represent an extreme event such as a tsunami wave. Moreover, since the lowest value of SF_S during the simulation is 0.87, the structure is well below the stability limit for the wave conditions tested, and is at risk of failing due to the caisson sliding. This is the main justification that drives the need to find new alternatives with which to achieve structural stability.

5.2.2 Global analysis of alternative A cases

Given the numerous cases and variables to analyse, the results for alternatives A, and later B, are represented as an array of radar plots in Figures 4 and 5. Each of the variables except SF_S are represented in terms of the percentage of variation with respect to the reference case values. The central line indicates no variation (0%), while the upper and lower limits are indicated in the top left panel. Variable SF_S is scaled differently: the central line corresponds to the limit state, $SF_S = 1$. Any values below 1 (outer region) are considered potentially unsafe, as the crown wall will be likely to slide. Therefore, for all variables, a point in the inner region indicates a safer condition with respect to the original geometry, while the outer region signals a more dangerous condition.

The results for alternative A (new armour layer) are displayed in Figure 4. The variation of all the indicators is quite limited, below 5%. Changes are low, around 1%, especially for those alternatives with large negative freeboard and narrow berm width (upper and left sections). At the lowest freeboard values (–0.8 and –0.6, top two rows), increasing the berm width does not produce significant changes, as the crest of the new layer is too deep to alter the incoming wave significantly. As the freeboard increases (–0.4 and –0.2), the effect of the berm width starts being noticeable, reducing both q_m and $\sum Q$ up to 5% and 2.5%, respectively. However, in all cases the SF_S is still below 0.9, thus, the stability of the crown wall continues to be compromised. The cases with the highest freeboard ($f = 0$, bottom row) show a consistent and noticeable decrease in the hydrodynamic forces and overtopping up to a 6% as the berm width increases, bringing SF_S closer to 1 progressively. The only case in which SF_S is just above 1 is ($f = 0$, $b = 1$). This is the most expensive alternative in terms of building cost, and will be analysed and compared with the original structure later in the text.

5.2.3 Global analysis of alternative B cases

The results for alternative B (detached breakwater) are shown in Figure 5. In this plot the range of variation of each indicator variable is different, as indicated in the top left panel. Similarly to what occurred for alternative A, the cases with the lowest freeboard (-0.8, -0.6) do not show significant differences with respect to the original alternative. Also, separation "s" does not have a major impact either, as the four lines represented are almost collapsed into one. For example, a detached breakwater with $f = -0.8$ will only reduce the variables studied by 2% at most, independent of its berm width and separation from the main breakwater. Nevertheless, for $f = -0.6$, the reduction in the total overtopping volume starts becoming noticeable (5%) for the largest berm widths. Also, the separation of the detached breakwater (s) starts having a noticeable impact in q_m for the longest berm widths (0.6–1) at $f = -0.4$.

As the freeboard continues to increase, the overtopping variables decrease significantly faster than forces. For example, at $f = -0.2$ the force reduction reaches 3–5% while $\sum Q$ decreases up to 29%. The main reason behind this is that the detached breakwater produces local wave transformations, but does not induce wave breaking for $f = -0.2$, thus, the amount of energy reflected and dissipated is limited. Moreover, since q_m decreases much less than $\sum Q$, it means that the time of the overtopping event ($\sum Q/q_m$) must be reduced too.

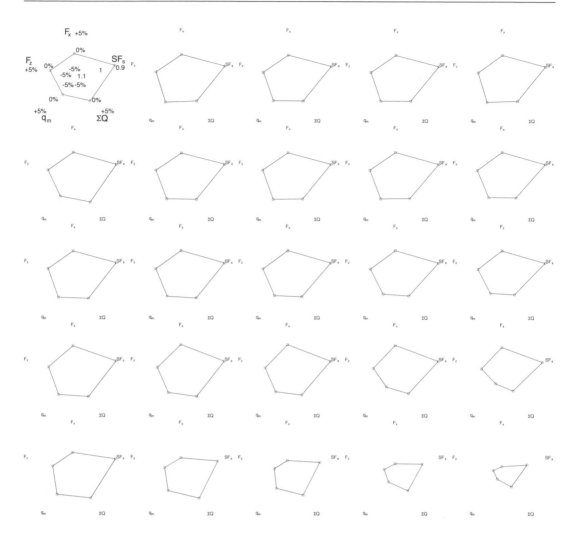

Figure 4: Percentage of variation for the horizontal and uplift forces, mean overtopping discharge rate and total overtopping volume and safety factor against sliding for the cases in alternative A. From top to bottom: f =(–0.8, –0.6, –0.4, –0.2, 0). From left to right: b =(0.2, 0.4, 0.6, 0.8, 1). The range of variation per variable is indicated in the top left panel.

It can be noted that there is no linear correlation between s and the value of q_m. While the maximum reduction of q_m corresponds to cases with $s = 0$, q_m magnitude can also increase with respect to the original case for a certain range of parameters. For example, the largest reduction of q_m is 9%, for $s = 0$ (square marker). Cases with $s = 1$ and $s = 3$ also show a reduction, although less significant. However, q_m increases slightly in cases with $s = 2$ (triangle marker). This irregular evolution can only be explained due to local effects on wave transformation.

Even though overtopping reductions can be significant for $f \leq -0.2$ in alternative B, no single case fulfils that $SF_S > 1$, hence, the structure would still be at risk of failing by sliding. The bottom row in Figure 5 ($f = 0$) includes all the cases in which $SF_S > 1$. The effect of the berm width plays an important role in reducing the overtopping volume, whereas the separation s does not produce significant differences for this variable. The separation (s)

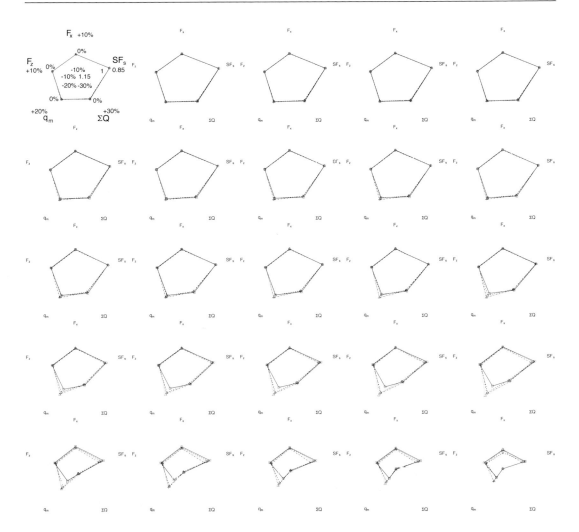

Figure 5: Percentage of variation for the horizontal and uplift forces, mean overtopping discharge rate and total overtopping volume and safety factor against sliding for the cases in alternative B. From top to bottom: $f = (-0.8, -0.6, -0.4, -0.2, 0)$. From left to right: $b = (0.2, 0.4, 0.6, 0.8, 1)$. The range of variation per variable is indicated in the top left panel. The separation of the structure (s) is: 0 for continuous line and square marker, 1 for dashed line and circumference marker, 2 for dotted line and triangle marker and 3 for dash-dot line and plus sign marker.

is, however, important when analysing SF_S, especially for the narrowest berms. In case of $f = 0$, $b = 0.2$, the two smallest separations ($s = 0$ and 1) fail, while cases with $s = 2$ and 3 are above $\mathrm{SF}_S = 1$. For cases with $b = 0.4, 0.6$, only the case with $s = 0$ would produce a $\mathrm{SF}_S < 1$. Above $b = 0.6$ all the cases, independent of s, would fulfil $\mathrm{SF}_S > 1$. The largest SF_S obtained is 1.08, for case ($f = 0$, $b = 1$, $s = 3$). Therefore, this case will be studied in depth below.

It must be remarked that there is no direct relation between the maximum reduction of horizontal and uplift forces and the increase in SF_S, because the maximum forces in both directions do not occur at the same instant.

5.2.4 Detailed analysis of the alternative A and B selected cases

In this subsection we perform a detailed comparison of the selective cases, alternative A ($f = 0$, $b = 1$), alternative B ($f = 0$, $b = 1$, $s = 3$) and the original reference case.

The evolution of the wave and velocity field before and during the overtopping phase are represented in Figures 6 and 7. At $t = 7.30$ s the additional layer in alternative A has not produced any significant effects yet, as the wave is still several water depths away from the toe of the breakwater, whereas the detached breakwater has induced wave breaking, thus changing the shape and kinematics of the incident wave. The most noticeable effects in alternative B include a larger particle velocity between the free surface and the top of the breakwater and a lower velocity inside the breakwater due to the porous medium drag force. Also, the maximum free surface elevation decreases as compared with the reference case, showing a flatter wave crest.

At $t = 8.65$ s the wave has just impacted the crown wall. Both alternatives show noticeable differences with respect to the original case. The most obvious difference is that overtopping is delayed, since it is incipient for the base case (bottom subpanel) but no water has surpassed the crown wall yet in either of the other two alternatives. In solution A, the deviation with respect to the initial case is limited, and localized near the crown wall because the new armour layer attached to the breakwater produces energy dissipation and a wake effect above it, effectively delaying the impact and overtopping phases. The water level at the front of the crown wall is significantly lower for alternative B at this stage, because of wave breaking occurrence. The detached breakwater also produces noticeable energy dissipation, inducing a wake after the structure that is intense and creating a clockwise vortex that remains in place for a long time.

Figure 7 shows two time steps during the overtopping phase. At $t = 9.15$ s, the overtopping jet in alternative A looks very similar to that in the original case, although in a previous stage of development, i.e., while the jet has recently impacted the esplanade of the crown wall and is flowing to the leeside, it has not yet been projected back towards the vertical wall,

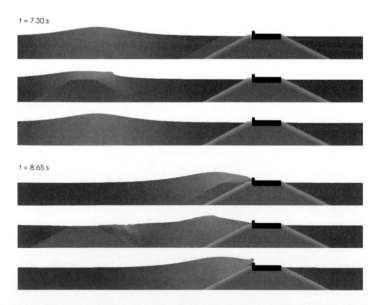

Figure 6: Evolution of the solitary wave before overtopping. For each instant, the top subpanel is case A ($f = 0$, $b = 1$), the middle subpanel is case B ($f = 0$, $b = 1$, $s = 3$) and the bottom subpanel is the reference case.

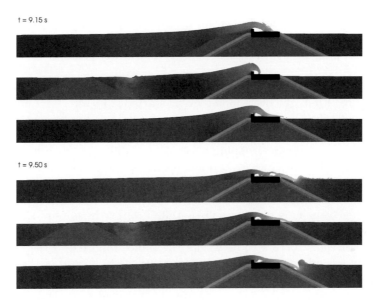

Figure 7: Evolution of the solitary wave during the overtopping phase. For each instant, the top subpanel is case A ($f = 0$, $b = 1$), the middle subpanel is case B ($f = 0$, $b = 1$, $s = 3$) and the bottom subpanel is the reference case.

as in the base case. This effect will eventually occur for all three cases, the only differences being the magnitude of the back-flow and the time of occurrence. The alternative B presents a significantly different jet shape, caused by a deficit in horizontal momentum as a result of wave breaking. Nevertheless, the depth of the overtopping jet is thicker at this instant.

At the final panel, $t = 9.50$ s, all three overtopping jets show similar features as recirculation behind the vertical wall and impact on the quiescent water body behind the breakwater. The discharge volume is so large that the disturbance created on the leeside of the breakwater is of the order of magnitude of the crown wall height. This observation is aligned with the reports in EurOtop, by which even much smaller discharge volumes will cause safety threats or even sink large vessels. Another feature to note is the vortex located on the leeside of the detached breakwater in alternative B. As mentioned before, this coherent structure persists a long time in that position. Although it has not appeared yet, another vortex is created on the seaside of the new berm in alternative A, which also persists for long time, although with a lower absolute velocity.

The reflection coefficient R is calculated at a wave gauge near the wave generation boundary, as the ratio between the maximum elevation of the reflected wave and the height of the incident wave. The range of variation of R among all the cases simulated is 28.0% to 40.1%, with $R = 32.9\%$ for the original alternative. The selected alternative A yields the lowest global reflection coefficient, 28.0%, whereas the selected alternative B produces a reflection coefficient of 31.6%. This 3.6% difference is not very significant, as it corresponds to less than 1.5 cells, but indicates that since overtopping has also been reduced in both cases, the new structural elements produce additional energy dissipation.

The time evolution of the overtopping variables is represented in Figure 8. The instantaneous overtopping rate (q) shows that the alternative A causes a delay of 0.11 s in the start of overtopping, and the slope of the curve is also steeper initially. The rest of features, such as the maximum value or ending time, remain practically unchanged, thus the total overtopping time is reduced. An almost identical evolution can be observed for the instantaneous overtopping water depth (central panel), except that the alternative A produces a 15% reduction with respect to the highest value. This translates into a small reduction of the total overtopping volume (3%, right panel) and of the mean instantaneous overtopping rate (5%).

Alternative B produces significant changes because the solitary wave breaks before reaching the structure. First, the overtopping event starts 0.25 s after and finishes almost at the same instant as in the reference case, therefore, the duration of the overtopping event is reduced by almost 20%. The initial slope of q is almost identical to that of the reference case. However, the shape of the overtopping curve is very different, with a marked peak instead of a smooth crest. The variable h_o also follows the same trend, with a maximum at 30 cm water depth. The maximum value of q is larger than in the reference case by 9%, yet the area under the curve is significantly smaller, as can be seen in the right panel, in which the total overtopping volume is reduced substantially (32%). In spite of slightly increasing the maximum q value, alternative B reduces its mean value by 13%. Nevertheless, this reduction is not enough for the structure to be within the values reported in the EurOtop manual, as the original wave condition is extreme compared to the freeboard of the structure.

The wave-induced forces acting on the structure are presented in Figure 9. Some of the previous observations for overtopping are also applicable to the forces time series. The main difference is that forces increase very slowly at first, and start growing faster prior to the overtopping event. This is because the wave progressively impacts the vertical wall, increasing the water level steadily and building up the pressure prior to surpassing it. After overtopping starts, the growth rate decreases, and forces reach their maximum value. At

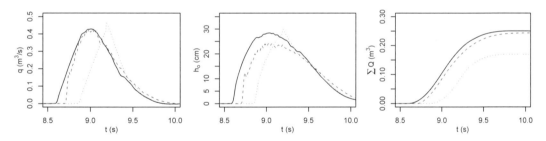

Figure 8: Overtopping analysis of the selected alternatives. The left panel represents the instantaneous overtopping rate (q). The central panel represents the instantaneous overtopping water depth (h_o). The right panel represents the cumulative overtopping volume ($\sum Q$). The continuous line denotes the reference case, the dashed line is for the selected alternative A ($f = 0$, $b = 1$) and the dotted line represents the selected alternative B ($f = 0$, $b = 1$, $s = 3$).

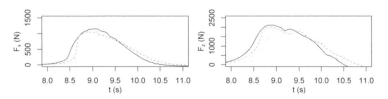

Figure 9: Wave-induced force analysis. The left panel represents the horizontal force and the right panel represents the uplift force. The continuous line denotes the reference case, the dashed line is for the selected alternative A and the dotted line represents the selected alternative B.

this stage the total maximum force reduction is 5% for case A and 10% for case B, both for F_x and F_z. In view of the time series, it can be concluded that the wave breaking event in alternative B only has a significant effect on the initial part of the F_x curve. The evolution of the trailing edge is quite similar in all cases, with forces that start to decrease progressively, following the trend of the overtopping water depth. However, small differences in F_z, in terms of a slight delay in the end of the uplift force time series can also be observed. In any case, not only the magnitude of the maximum forces is reduced, but also the total duration of the impact event is shortened for case B.

As mentioned before, these cases were selected because the new design prevents the structural failure in terms of a quasi-static analysis. The reduction in the force magnitudes translates into a SF_S that is just above 1 for alternative A and a SF_S equals to 1.08 in case B.

5.3 Concluding remarks

This example portraits the capabilities of VARANS modelling to evaluate different structural alternatives for a given scenario. The numerical model has proven to be a useful tool to calculate the overtopping and forces acting on a structure, therefore, to establish its functionality and stability regimes. However, numerical modelling is complementary to experimental modelling, as validation is a necessary step to check whether the model (in combination with the numerical schemes and mesh used) is able to simulate accurately all the physical processes involved in a specific case.

6 Applications: Wave and sediment grain interaction by a nonbreaking solitary wave on a steep slope

The second application case reproduces the interaction of a solitary wave with solid particles representing the sediment grains of a beach, modelled by means of the Discrete Element Method (DEM). Particles can collide and interact between them and with the walls, and are coupled with the flow via the individual drag forces and dynamic porosity terms included in the VARANS equations. The base setup is similar to the one depicted in Higuera et al. (2018). The goal of this example is to observe the evolution of the sediment grains in the swash zone and compare the numerical results qualitatively with observations available in literature.

6.1 Introduction to DEM

The purpose of this subsection is to serve as a brief introduction to the physics and implementation of DEM modelling coupled with CFD. Due to practical space limitations, only the basics of DEM will be covered. The reader is referred to Kloss et al. (2012); Smuts (2015); Jing et al. (2016); Norouzi et al. (2016) for further details concerning the underlying theory.

6.1.1 Overview

The Discrete Element Method, also known as Distinct Element Method, is a numerical method that has been widely used in literature to solve discontinuous mechanics problems such as granular flow or rock mechanics simulations since Cundall and Strack (1979). More recent advances include the so-called extended discrete element method (XDEM) (Peters, 2013), which enhances DEM, allowing solving multi-physics problems such as chemical reactions (e.g., pellets acting as catalysts), heat transfer or electro-magnetic forces (e.g., magnetic particle sorting and separation). Although DEM particles are often spherical in shape, this is not a restriction, as complex shapes can be built from overlapping spheres of different sizes (Ferellec and McDowell, 2010). There are two main approaches for sphere interaction. In the inelastic hard sphere method, the spheres are rigid and particle collisions produce an exchange of momentum through impulsive forces. This method has a limitation, it does not allow sustained contact between particles, therefore, it is suitable for the so-called collisional particle regime only. The other method is the soft sphere approach, which is more common in DEM. In it, particles, although still assumed rigid, are allowed to overlap slightly. This way, the interaction between particles takes a finite time and allows simulating a sustained contact between particles (i.e., frictional forces) as well as regular collisions (Mitarai and Nakanishi, 2003). The soft sphere method will be used in the present work.

In a more flexible approach, DEM has also been coupled with Finite Element Method (FEM) (Munjiza, 2005), introducing tetrahedral meshes to represent arbitrary shapes and enabling calculating internal stresses within the elements (Latham et al., 2008). The same research group applied the FEM-DEM to model rubble mound coastal structures in Latham et al. (2013); Anastasaki et al. (2016), although their model was not coupled with CFD at that time, and their investigations were limited to seismic-type loading, as they were exciting armour units using a global acceleration field.

DEM coupling with CFD has been recently gaining momentum for research and industrial applications alike, propelled by open source models. For example, Klossr and Goniva (2010) developed CFDEM, a coupling between the CFD solver OpenFOAM® and the DEM solver LIGGGHTS (originally derived from LAMMPS), which has been widely used in literature ever since (e.g., Zhao and Shan (2013); Schmeeckle (2014)). Other references with a specific focus on modelling multiphase flows with the VOF technique include Xiang et al. (2012), Jing et al. (2016) and de Lataillade et al. (2017).

Two approaches exist for coupling CFD and DEM. In the resolved coupling, the particles are larger than the underlying CFD mesh cell size, allowing the detailed flow around the particles to be simulated. This technique requires the immersed boundary or fictitious domain method to be implemented in the CFD solver, so that the cells inside the large particles are blocked, acting like the solid part of a moving obstacle, and the cells intersected by particle boundaries act as moving walls (Guo et al., 2013). Alternatively, in the unresolved coupling, particles are smaller than the cell size and the detailed flow around them is not simulated. Instead, the particles are assumed to be a point with a unique set of flow magnitudes (i.e., VOF, velocity, pressure...)

The benefits of using DEM for sediment transport is that loose granular flow can be represented in a more physical way, rather than as a continuum, provided that particle sizes are large enough. Also, given the physical properties of the material and grain shape, the critical angle of repose and the avalanching processes are captured implicitly. It must also be noted that the main disadvantages of DEM are the large computational costs because of the large number of particles that may be involved in simulations and the challenges posed to parallelization when coupled with mesh decomposition techniques. Modern models are parallelized or even GPU-accelerated (Qi et al., 2015), thus reducing some overheads, but there is still room for improvement.

In this work, the DEM model developed in OpenFOAM®, consisting of solid spheres, will be used to simulate individual sediment (sand) particles. Sediment transport studies are common in the DEM literature. Although most of them consider a single phase flow (Schmeeckle, 2014), they cannot represent a moving shoreline, but they can include various types of particle shapes (Sun and Xiao, 2017). Moreover, the native DEM model in OpenFOAM® has previously been validated in a backward-facing step setup for a single phase flow (Greifzu et al., 2016).

6.1.2 Particle physics

In DEM every particle is defined by its physical properties (e.g., shape, inertia, material...) and its position, orientation and linear and angular momentum (i.e., velocity and rotation velocity). Under certain conditions, some of the previous information can be neglected, as for example the orientation of the particles when considering the soft sphere approach.

In this particular case in which the particles are considered rigid spheres, the shape will not change in time and the contact points with their neighbours and the walls will be straightforward to calculate. In spite of being considered rigid, the particles are allowed to overlap, thus through contact models, the forces acting on each particle can be calculated. Particles follow Newton's 2nd law, therefore, acceleration (linear and angular) can be easily calculated given the forces and the properties of the particle, as shown in the following equations:

$$m\frac{dv}{dt} = \sum F \tag{20}$$

$$I\frac{d\omega}{dt} = \sum M \tag{21}$$

in which m is the mass and I is the moment of inertia of the particle, v is velocity, ω is the angular velocity. $\sum F$ represents the sum of all forces acting on the particle, which include the gravitational force, buoyant force, drag force, contact forces (normal and tangential), added mass force and/or Magnus (lift due to rotation) and Saffman (lift due to shear) forces. $\sum M$ represents the sum of all moments (torques) acting on the particle.

Given the previous equations, the velocity and position of the particle in the new time step can be obtained directly by numerical integration.

6.1.3 Numerical implementation

In this subsection we will first introduce the particulars regarding the numerical implementation of the unresolved DEM model in OpenFOAM® and the coupling of DEM with the two-phase flow solver.

The DEM physics has been integrated in a solver derived from *olaFlow*, in which the particle evolution loop has been included within the main solving loop before advecting the VOF function. Since this is an unresolved method, flow properties of a particle are taken from the cell in which the centre of the particle is located.

The particle effects have been added to the VOF, conservation of mass and conservation of momentum equations as explained in Section 3.3. In this case the contribution to the different cells, in the form of porosity and source terms, is proportional to the volume of the particle inside each cell over the volume of the cell.

Regarding the forces acting on the particles, gravity and buoyancy are straightforward to consider, as they only depend on the mass and volume of the particle and the density of the fluid in which it is submerged. There are two types of forces that are modelled: drag and contact forces, which are the most significant. The rest of the forces, i.e., added mass, Magnus and Saffman, are not taken into consideration due to the unresolved approach used, although they are expected to be of second order importance in this case.

The drag coefficient (C_D) is modelled according to a complex expression that fits the experimental data for a single smooth sphere by Schlichting (1979). The equation, presented in Morrison (2013), is as follows:

$$C_D = \frac{24}{Re} + \frac{2.6\left(\frac{Re}{5}\right)}{1+\left(\frac{Re}{5}\right)^{1.52}} + \frac{0.411\left(\frac{Re}{263000}\right)^{-7.94}}{1+\left(\frac{Re}{263000}\right)^{-8}} + \frac{0.25\left(\frac{Re}{10^6}\right)}{1+\left(\frac{Re}{10^6}\right)} \tag{22}$$

where Re is the Reynolds number. This model can be used from creeping flow ($Re \sim 10^{-1}$) up to a turbulent value of $Re \sim 10^6$.

Contact forces account for continuous contacts (e.g., particle resting or rolling on a wall) and for instantaneous contacts (e.g., collisions). These forces are modelled with the spring-slider-dashpot model, which takes into account the normal force, the friction force and the damping of energy, respectively. The inputs for the model are the Young's modulus and shear modulus, the coefficients of friction (tangential sliding) and restitution and the cohesion energy density, to model cohesive materials. For further reference the reader is referred to Jiang et al. (2009).

Solving the equations for particle motion (20–21) has a lower complexity as compared with solving Navier-Stokes equations. However, the time step of the DEM model needs to be much smaller than that of the CFD to limit the amount of overlapping between particles. Otherwise, large overlapping would yield excessively large forces. Moreover, short time steps ensure that the energy is transferred at a maximum of the Rayleigh wave speed (Otsubo et al., 2017). Therefore, for every CFD time step, which is on the order of 10^{-3}, the DEM solver can calculate up to thousands of sub-time steps.

6.2   Numerical setup

This simulation reproduces a two-dimensional wave flume with a constant-depth region and a steep slope (1 on 3). The two-dimensional setup, shown in Figure 10, is similar to the one used in Higuera et al. (2018). The numerical domain is 2 m long in the wave propagation direction and 0.187 m in the vertical direction. The solitary wave height is $H = 2.8$ cm and the water depth is $h = 8$ cm. The wave is generated with the wave generation boundary conditions in *olaFlow* (Higuera, 2015) on the left boundary, applying Grimshaw's third order theory (Grimshaw, 1971). The wave propagates over a 1.5 m-long smooth and flat bottom before arriving at the toe of the slope. The slope is made from sand grains, which are contained in a sandbox that extends 5 mm vertically.

The mesh has been created to minimise the computational cost, allowing lower resolution in areas where mesh quality is not critical (e.g., away from the water, near the top boundary). Cell size variations have been performed by geometric progression to ensure smooth transitions. The smallest cell size is 1 x 1 mm, over the slope, in the vicinity of the beach. The horizontal cell size increases gradually from 1 mm at the toe of the slope to 5 mm at the wave generation boundary. The vertical cell size also increases from 1 mm at $Z = h + H$ to 5 mm at the top boundary. The mesh which is unstructured but hexahedral-cell dominant, totals 83,500 cells. Unlike the mesh used in Higuera et al. (2018), no boundary layer refinement has been performed near the bottom walls in order to accommodate the sediment particles.

Sediment grains are represented by 32,000 individual spherical particles. Each sphere has a diameter $D_{50} = 0.25$ mm, a density of 2600 kg/m^3, a Young's modulus of $E = 75$ GPa and

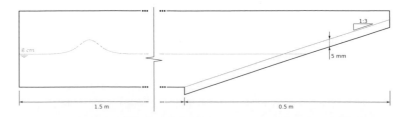

Figure 10: Sketch of the beach flume. Not to scale.

a Poisson's ratio of $\nu = 0.17$, corresponding to typical values for fine sand (Daphalapurkar et al., 2011; Román-Sierra et al., 2014). The initial positions of the sediment particles are initialized randomly, thus preventing an artificial perfect packing. The cell porosity has been limited numerically to a minimum value of 40%, also to maintain typical values.

Turbulence has been modelled with the volume-averaged $k - \omega$ SST turbulence model, modified according to Larsen and Fuhrman (2018) and Devolder et al. (2017). No turbulence enhancement due to the sediment particles has been considered.

The simulation time is 4 seconds and takes 20 days in a single Xeon core (2.50 GHz), which points out the needs for optimisation and parallelization of DEM techniques to achieve reasonable computational times.

6.3   Numerical results

For consistency with Higuera et al. (2018), the zero time reference is chosen when the crest of the solitary wave is located directly above the toe of the slope ($x = 0$ m). During the simulation five phases can be distinguished. Initially, during the propagation phase, the wave travels over the horizontal bottom with a constant form, subjected to frictional effects induced by the bottom. The second stage, the shoaling phase, starts as the wave propagates over the slope. The shoaling wave changes its shape, losing its symmetry and developing a steeper front, but it does not break for the selected nonlinearity and steep slope values. It can be noted that up to this stage, there are no evident hydrodynamic changes when compared to the simulation in Higuera et al. (2018).

The next phase is runup, in which the wave-driven uprush flow over the slope is decelerating and thinning due to gravity, until the maximum runup takes place and the flow tip stops before moving down. The starting point of the runup phase is defined when the flow tip (i.e., shoreline) becomes the highest point in the water domain. This happens at $t = 0.31$ s, slightly behind $t = 0.27$ s in the original experiments and $t = 0.29$ s in the numerical simulation in Higuera et al. (2018). This difference is probably caused by the additional friction imposed by the sand grains.

As the water tongue flows up the slope, the flow tip advances over the beach, slowly percolating and trapping pockets of air between the solid bottom and the beach surface. As presented in Figure 11, the number of particles dragged initially by the uprush (left panel) is limited. The right panel in Figure 11 shows a latter stage, before maximum runup. As it can be observed, the water tongue mobilises a larger number of particles from the top layers of the beach, producing sheet-flow-like sediment transport, which was also observed in similar experiments by Sumer et al. (2011). In both panels, the velocities inside the porous beach are significantly lower than in the free-flow region.

The comparison of free surface elevation (FSE) profiles during the runup phase is shown in the top panel of Figure 12. Generally, the shape of the free surface over the sediment is very similar to that of the glass-bottom experiments, especially away from the flow tip. In that area, results are comparable with those reported in Higuera et al. (2018) too, except for $t = -0.14$ s and $t = 0.06$ s, in which the profile elevation is slightly above the smooth bottom simulation results. The major FSE differences between the present simulation and the experimental measurements take place during the latter runup phase ($t = 0.45 - 0.65$ s) at the flow tip, where significant decrease in the runup extent can be observed, despite the good agreement in the offshore region. For example, maximum runup takes place at $t = 0.66$ s, almost at the same instant ($t = 0.65$ s) in Higuera et al. (2018). However, the present runup height is noticeably lower: $z_{\mathrm{ru}} = 7.1$ cm, as compared to $z_{\mathrm{ru}} = 8.4$ cm and 8.0 cm of the experiment and numerical simulation in Higuera et al. (2018), respectively. This difference can be explained by the additional friction and the percolation that the flow

Figure 11: Flow and sediment velocities near the shoreline during the initial (left panel) and mid (right panel) runup phase. The horizontal component of velocity is represented for the flow, whereas the velocity magnitude is presented in coloured arrows for the sand grains.

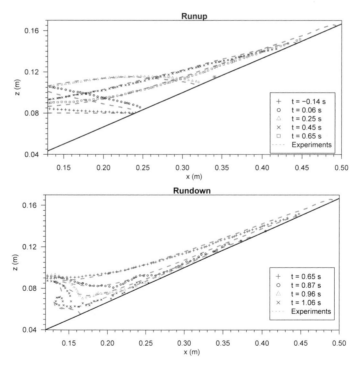

Figure 12: Comparison of experimental and numerical free surface elevation profiles during the runup (top panel) and rundown phases (bottom panel).

experiences during runup because of the porous beach, which produces additional energy dissipation.

At the start of the rundown phase the hydrodynamics have already transitioned from being wave-driven to gravity-driven. The potential energy of the flow is progressively converted into kinetic energy as the flow accelerates due to gravity. The increasing momentum produces a fast and shallow flow (supercritical flow regime) which moves into the quiescent deeper waters (subcritical regime), eventually producing a hydraulic jump that overturns.

Some of the pockets of air that were trapped during the runup phase start playing an important role at this stage. The air tries to flow up towards the free surface but the friction of the sand grains prevents it from ascending as fast as a free bubble. This creates a "fluidised bed" effect, which induces lift forces on the particles, thus easing the transport by the retreating swash flow. Furthermore, the flow velocities during the rundown phase are significantly larger than during the runup phase and the sediment transport in the offshore direction is favoured by gravity, therefore, this induces a sheet-like sediment flow. The amount of particles transported is massive, as shown in Figure 13.

The comparison of free surface elevation (FSE) profiles during the rundown phase is presented in the bottom panel of Figure 12. Since the starting point of this phase, the maximum runup, was lower than in the experiments, the retreating flow tip is at all times below the experimental mark. Furthermore, undulations in the FSE start to develop at the lower part of the slope. This effect is caused by the sediment as it does not appear in the numerical simulation in Higuera et al. (2018). Apart from these discrepancies, the overall agreement of the initial FSE profiles follows the same trend.

Before the hydraulic jump occurs, the difference in water depth between the supercritical and subcritical regions induces a large pressure gradient, which overcomes the relatively low momentum in the boundary layer near the beach interface, reversing the flow in the area of transition, as can be seen in the top left panel in Figure 13. Flow reversal starts taking place at $t = 0.93$ s in the present simulation, whereas in the experiments and numerical simulation for the smooth bottom in Higuera et al. (2018), flow reversal occurs at $t = 0.91$ s and $t = 0.81$ s, respectively. As an immediate result of flow reversal, an anti-clockwise vortex, which is perfectly visible in the LIC pattern, is created and evolves together with the hydraulic jump, which starts overturning at $t = 0.98$ s. This eddy picks sediment up from the beach surface and puts it in suspension. This phenomenon was already observed in Matsunaga and Honji (1980), in which they reported that the sediment was lifted up from the bed by the backwash vortex, and not by the plunging wave. Similar observations were made in Sumer et al. (2013). The vortex also acts like a trampoline, projecting the incoming sediment upwards, while it is being advected in the offshore direction by the downrush flow, from which the eddy gains energy to grow. At this stage ($t = 0.98$ s) there is a smaller second anti-clockwise vortex up the slope from the main vortex, which is more noticeable in the following panels. By $t = 1.03$ s the hydraulic jump is plunging onshore. The second vortex becomes more visible, and between the two vortices mentioned, a new vortex is created, which rotates in the clockwise direction for compatibility reasons. Although it is partially covered by the sediment, half of it is above the beach interface. Another difference of this simulation when compared with Higuera et al. (2018) is the shape of the overturning wave (see $t = 1.06$ s in the bottom panel of Figure 12), which initially traps less air when breaking. The system of three vortices continues to be advected downstream, growing and deforming, as it can be appreciated perfectly at $t = 1.11$ s, when the hydraulic jump has overturned and broken, projecting the wave lip shoreward. Up to this stage, prior to wave breaking, the evolution of the system of vortices is extremely similar to that under smooth bed conditions (Higuera et al., 2018).

Wave breaking creates very complex flow patterns ($t = 1.23$ s), trapping significant amounts of air and transporting plumes of sediment towards the surface. Nevertheless, it must be noted that this is a 2D simulation, therefore, breaking cannot be characterised completely, as it is a 3D process predominantly. For example, when air is trapped in a 2D simulation, it will disturb and "cut" the flow as it moves up to the surface since 2D air entrainment occurs in infinitely wide pressurised rollers, whereas in 3D simulations the disturbance might be localised in certain areas only, as air and water can mobilise sideways and create the low pressure tubes and bubbles typical in air entrainment processes during

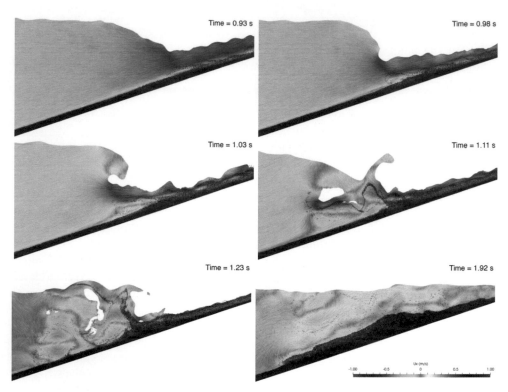

Figure 13: Flow velocities and sediment grains over the slope during the rundown phase and the development of the hydraulic jump. The horizontal component of velocity is represented for the flow with Line Integral Convolution technique (LIC, i.e., showing instantaneous streamlines). Sand grains are depicted as black dots.

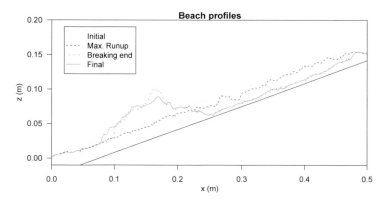

Figure 14: Beach profiles during the simulation. The differences in the vertical direction with respect to the initial profile (dotted line) have been amplified by a factor of 5. The straight solid line represents the bottom of the sand pit.

wave breaking. The situation towards the end of the simulation ($t = 1.92$ s), when the system is almost at rest, shows some particles that are still in suspension due to the remaining vortices, but most of the sediment has settled down already.

The time evolution of the beach profile is presented in Figure 14. The reader should note that the differences in the elevation of the sediment bed with respect to the original profile have been amplified by a factor of 5 to ease comparisons.

The initial stage depicts an almost perfect straight line (blue dotted line) following the 1:3 slope, parallel to the physical bottom of the tank (black continuous line). At the maximum runup instant most of the profile (red dashed line) shows a lower level, up to a maximum of 2.9 mm below the original line. This is caused because the initial packing is random and, as can be seen in Figure 11, some gaps exist within the internal structure, therefore, the sediment packing compacts due to the wave-induced velocities and pressures during the runup phase. Moreover, at this instant the flow has already started moving down the slope except at the tip. The sediment is also dragged down and some bumps appear and possibly induce the free surface oscillations that were previously observed in Figure 13. The beach profile when wave breaking has ended (green dash-dot line) shows that the sediment has piled up. The largest peak appears at a location between the downrush flow reversal and the hydraulic jump, 7.5 cm offshore from the shoreline, and is 8.6 mm above the original beach profile, which is consistent with the observations in Sumer et al. (2011) and Li et al. (2019). Nevertheless, the sediment on the bottom is still in movement and avalanches, which are captured by the DEM method, are ongoing at this instant. The final beach profile, after the sediment has come to a stop (grey solid line) shows the results of the avalanches, reducing the height of the peak by extending the mound down the slope and producing smoother slopes. The sediment uphill and downhill from the new berm remains almost unchanged in the last 2 snapshots. Finally, it can be noted that two areas of the slope remain unaltered for longer time than others. There is a small area within 2.3 cm from the toe of the slope that presents no erosion or accretion during the whole simulation. Additionally, the area above $x = 0.478$ m (maximum runup reaches $x = 0.453$ m) shows no movement after the initial settling since maximum runup occurs at this stage.

6.4 Concluding remarks

This case has shown the potential of RANS modelling for simulating systems in which sediment is represented particle by particle. This simulation must be viewed as a proof of concept, pointing out the benefits of this method, and can in principle be applied in diverse applications such as to estimate scour around coastal or offshore structures.

There are two main challenges that must be overcome in order to use this technique for real applications (i.e., consultancy cases). The first factor is performing three-dimensional simulations, which are required to represent important physical processes such as wave breaking. For example, 3D effects were found negligible prior to wave breaking in Higuera et al. (2018), whereas in the present simulation the air pockets trapped in the sediment bed flow up towards the free surface, eventually cutting through the downwards flow. This is an obvious side effect of the 2D simulation, as in a 3D simulation air could also escape sideways. However, running a 3D simulation with particles would increase the simulation time exponentially. In view of this, the second factor is computational cost, which in this case is extremely large even for the limited number of particles in 2D, due to the particular implementation of DEM in OpenFOAM®. Fortunately, more efficient DEM solvers exist (Klossr and Goniva, 2010) and can be coupled with OpenFOAM®.

7 Final remarks

In this chapter the interaction of wave and porous structures has been discussed. The framework for simulating porous materials was introduced by means of the Volume-Averaged

RANS equations, a set of equations in which porosity and other geometrical properties of the porous media, plus linear and nonlinear friction factors representing the force induced by the porous geometry come into play. Turbulent equations need to be volume-averaged as well and closure terms are required to model a set of terms that cannot be simulated. This closure is currently only available for $k - \varepsilon$ turbulence model, although it can be argued that the enhancement of turbulent dissipation may be included in the friction terms in the lack of turbulent closure.

Two applications of CFD involving porous media have been modelled with the open source model *olaFlow*, within the OpenFOAM® framework. In the first, the model has been proven to be a useful tool to test new structural typologies. In this sense, the original structure, which was not ready to withstand a certain wave condition has been modified to continue fulfilling its role. This type of problem will be of importance in the near future, when enhancements to existing structures might be needed to mitigate the risks of raising sea levels and higher-magnitude and more frequent storms (Knutson et al., 2010).

In the second application sediment grains have been represented as moving particles with a DEM model. The movement and interaction of such particles induces local time changes in porosity, which are also incorporated in the equations. Differences with respect to a fixed smooth bottom simulation have been highlighted. The simulation presents interesting swash-sediment interaction processes that agree qualitatively with previous similar references on experiments and real condition observations. Although this application required a very high computational cost, it highlights the advantages of treating sediment with DEM.

Future research needs to increase the efficiency of the DEM simulations to allow a practical use of the model in consultancy companies. Moreover, in the future, the simulation of solid elements larger than the mesh (e.g., concrete units of the primary armour layer) via the DEM and immerse boundary method (Guo et al., 2013), in conjunction with the VARANS equations for the internal layers and core of the structure should be explored. With this method, the stability of coastal structures, especially in challenging areas such as curved layers and in locations of element size transition could be analysed computationally. Furthermore, variables that are difficult or not possible to measure in the laboratory, as for example shear stresses or forces on individual units, may be obtained with the numerical model without perturbing the flow.

References

Altomare, C., Crespo, A. J. C., Rogers, B. D., Dominguez, J. M., Gironella, X. and Gómez-Gesteira, M. (2014). Numerical modelling of armour block sea breakwater with smoothed particle hydrodynamics. Computers & Structures 130: 34–45.

Anastasaki, E., Latham, J. -P. and Xiang, J. (2016). Numerical test for single concrete armour layer on breakwaters. Proceedings of the Institution of Civil Engineers-Maritime Engineering 169(4): 174–187.

Brackbill, J. U., Kothe, D. B. and Zemach, C. (1992). A continuum method for modeling surface tension. Journal of Computational Physics 100(2): 335–354.

Burcharth, H. and Andersen, O. (1995). On the one-dimensional steady and unsteady porous flow equations. Coastal Engineering 24(3-4): 233–257.

Cundall, P. A. and Strack, O. D. L. (1979). A discrete numerical model for granular assemblies. Géotechnique 29(1): 47–65.

Dalrymple, R. A., Losada, M. A. and Martin, P. (1991). Reflection and transmission from porous structures under oblique wave attack. Journal of Fluid Mechanics 224: 625–644.

Daphalapurkar, N. P., Wang, F., Fu, B., Lu, H. and Komanduri, R. (2011). Determination of mechanical properties of sand grains by nanoindentation. Experimental Mechanics 51: 719–728.

de Lataillade, T., Dimakopoulos, A., Kees, C., Johanning, L., Ingram, D. and Tezdogan, T. (2017). Cfd modelling coupled with floating structures and mooring dynamics for offshore renewable energy devices using the proteus simulation toolkit. Proceedings of the European Wave and Tidal Energy Conference, 2017.

de Lemos, M. J. S. and Pedras, M. H. J. (2001). Recent mathematical models for turbulent flow in saturated rigid porous media. Journal of Fluids Engineering 123: 935–941.

del Jesus, M. (2011). Three-dimensional interaction of water waves with maritime structures. Ph.D. thesis, University of Cantabria.

del Jesus, M., Lara, J. L. and Losada, I. J. (2012). Three-dimensional interaction of waves and porous structures. Part I: Numerical model formulation. Coastal Engineering 64: 57–72.

Dentale, F., Donnarumma, G. and Carratelli, E. P. (2014). Simulation of flow within armour blocks in a breakwater. Journal of Coastal Research 295: 528–536.

Dentale, F., Reale, F., Di Leo, A. and Carratelli, E. P. (2018). A CFD approach to rubble mound breakwater design. International Journal of Naval Architecture and Ocean Engineering 10(5): 644–650.

Devolder, B., Rauwoens, P. and Troch, P. (2017). Application of a buoyancy-modified komega SST turbulence model to simulate wave run-up around a monopile subjected to regular waves using OpenFOAM. Coastal Engineering 125: 81–94.

Engelund, F. (1953). On the laminar and turbulent flow of ground water through homogeneous sand. Transactions of the Danish Academy of Technical Sciences, 3.

Espinoza-Andaluz, M., Andersson, M. and Sundén, B. (2017). Computational time and domain size analysis of porous media flows using the lattice boltzmann method. Computers & Mathematics with Applications 74(1): 26–34.

Ferellec, J. -F. and McDowell, G. R. (2010). A method to model realistic particle shape and inertia in DEM. Granular Matter 12(5): 459–467.

Forcheimer, P. (1901). Wasserbewegung durch Boden. Z. Ver. Deutsch. Ing. 45: 1782–1788.

Gray, W. G. (1975). A derivation of the equations for multi-phase transport. Chemical Engineering Science 30: 229–233.

Greifzu, F., Kratzsch, C., Forgber, T., Lindner, F. and Schwarze, R. (2016). Assessment of particle-tracking models for dispersed particle-laden flows implemented in OpenFOAM and ANSYS FLUENT. Engineering Applications of Computational Fluid Mechanics 10(1): 30–43.

Grimshaw, R. (1971). The solitary wave in water of variable depth. Part 2. Journal of Fluid Mechanics 46: 611–622.

Guanche, R., Losada, I. J. and Lara, J. L. (2009). Numerical analysis of wave loads for coastal structure stability. Coastal Engineering 56(5-6): 543–558.

Guo, Y., Wu, C. -Y. and Thornton, C. (2013). Modeling gas-particle two-phase flows with complex and moving boundaries using DEM-CFD with an immersed boundary method. Particle Technology and Fluidization 59(4): 1075–1087.

Higuera, P., Lara, J. L. and Losada, I. J. (2013a). Realistic wave generation and active wave absorption for Navier-Stokes models: Application to OpenFOAM. Coastal Engineering 71: 102–118.

Higuera, P., Lara, J. L. and Losada, I. J. (2013b). Simulating coastal engineering processes with OpenFOAM. Coastal Engineering 71: 119–134.

Higuera, P., Lara, J. L. and Losada, I. J. (2014a). Three-dimensional interaction of waves and porous coastal structures using OpenFOAM. Part I: Formulation and validation. Coastal Engineering 83: 243–258.

Higuera, P., Lara, J. L. and Losada, I. J. (2014b). Three-dimensional interaction of waves and porous coastal structures using OpenFOAM. Part II: Applications. Coastal Engineering 83: 259–270.

Higuera, P. (2015). Application of computational fluid dynamics to wave action on structures. Ph.D. thesis, University of Cantabria.

Higuera, P., Losada, I. J. and Lara, J. L. (2015). Three-dimensional numerical wave generation with moving boundaries. Coastal Engineering 101: 35–47.

Higuera, P. (2017). olaflow: CFD for waves.

Higuera, P., Liu, P. L. -F., Lin, C., Wong, W. -Y. and Kao, M. -J. (2018). Laboratory-scale swash flows generated by a non-breaking solitary wave on a steep slope. Journal of Fluid Mechanics 847: 186–227.

Hirt, C. W. and Nichols, B. D. (1981). Volume of fluif (VOF) method for the dynamics of free boundaries. Journal of Computational Physics 39: 201–225.

Hsu, T., Sakakiyama, T. and Liu, P. (2002). A numerical model for wave motions and turbulence flows in front of a composite breakwater. Coastal Engineering 46(1): 25–50.

Hur, D. -S., Lee, K. and Yeom, G. (2008). The phase difference effects on 3D structure of wave pressure acting on a composite breakwater. Ocean Engineering 35(17-18): 1826–1841.

Jacobsen, N. G., Fuhrman, D. R. and Fredsøe, J. (2012). A wave generation toolbox for the open-source CFD library: OpenFOAM. International Journal for Numerical Methods in Fluids 70(9): 1073–1088.

Jasak, H. (1996). Error analysis and estimation for the finite volume method with applications to fluid flows. Ph.D. thesis, Imperial College of Science, Technology and Medicine.

Jensen, B., Jacobsen, N. G. and Christensen, E. D. (2014). Investigations on the porous media equations and resistance coefficients for coastal structures. Coastal Engineering 84: 56–72.

Jiang, M., Leroueil, S., Zhu, H., Yu, H. -S. and Konrad, J. -M. (2009). Two-dimensional discrete element theory for rough particles. International Journal of Geomechanics 9(1): 20–33.

Jing, L., Kwok, C. Y., Leung, Y. F. and Sobral, Y. D. (2016). Extended CFD-DEM for free-surface flow with multi-size granules. Numerical and Analytical Methods in Geomechanics 40(1): 62–79.

Kaviani, M. (1995). Principles of Heat Transfer in Porous Media. Mechanical Engineering Series. Springer-Verlag New York, 2nd edition.

Kloss, C., Goniva, C., Hager, A., Amberger, S. and Pirker, S. (2012). Models, algorithms and validation for opensource DEM and CFD-DEM. Progress in Computational Fluid Dynamics 12(2-3): 140–152.

Klossr, C. and Goniva, C. (2010). LIGGGHTS: a new open source discrete element simulation software. In 5th International Conference on Discrete Element Methods, London, UK, pp. 25–26.

Knutson, T. R., McBride, J. L., Chan, J., Emanuel, K., Holland, G., Landsea, C., Held, I., Kossin, J. P., Srivastava, A. K. and Sugi, M. (2010). Tropical cyclones and climate change. Nature Geoscience 3: 157–163.

Lara, J. L., Garcia, N. and Losada, I. J. (2006). RANS modelling applied to random wave interaction with submerged permeable structures. Coastal Engineering 53: 395–417.

Lara, J. L., Losada, I. J. and Guanche, R. (2008). Wave interaction with low mound breakwaters using a RANS model. Ocean Engineering 35: 1388–1400.

Lara, J. L., del Jesus, M. and Losada, I. J. (2012). Three-dimensional interaction of waves and porous structures. Part II: Model validation. Coastal Engineering 64: 26–46.

Larese, A., Rossi, R. and Oñate, E. (2015). Finite element modeling of free surface flow in variable porosity media. Archives of Computational Methods in Engineering 22(4): 637–653.

Larsen, B. E. and Fuhrman, D. R. (2018). On the over-production of turbulence beneath surface waves in RANS models. Journal of Fluid Mechanics 853: 419–460.

Latham, J. -P., Munjiza, A., Garcia, X., Xiang, J. and Guises, R. (2008). Three-dimensional particle shape acquisition and use of shape library for DEM and FEM/DEM simulation. Minerals Engineering 21(11): 797–805.

Latham, J. -P., Anastasaki, E. and Xiang, J. (2013). New modelling and analysis methods for concrete armour unit systems using FEMDEM. Coastal Engineering 77: 151–166.

Li, J., Qi, M. and Fuhrman, D. (2019). Numerical modeling of flow and morphology induced by a solitary wave on a sloping beach. Applied Ocean Research 82: 259–273.

Lin, P. and Liu, P. -F. (1998). A numerical study of breaking waves in the surf zone. Journal of Fluid Mechanics 359: 239–264.

Liu, P. L. -F., Lin, P., Chang, K. and Sakakiyama, T. (1999). Numerical modeling of wave interaction with porous structures. Journal of Waterway, Port, Coastal, and Ocean Engineering 125: 322–330.

Losada, I. J., Losada, M. A. and Baquerizo, A. (1993). In analytical method to evaluate the efficiency of porous screens as wave dampers. Applied Ocean Research 15(4): 207–215.

Losada, I. J., Lara, J. L. and del Jesus, M. (2016). Modeling the interaction of water waves with porous coastal structures. Journal of Waterway, Port, Coastal, and Ocean Engineering 142(6).

Madsen, O. S. (1974). Wave transmission through porous structures. Journal of Waterway, Port, Coastal and Ocean Engineering 100(3): 169–188.

Márquez, S. (2013). An extended mixture model for the simultaneous treatment of short and long scale interfaces. Ph.D. thesis, Universidad Nacional del Litoral.

Matsunaga, N. and Honji, H. (1980). The backwash vortex. Journal of Fluid Mechanics 99(4): 813–815.

Mitarai, N. and Nakanishi, H. (2003). Hard-sphere limit of soft-sphere model for granular materials: Stiffness dependence of steady granular flow. Physical Review E 67: 021301.

Morrison, F. A. (2013). An Introduction to Fluid Mechanics. Cambridge University Press, New York.

Munjiza, A. (2005). The Combined Finite-Discrete Element Method. John Wiley & Sons, Ltd.

Nakayama, A. and Kuwahara, F. (1999). A macroscopic turbulence model for flow in a porous medium. Journal of Fluids Engineering 121: 427–435.

Nikora, V., Ballio, K., Coleman, S. and Pokrajac, D. (2013). Spatially averaged flows over mobile rough beds: Definitions, averaging theorems, and conservation equations. Journal of Hydraulic Engineering 139(8): 803–811.

Norouzi, H. R., Zarghami, R., Sotudeh-Gharebagh, R. and Mostoufi, N. (2016). Coupled CFD-DEM Modeling: Formulation, Implementation and Application to Multiphase Flows. John Wiley & Sons.

Otsubo, M., O'Sullivan, C. and Shire, T. (2017). Empirical assessment of the critical time increment in explicit particulate discrete element method simulations. Computers and Geotechnics 86: 67–79.

Oumeraci, H. and Kortenhaus, A. (1997). Wave impact loading—tentative formulae and suggestions for the development of final formulae. Proceedings 2nd task 1 Workshop, MAST III, PROVERBS-Project: Probabilistic Design Tools for Vertical Breakwaters, Edinburgh, U.K., pages Annex 1.0.2., 13 pp.

Pedersen, J. (1996). Wave forces and overtopping on crown walls of rubble mound breakwaters: an experimental study. Aalborg Universitetsforlag (Series Paper; No. 12).

Peters, B. (2013). The extended discrete element method (XDEM) for multi-physics applications. Scholarly Journal of Engineering Research 2: 1–20.

Polubarinova-Kochina, P. Y. (1962). Theory of Ground Water Movement. Princeton University Press.

Qi, J., Li, K. -C., Jiang, H., Zhou, Q. and Yang, L. (2015). GPU-accelerated DEM implementation with CUDA. International Journal of Computational Science and Engineering 11(3): 330–337.

Ren, B., Wen, H., Dong, P. and Wang, Y. (2016). Improved SPH simulation of wave motions and turbulent flows through porous media. Coastal Engineering 107: 14–27.

Román-Sierra, J., Muñoz-Perez, J. J. and Navarro-Pons, M. (2014). Beach nourishment effects on sand porosity variability. Coastal Engineering 83: 221–232.

Rusche, H. (2002). Computational fluid dynamics of dispersed two-phase flows at high phase fractions. Ph.D. thesis, Department of Mechanical Engineering, Imperial College of Science, Technology & Medicine, London.

Sakakiyama, T. and Kajima, R. (1992). Numerical simulation of nonlinear waves interacting with permeable breakwaters. Proceeding, 23rd International Conference on Coastal Engineering, Reston, Virginia, pp. 1517–1530.

Schlichting, H. (1979). Boundary Layer Theory. McGraw-Hill, New York., 7 edition.

Schmeeckle, M. W. (2014). Numerical simulation of turbulence and sediment transport of medium sand. Journal of Geophysical Research: Earth Surface 119(6): 1240–1262.

Shao, S. (2010). Incompressible SPH flow model for wave interactions with porous media. Coastal Engineering 57(3): 304–316.

Slattery, J. C. (1967). Flow of viscoelastic fluids through porous media. American Institute of Chemical Engineers Journal 13(6): 1066–1071.

Smuts, E. M. (2015). A Methodology for Coupled CFD-DEM Modelling of Particulate Suspension Rheology. Ph.D. thesis, University of Cape Town.

Sumer, B. M., Sen, M. B., Karagali, I., Ceren, B., Fredsøe, J., Sottile, M., Zilioli, L. and Fuhrman, D. R. (2011). Flow and sediment transport induced by a plunging solitary wave. Journal of Geophysical Research 116: C01008.

Sumer, B. M., Guner, H. A. A., Hansen, N. M., Fuhrman, D. R. and Fredsøe, J. (2013). Laboratory observations of flow and sediment transport induced by plunging regular waves. Journal of Geophysical Research 118: 6161–6182.

Sun, R. and Xiao, H. (2017). Realistic representation of grain shapes in CFD-DEM simulations of sediment transport: A bonded-sphere approach. Advances in Water Resources 107: 421–438.

Sun, X., Sun, M., Takabatake, K., Pain, C. and Sakai, M. (2018). Numerical simulation of free surface fluid flows through porous media by using the explicit MPS method. Transport in Porous Media, pp. 1–27.

Troch, P. and De Rouck, J. (1998). Development of 2D numerical wave flume for simulation of wave interaction with rubble mound breakwaters. Proceedings 26th International Conference on Coastal Engineering (ICCE), Copenhagen.

Troch, P., De Rouck, J. and Burcharth, H. F. (2002). Experimental study and numerical modelling of pore pressure attenuation inside a rubble mound breakwater. In Proceedings of the 28th International Coastal Engineering Conference (ICCE), Cardiff, United Kingdom.

van der Meer, J. W., Allsop, W., Bruce, T., De Rouck, J., Kortenhaus, A., Pullen, T., Schüttrumpf, H., Troch, P. and Zanuttigh, B. (2018). EurOtop, 2018. Manual on Wave Overtopping of Sea Defences and Related Structures. An Overtopping Manual Largely based on European Research, but for Worldwide Application., second edition. www.overtopping-manual.com.

van Gent, M. R. A. (1995). Porous flow through rubble-mound material. Journal of Waterway, Port, Coastal and Ocean Engineering 3(121): 176–181.

Vanneste, D. (2012). Experimental and Numerical Study of Wave-induced Porous Flow in Rubble-mound Breakwaters. Ph.D. thesis, Ghent University.

Vanneste, D. and Troch, P. (2012). An improved calculation model for the wave-induced pore pressure distribution in an rubble-mound breakwater core. Coastal Engineering 66: 8–23.

Wellens, P., Borsboom, M. J. A. and van Gent, M. R. A. (2010). 3D simulation of wave interaction with permeable structures. Proceedings of the 32nd International Conference on Coastal Engineering, Shanghai, China.

Weller, H., Tabor, G., Jasak, H. and Fureby, C. (1998). A tensorial approach to computational continuum mechanics using object-oriented techniques. Computers in Physics 12(6): 620–631.

Whitaker, S. (1967). Diffusion and dispersion in porous media. American Institute of Chemical Engineers Journal 13(3): 420–427.

Whitaker, S. (1996). The Forcheimer equation: a theoretical development. Transport in Porous Media 25: 27–61.

Xiang, J., Latham, J. -P., Vire, A., Anastasaki, E. and Pain, C. C. (2012). Coupled Fluidity/Y3D technology and simulation tools for numerical breakwater modelling. In Proceedings of the 33rd International Coastal Engineering Conference (ICCE), Santander, Spain.

Zhao, J. and Shan, T. (2013). Coupled CFD-DEM simulation of fluid-particle interaction in geomechanics. Powder Technology 239: 248–258.

Chapter 7

CFD Modelling of Scour in Flows with Waves and Currents

Nicholas S Tavouktsoglou,[1,*] *David M Kelly*[2] and *John M Harris*[1]

1 Introduction

When it comes to sediment transport and scour modelling, the science must precede the engineering practice: simply measuring waves and currents and employing back of the envelope formulae is seldom adequate. Instead, performing detailed numerical modelling of multiple scenarios can offer real insight into a design problem. Moreover, in order to predict the sediment budget or the sediment transport, in relation to coastal works, a wide range of conditions need to be considered; thus, numerical models cannot be ignored. This begs the fundamental question: are accurate sediment transport predictions possible?

One reason sediment transport calculations are difficult is because the sediment transport is highly dependent on the hydrodynamics. To obtain reliable estimates of erosion rates we need to accurately predict the hydrodynamics of waves, currents and turbulence before the sediment flux can be determined. Importantly, the currents could result from processes at different time scales (for example, winds and cyclic loading etc.) developed within different generating locations (remote, fairly local or immediate vicinity) and can be highly variable in both space and time. The waves can also be driving currents, and they may initiate sediment motion themselves. This results in a complex fully three dimensional problem. In sediment transport, we are primarily dealing with a boundary layer of 1–2cm, but within this zone there are numerous factors that may contribute to high uncertainty in the final calculations.

Further to the effects the local hydrodynamics have, it has been observed that the bed has different responses depending on its local composition (sand, clay or silt) which in turn effects the movement of the sediment (Soulsby, 1997). The main sediment transport mechanisms can be divided into two categories:

- Bedload, and;
- Suspended load

These modes of sediment transport are in turn characterised by other processes and properties such as the bed pore pressure, median sediment size intra-particle forces etc. The result of all these properties, in conjunction with the local hydrodynamics and turbulence, are used to determine the sediment flux. To better understand and determine how sediment

---

[1] HR Wallingford, Wallingford, Oxon, OX10 8BA, UK.

[2] Centre for Environment Fisheries and Aquaculture Science (CEFAS), Pakefield Rd, Lowestoft, NR33 0HT, UK.

* Corresponding author: N.Tavouktsoglou@hrwallingford.com

moves in oceanic, coastal and fluvial environments it is of essential importance to understand fundamentally how sediment particles move under the forcing of the hydrodynamics. The latest research is moving towards such solutions and by taking into consideration the increasing capabilities of computers, more direct methods to calculate sediment transport are being developed. These methods are based on Lagrangian approaches where each sediment grain is modelled as a particle or, for computational efficiency, a number of grains are collected together and modelled as a sediment parcel. Better measurements have led to better theory and better computer simulations. Sediment transport still remains challenging, but major advances in hardware, modelling and science have made it more and more accurate.

This chapter examines some of the authors' research into numerical models for sediment transport using computational fluid dynamics (CFD) in non-cohesive sediments. In addition, this chapter will provide a summary of key studies that lead us to the current state of sediment transport numerical modelling along with a summary of the current methods used to model sediment transport using CFD.

2 Types of sediment transport models in CFD

In recent years, with the rapid development of CFD and computational power, numerical studies of local scour have become increasingly popular. Numerical models for the simulation of scour usually comprise two distinct elements:

- A hydrodynamic module to solve the fluid flow; and

- A morphological module to deal with the sediment transport and bed erosion.

The hydrodynamic module solves the (Reynolds Averaged) Navier-Stokes equations for the flow field. The sediment-laden flow can be treated as a two-phase flow, which includes the water phase and the sediment phase, or as a single-phase flow, in which the two phases are modelled as a mixture. The two-phase flow model can be categorized into an Euler-Euler type (Chen et al., 2011), Lagrange-Lagrange type (Zubeldia et al., 2018) and an Euler-Lagrange type (Li et al., 2014). The Euler-Lagrange type of model treats the sediment phase as representing the motion of a certain number of individual particles. This approach succeeds in capturing the individual and collective dynamics of natural sand grains and many very good numerical models of the Euler-Lagrange type have been proposed recently such as those by Li et al. (2014), Finn et al. (2016) and Sun & Xiao (2016). However, this approach requires a large number of particles to simulate practical problems which translates into a huge demand on computational resources. The Euler-Euler approach describes the dispersed phase in a similar manner to that used for the continuous phase, and efforts have been made to develop such sediment transport models; representative examples are the models of Jha & Bombardelli (2010) and Chen et al. (2011). More recently a new type of Euler-Euler approach has been developed that makes use of two different intergranular stress models, the kinetic theory of granular flows and the dense granular flow rheology (Chauchat et al., 2017 and Mathieu, 2019). However, the Euler-Euler type of two-phase models are typically complicated (Chen et al., 2011). The two-phase model needs to solve the continuity and momentum equation for both of the phases and the most difficult part is to properly describe the turbulence characteristics of the two phases. For this reason, in hydraulic engineering applications, including scour estimations, single phase models are typically used.

The most popular single-phase approach neglects the effects of the dispersed phase on the continuous phases and solves the volumetric concentration of the dispersed phase using an advection-diffusion model. The advection-diffusion model has been successfully applied to many numerical studies of scour problems; see, for example, the work of Liang et al. (2005)

and Roulund et al. (2005). However, it should be noted that the advection-diffusion model, which neglects the effects of the dispersed phase, is only valid for dilute problems where the suspended load concentration is not large enough to influence the principle properties of the flow field. Another issue concerning the hydrodynamic module is whether to include free surface effects in the model, or not. The air-water interface is affected when the flow interacts with obstacles—this causes local variations in the water depth. If the variations are comparable with the water depth, they may have a significant influence on the flow field and thus on the scour formation. Free surface effects are very common in scour problems, as the obstacles causing scour can have significant dimensions and thus the variation of the flow depth can be large. Recently, many numerical models for scour, or bed erosion, problems have included the free surface effect in the model; see, for example, the model of Liu and Garcia (2008).

The final important issue concerning the hydrodynamic module is to properly choose the turbulence closure for solving the (Reynolds Averaged) Navier-Stokes equations. The flows in scour problems are often turbulent. A direct numerical simulation (DNS) Orszang (1970) of the full Navier-Stokes equations will resolve the whole range of spatial and temporal scales of the turbulence, but the computational meshes must be fine enough to ensure resolution of the smallest dissipative scale (Kolmogorov microscales); for many problems this can be $\mathcal{O}(10^{-5})$ m. The computational resources required to solve this sort of problem exceed what is currently practical. The most commonly used turbulence closure conditions are based on Reynolds Averaged Navier-Stokes (RANS) and Large Eddy Simulation (LES) models. Various studies have investigated the influence of the choice of turbulence model on the results of scour simulations. Aghaee and Hakimzadeh (2010) simulated the scour caused by bridge piers with both RANS and LES turbulence models and concluded that although the LES model simulates the flow field more accurately, especially for the periodic behaviour of the vortex shedding, a RANS model was generally sufficient to give a satisfactory estimate of scour development. The morphological module for scour simulations generally includes a sediment transport model and a sediment conservation equation (the Exner equation) to calculate the bed level change. There are many possible choices for the sediment transport model which usually includes both suspended load and bed load. As mentioned above, the suspended load in scour problems is often solved by the advection-diffusion model with a reference concentration given at a certain reference height. The reference concentration is usually given by empirical models of sediment entrainment. Two entrainment models commonly used can be found in the work of van Rijn (1984) and Garcia and Parker (1991).

To determine the bed load, many empirical equations from laboratory flume data have been given by previous studies. Most of them depend on the bed shear stress τ_b, i.e., the Shields parameter θ, the density ratio between the sediment and the flow ($s = \rho_s/\rho_w$) and the nominal sediment particle diameter d_{50}. In order to facilitate the expression, a non-dimensional form of transport rate which is referred as the Einstein number in some papers Φ_b is defined.

$$\Phi_b = \frac{|q_b|}{\sqrt{(s-1)\,gd_{50}^3}} \tag{1}$$

An extensive summary of the main methods to determine Φ_b is found in Soulsby (1997).

3 The scourFOAM model

The scourFOAM model was born out of the need for a better scour prediction tool for certain engineering projects at HR Wallingford, UK. Since the inception of the original scourFOAM model by Kelly (2013) the code has been in continuous development especially with regards to the parametrization of turbulence (Li et al., 2014, Tavouktsoglou et al. 2018). As such, scourFOAM is a state-of-the-art three-phase model for free surface flow over a mobile sediment bed, the air and water phases are combined together as a mixed continuum phase and solved by the existing OpenFOAM® interFOAM solver using a single momentum equation over the whole computational domain (Jasak 1996, Higuera, 2013). It should be noted that, within this framework, the air-phase is assumed to be incompressible. The Volume of Fluid (VOF) method (Hirt and Nichols, 1981) is employed to track the water-air interface and determine the fluid properties. The sediment phase is introduced as Lagrangian particle parcels, whose motions are coupled to the local flow field. The size of the particle parcel is assumed to be small in comparison with the mesh size and hence the presence of the particle is considered at a sub-grid scale.

3.1 Governing equations

3.1.1 Fluid phases (Eulerian)

The dynamics of the fluid phase, i.e., the mixture of water and air, is resolved via the incompressible Navier-Stokes equations, which are modified with the two-fluid method as described in Rusche (2003). The governing equations for the fluid phase are written as:

$$\nabla \cdot \left(\overrightarrow{U_f} \right) = 0 \tag{2}$$

$$\frac{\partial \rho \overrightarrow{U_f}}{\partial t} + \nabla \cdot \left(\rho_f \overrightarrow{U_f} \overrightarrow{U_f} \right) - \nabla \cdot \left(\mu_{eff} \nabla \overrightarrow{U_f} \right) - \left(\nabla \cdot \overrightarrow{U_f} \right) \cdot \nabla \mu_{eff} = -\nabla p + S_{imt} \tag{3}$$

where subscripts f denotes the fluid phase, $\overrightarrow{U_f}$ is the flow velocity vector, ρ is the density, μ_{eff} is the efective dynamic viscosity, p is the total pressure, S_{imt} is the inter-phase momentum transfer term, accounting for the influences of the solid phase on the fluid phase. A phase volume fraction α, is introduced by the VOF method. It represents the volume fraction occupied by water within a cell. With the aid of α, the properties of the fluid phase can be expressed as a weighted sum of that from air and water. Thus, $\overrightarrow{U_f} = \alpha \overrightarrow{U_a} + (1-\alpha) \overrightarrow{U_w}$ etc. It should be noted that the particles are represented at the sub-grid scale and, as such, since material elasticity effects do not play an important role, the solid phase is considered as an incompressible phase, hence the continuity for the fluid maintains its well known form. Thus, in formulating the pressure Poisson equation the fluid velocity field is considered to be divergence free.

The transport equation for α is given by:

$$\frac{\partial \alpha}{\partial t} + \nabla \cdot \left(\overrightarrow{U_f} \alpha \right) + \nabla \cdot \left(\overrightarrow{U_r} \alpha \left(1 - \alpha \right) \right) = 0 \tag{4}$$

where $\vec{U_r}$ is the relative velocity between air and water, $\vec{U_r} = \vec{U_w} - \vec{U_\alpha}$. The last term on the left hand side (l.h.s.) is an artificial compression term, which is introduced in order to reduce artificial diffusion of the air/water interface ('interface smearing') without the need to resort to an overly complex advection scheme (Jasak, 1996). With the definition of α as a phase fraction, this term vanishes in cells where there is only water or air and is therefore only active along the air-water interface.

3.1.2 Turbulence model

In order to describe the turbulent characteristics at the various scales involved in the scour process, the Large Eddy Simulation (LES) with the k-equation sub grid-scale model of

Yoshizawa (1993) is chosen for the present incarnation of scourFOAM. The momentum equation Equation (3) is filtered for performing LES simulation, with the additional sub-grid stress term appearing on the l.h.s., which can be modelled based on an eddy viscosity approximation as:

$$\tau_{ij} = -2\mu_{sgs}S_{ij} + \frac{1}{3}\tau_{ij}\delta_{ij} \tag{5}$$

Here, the sub-grid-scale eddy viscosity $\rho_{sgs} = C_k\rho_f k_{sgs}^{\frac{1}{2}}\Delta$, Δ is the cell length scale, and S_{ij} is the strain tensor rate, and C_k is a model constant. The transport equation for sub-grid-scale kinetic energy k_{sgs} can be written as:

$$\frac{\partial\rho_f k_{sgs}}{\partial t} + \frac{\partial(\vec{\rho_f}U_f k_{sgs})}{\partial x_i} = \frac{\partial}{\partial x_i}\left[(\mu + \mu_{sgs})\frac{\partial k_{sgs}}{\partial x_i}\right] - C_\epsilon\rho_f\frac{k_{sgs}^{3/2}}{\Delta} + 2\mu_{sgs}S_{ij}S_{ij} \tag{6}$$

The effective dynamic viscosity can then be computed $\mu_{eff} = \mu + \mu_{sgs}$ and C_ϵ is the model constant.

Similar to the treatment in the flow module, the assumption is made that the size of the particles is small enough such that it does not interfere with the turbulence eddies at the mesh size that the LES resolves directly. At the sub-grid scale, influences from particles on the turbulence dissipation can be significant, particularly where the suspended sediment concentration is high. Such a process is taken into account through the mixture viscosity concept, rather than directly modifying the sub-grid turbulence viscosity, which can introduce large uncertainties in the scour simulation.

3.1.3 Particle phase (Lagrangian)

The sediment phase is treated as individual discrete particles, and their motion is solved based on Newton's Law of motion. The governing equations for the motion of an individual particle are given as:

$$\frac{dx_p}{dt} = \vec{U_p} \tag{7}$$

$$\frac{d\vec{U_p}}{dt} = D_p\left(\vec{U_f} - \vec{U_p}\right) - \frac{\nabla p}{\rho_p} + \left(1 - \frac{\rho_f}{\rho_p}\right)g - \frac{1}{\theta_s\rho_p}\nabla\tau \tag{8}$$

Here the subscript p denotes a particle property and subscript f refers to a fluid property interpolated to the particle position. The terms on the right hand side (r.h.s) account for particle acceleration due to the drag force, the pressure gradient force, the net buoyant force (gravity minus buoyant force), and the inter-particle stress gradient, respectively, and are explained in more detail below.

3.1.4 Drag model

The model by Andrews and O'Rourke (1996) is selected to represent the drag force from the fluid on each particle, shown as the first term on the r.h.s of Equation 8. The parameter D_p is given as:

$$D_p = C_d\frac{3}{8}\frac{\rho_f}{\rho_p}\frac{\left|\vec{U_f} - \vec{U_p}\right|}{r_p} \tag{9}$$

In which:

$$C_d = \frac{24}{Re_p} \left(\theta_f^{-2.65} + \frac{1}{6} Re_p^{\frac{2}{3}} \theta_f^{-1.78} \right) \tag{10}$$

and

$$Re_p = \frac{2\rho_f \left| \overrightarrow{U_f} - \overrightarrow{U_p} \right| r_p}{\mu_f} \tag{11}$$

in which r_p is the radius of the particle, θ_f is the volume fraction for the mixture fluid phase (air-water), i.e., $\theta_f = 1 - \theta_s$, and θ_s is the volumetric fraction of solid particles in a cell.

3.1.5 Inter-particle stress

When the solid volume fraction (θ_s) exceeds about 5%, frequent particle collisions will take place (Patankar and Joseph, 2001), and this should be accounted for in the model. In the MP-PIC method, an isotropic particle collisional pressure based on the Eulerian grid is usually adopted to represent particle collisions and prevent the solid volume fraction from exceeding its critical value. Such continuum models have been proven to be suitable and efficient in Eulerian-Lagrangian models (Patankar and Joseph, 2001; Snider, 2001). Following Snider (2001), a continuum particle stress model is employed here to take into account the particle interactions at high volume fractions:

$$\tau = \frac{P_s \theta_s^\beta}{max\left(\theta_{cs} - \theta_s, \epsilon\left(1 - \theta_s\right) \right)} \tag{12}$$

where P_s is a constant with the unit of pressure, and $2 \leq \beta \leq 5$. ϵ is a small number of the order 10^{-4} which is introduced to remove the sudden accelerations at very high packing densities. In the case of sandy particles, the porosity determines that the maximum volume concentration for fully packed bed is approximately 60–65%. Therefore, a critical solid volume fraction θ_{cs} is employed with an assigned value of 0.6.

This normal particle stress is represented as a continuum calculation of the "particle pressure" on the Eulerian grid, which is in turn interpolated back to the discrete particle's positions to calculate the normal stress due to motion and inelastic collision of particles.

3.1.6 Influence of particles on the flow

After the information is updated for all particles, the impacts to the fluid flow are then evaluated by integrating the particles' information within the cell they occupy. For such a purpose, the solid volume fraction (θ_s) is an essential variable which accounts for the volume occupied by the solid particles within a cell:

$$\overrightarrow{\theta_s} = V_c^{-1} \sum_{p=1}^{np} V_p \delta_s \tag{13}$$

Here, V_c is the volume of the computational cell, V_p is the particle volume, np is the number of particles in the cell, and δ_s is a binary function that is equal to 1 for the cell containing a given particle, and 0 for all other cells.

In the scour process, particles are influenced by the flow through the drag force and pressure gradient force, and spontaneously gain a certain amount of momentum from the fluid phase. To model their equal and opposite reaction to the flow, the interphase momentum transfer, S_{imt} as in Equation 3, is computed as in Snider (2001); Patankar and Joseph (2001)

$$\overrightarrow{S}_{imt} = V_c^{-1} \sum_{p=1}^{np} \left[\rho_p V_p \left(D_p \left(\overrightarrow{U_f} - \overrightarrow{U_p} \right) - \frac{\nabla p}{\rho_p} \right) \bigg|_p \right] \tag{14}$$

3.1.7 Mixture velocity

Previous studies show that in dilute suspensions, concentration and viscosity are linearly related, but as the concentration approaches the maximum packing status, the viscosity becomes infinite. In order to account for this effect the effective viscosity of the suspension is modelled using a equation as follows:

$$\mu_{mix} = \mu_{eff} \left[1 + \frac{0.5\mu_0 \theta_s}{\left(1 - \frac{\theta_s}{\theta_{cs}} \right)} \right]^2 \tag{15}$$

Here μ_0 is the intrinsic viscosity which accounts for the shape of particles. For roughly spherical sediment particles, the value $\mu_0 = 2.5$ is used as recommend in Penko et al. (2009).

3.1.8 Enforcing the maximum sediment fraction

When using MP-PIC type solvers the maximum packing fraction is usually enforced by employing a particle acceleration term based on inter-particle stress gradients at particle locations, e.g., (12). These stress gradients provide a pressure type force that prevents over-packing. The inter-particle stress gradients are usually computed on the Eulerian mesh as a function of the solid fraction of a cell. Whilst this approach is computationally efficient, it does not always prevent over packing and can lead to numerical instabilities (Apte et al., 2003). For example, when considering a particle inside the packed bed, the concentration gradient is small, which will not prevent the particle from entering a fully packed cell. To resolve the motion of particles under high concentration values, O'Rourke and Snider (2010) and O'Rourke et al. (2009) proposed a new collision term to improve the performance of the MP-PIC method in a high particle fraction regime. However, the uncertainty in estimating the collision time-scale could lead to noticeable errors in the scour simulation where the particle-fluid interaction time scale is very important. An alternative is to employ a Lagrangian type particle based collision model or the Discrete Element Method (DEM) of Cundall and Strack (1979). But these methodologies are computationally very expensive. In the current work, a Eulerian type concentration gradient approach is employed. Following the approach of Kelly (2013), if a parcel entering a cell causes the sediment fraction of that cell to exceed maximum pack the neighbouring cell with the highest sediment concentration gradient is identified and the parcel moved into it. If the presence of the particle causes the sediment to exceed maximum pack in the new cell as well, the procedure is repeated. This approach has proven to be both stable and computationally efficient. The approach is able to maintain a quiescent condition for densely packed sediment as well as allowing sediment to settle out of suspension to maximum pack.

4   Numerical solution technique

4.1   The solver

The Finite Volume Method (FVM) is employed to solve the governing equations for the fluid mixture (air-water) phases. The computational domain is discretised into control volumes

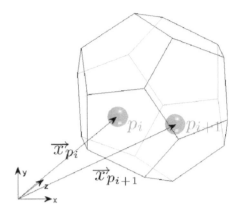

Figure 1: Sketch of particle parcels within a typical computational control volume.

(cells) without overlapping (Figure 1). Unstructured meshes can be used to resolve complex geometry and achieve local grid refinement. In OpenFOAM®, all dependent variables are collocated at the centroid of the control volume. To overcome the checkerboard oscillation, Rhie-Chow interpolation (Rhie and Chow, 1983) is used. The velocity values at cell faces are evaluated via the discretised momentum equation with locally linearised convective terms, which are interpolated from adjacent grid nodes. With the exception of the sediment-fluid momentum transfer term as the additional source term in Equation 3 are incorporated into the L.H.S of the discretised momentum equation. In addition, a face limiting procedure in the OpenFOAM® library is also used to eliminate the localised over and under shoots during the reconstruction of gradients at cell centres.

For the solid phase, computational particles within a control volume are aggregated to calculate the solid fraction in each volume (see Figure 1). Particles are regarded to be inside a control volume as long as their centroids lie within the control volume boundary. As there are a large number of particles involved in scour processes, it is essential to introduce the concept of a *parcel*, which is assumed to be a group of particles with the same properties such as size, velocity and others. In this hybrid Eulerian-Lagrangian technique, parcels are used within the Lagrangian framework in order to reduce the computational expense to a level that is practicable.

4.2   Boundary and initial conditions

For a typical case, a rectangular domain is used as shown in Figure 2. An inlet boundary is specified at the upstream end with a prescribed water level, flow velocities and turbulence quantities. An outlet is used at the downstream boundary with all variables being linearly interpolated from internal node points to the boundary. The two cross-stream side (lateral) boundaries are specified as periodical boundaries. The structure is typically specified at the centre of the domain near the bottom with the surface being specified as a wall. The top boundary is treated as an inlet with an atmospheric pressure (zero pressure).

The initial conditions for the fluid mixture is typically specified at zero velocity. The volume fraction of water is prescribed as unity at levels below the free surface in the domain.

The requisite number of sediment particles are placed at the bottom of the domain within a pit with zero velocity initially (see Figure 2), to make up the prescribed total volume fraction. When a particle reaches the downstream boundary, it will no longer remain in the solution domain. If a periodic boundary condition is assigned, particles will enter from the corresponding boundary again into the computational domain.

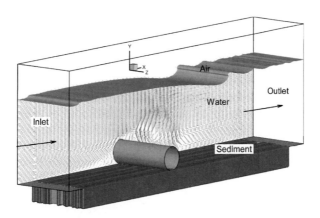

Figure 2: Computational model set-up.

4.3 Solution procedure

The solution of the fluid mixture phase on the Eulerian grid follows the multiphase solver interFOAM in OpenFOAM®. The pressure-velocity coupling of the Navier-Stokes equations follows the PIMPLE algorithm of Issa (1986) for transient flows. In these algorithms, the equations are all solved in a segregated approach. The calculation starts with prescribed motion initially until the flow converges to steady or dynamic steady state. Once the fluid phase motion has been resolved, the results are then interpolated onto the particle position, and the position and velocity of each particle can be updated accordingly. In particular, the motion of each sediment parcel can be tracked through time using the following procedure:

1. Interpolate relevant quantities (fluid density, fluid pressure gradient and ensemble-averaged fluid velocity) from the mesh to the particle location;

2. Set a Lagrangian sub time step Δt_p of the Eulerian time step Δt based on the $\overrightarrow{U_p}$ value from the previous time-step;

3. Update $\overrightarrow{U_p}$ using the Euler difference form of Equation 8 with time step Δt_p; and

4. Repeat until the Eulerian time step Δt has been reached.

This procedure effectively integrates Equation 6 for each parcel to determine its new position at the end of the Eulerian (fluid) time step. A sub-time stepping procedure is used for the Lagrangian parcels to enable the code to run in parallel.

As one would expect, it is essential to know where the discrete Lagrangian particles are on the Eulerian grid, so that the correct particle-fluid interaction can be realised. The algorithm of Nordin (2000) and Macpherson and Weller (2009) is used in searching for the particles to reduce the total computational cost.

5 Model applications

This section provides examples of the application of the present model (scourFoam) and comparisons with studies to demonstrate its prediction capabilities against experimental data. Firstly, the model's predictive capability is tested in 2D using the data of Alper Oner et al. (2008) and Mao (1986) and then it is applied to simulate the 3D scour around a complex foundation using the data of Tavouktsoglou et al. (2016).

5.1   2D scour application

The model is applied to simulate scour under a horizontal pipeline. Firstly, the study of Alper Oner et al. (2008) was used to test the model's prediction in hydrodynamics around the pipeline. To verify the ability of scourFOAM to predict current induced scour, the benchmark test of Mao (1986) has been simulated.

5.1.1   Hydrodynamics

The experiment of Alper Oner et al. (2008) involves detailed measurements of flow over a cylinder at different elevations above a fixed bed. The model was applied to one of the tests in which the cylinder was placed above the bed without any gap. The water depth was kept at 0.32 m and a steady current with a speed of 0.197 m/s was introduced at the inlet. The cylinder has a diameter of 50 mm The model is set up as shown in Figure 2 with the same parameters in the laboratory experiment. To simulate the flow-structure-sediment interaction, an 8 cm deep sediment rectangular sediment pit was introduced at the bottom while forcing a zero velocity condition on the bed particles to ensure the bed is immobile. The time step was set at 5×10^{-4} s and a mesh size of 5 mm was used in all three directions. The computed vertical profile of the stream-wise flow velocity was compared with the measured data at –1.5D, –1D, –0.5D, OD, 0.5D, 1D and 1.5D in Figure 3 where the zero level corresponds to the bed surface. It can be seen clearly that the agreement between the model results and measurements are very good at all sites. The computed velocity vertical profile inside the bed ($z < 0$) also shows the effects of particle on fluid motion, which shows that the interaction produces a good resistance to the flow dynamics near the bed surface.

5.1.2   Scour around 2D pipe

The model is applied to steady current-induced live-bed scour beneath a pipeline, using a benchmark laboratory test carried out by Mao (1986). According to Mao (1986), the water

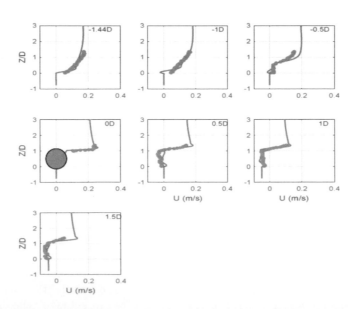

Figure 3: Comparison of computed and measured horizontal flow velocity in Alper Oner et al. (2008) test.

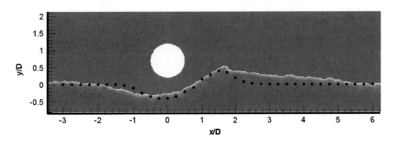

Figure 4: Comparison of computed and measured bed level at t = 1.5 min.

depth was set as 0.35 m and pipeline diameter as 0.1 m. The computational domain is 1.6 m long, 0.5 m deep and one cell wide. At the upstream inlet boundary the prescribed flow velocity of 0.5 m/s is distributed across the depth. At the downstream outlet boundary a zero gradient condition is assigned to the variables in the flow module. The particles were allowed to leave the domain when they moved out of the computational area. The bottom of the channel was specified as a wall with a sediment pit in the middle of the domain, taking up 14D in length and 8 cm in depth. The sediment particles with median diameter (d_{50}) of 36 mm were distributed inside the pit initially with the pipeline lying on the surface at the middle of this section (see Figure 2). The top boundary is specified as a zero pressure boundary open to the atmosphere. The two side boundaries in the cross-flow direction are treated as periodic to eliminate any side boundary effects. The time-step was specified as 2×10^{-4} s and a uniform mesh size of 5 mm was used in all three directions. Similar to the hydrodynamic test, the mesh around the pipeline is split to resolve the structure. In total the computational mesh comprised of 11,000 cells. The total number of particles exceeds 10,000 in the simulation.

Figure 4 shows the computed scour profile with the dots denoting the measured data at 1.5 min. The shape of the scour hole, the maximum scour depth, and the maximum deposition point in the downstream are all captured by the model. The downstream deposition area at x/D = 1.75 is slightly over predicted, which can be partly attributed to the complex flow structures of the wake vortices that are not strong enough near the bed surface to transport the particles away. Overall, the agreement is very good, which demonstrates the capability of scourFOAM in simulating the scour process over a live bed condition.

It should be noted that in the previous studies, e.g., Yeganeh-Bakhtiary et al. (2013), Zanganeh et al. (2012), Yeganeh-Bakhtiary et al. (2011), Liang and Cheng (2005b), Zhao and Fernando (2007), a gap between the pipeline and bed together with an arbitrary bed profile just underneath the structure are needed in the model in order to initiate the scouring process, due to the inability in dealing with particle-fluid interactions. However, in the present study, the pipeline is placed directly onto the particles on the bed surface without any gap. The initiation of the onset of scour is through the motion of individual particles under the influence of the flow hydrodynamics and the pressure difference between the upstream and downstream sides of the pipeline. Figure 5 details the onset of scouring and the surrounding flow dynamics for the first 1.2 s period at every 0.1 s interval. It can be seen that in the initial 0.5 s there is almost no disturbance in the bed upstream of the structure due to the low flow speed above the bed surface. On the contrary, the flow downstream is marked by the growing vortices which start to erode the bed in the immediate wake of the pipeline from 0.2 s onwards. With the strong flow velocity and vortex shedding, the deformation in the bed surface downstream of the pipeline (x>1D) has already become visible in the initial 0.4 s although the magnitude is small. The bed region between 0.1D and 0.2D also

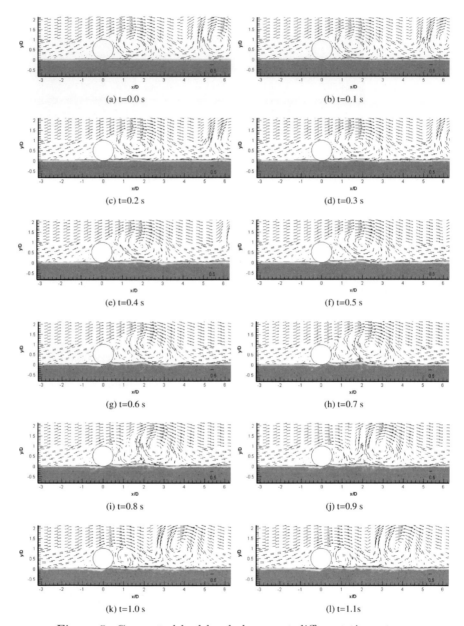

Figure 5: Computed bed level change at different time steps.

rises initially with piping taking place. Subsequently, the particles underneath the pipeline between –0.1D and 0.1D start moving along the pipeline perimeter and then towards the downstream side. It is noted that at the upstream of the pipeline, the flow decelerates and forms small vortices beneath the pipeline and the bed surface, which pushes the flow going into the bed. Downstream of the pipeline, the coherent flow structure is generated behind the pipeline and propagates along the bed surface, interacting with the sediment at the bed surface as they touch down and advect downstream.

The results clearly demonstrate the effectiveness of the method used in scourFOAM in dealing with particle motion within the high concentration region, i.e., a nearly fully packed bed. At the same time, these results also suggest that the initiation of the onset of scour is largely driven by the seepage flow within the bed. The bed liquefaction under these fluid-

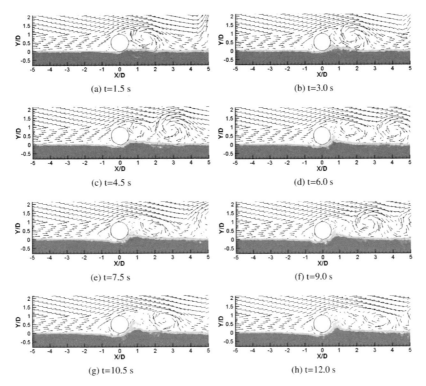

Figure 6: Computed tunnel erosion at various stages.

particle interactions are the primary driver for further scour, rather than the amplified shear stress. To a large extent, this indicates the necessity to include sub-surface flow information into practical modelling of sediment transport and scour.

Figure 6 shows the subsequent development of the scouring from 1.5 s to 12 s. At t = 1.5 s, a layer with low sediment concentration is visible at the bed surface connecting the upstream and downstream side of the pipeline, forming a very narrow pathway for the particles underneath the pipe. Subsequently, a slope underneath the pipe at the upstream side gradually develops from about 3 s, which encourages flow separation and vortice formation in front of the pipeline. Meanwhile, the transported particles also form a mound of deposition immediately behind the pipeline. From 4.5 s, the gap between the bed surface upstream and the pipeline increases (–1D < x < 0D), which allows water to flow under the structure and accelerate in the narrow pathway. The mixture of water and sand can move via this channel towards the downstream side more easily and feed into the mound on the immediate downstream side. Gradually the tunnel erosion stage starts. A jet of water and sand mixture is observed in the pathway underneath the pipe at 6 s, and the vortices in the lee-wake side are affected due to the presence of the mound. After that, the pathway keeps developing (see Figure 6), and the bed underneath the pipe is being eroded and transported downstream. In the later tunnel erosion stage, the pathway develops into a scour hole and the mound downstream grows significantly as shown in the figure.

Figure 7 presents the computed flow velocity vector distribution at the later stage of tunnel erosion and during the development phase of the scour hole. At the later stage of tunnel erosion (15 s), the gap between the pipeline and the underlying bed is still small, and the flow is largely deflected over the structure, generating a small vortex in front of the pipeline and larger lee-wake shedding behind. The gap expands, part of the flow diverges under the pipeline and accelerates as a jet, shooting out on the lee side, causing a vortex to

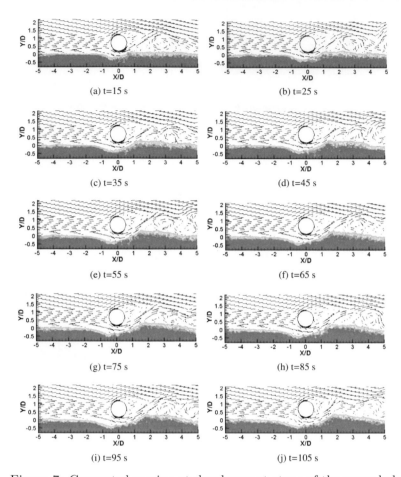

Figure 7: Computed erosion at development stage of the scour hole.

form higher up in the water column, immediately behind the pipeline increasing in size and leading to the generation of a vortex close to the bed at x = 2–3D. As the scour develops further, the jet increases in strength leading to lee-wake vortex shedding higher in the water column. The mound of sediment deposit also gradually increases in size and eventually suppresses the near-bed vortex behind the structure within a distance of 2D.

The computed instantaneous flow velocity vector distribution at lee-wake erosion, t = 1.5 min is presented in Figures 6–8. In general, the flow field at the upstream side of the pipeline remains undisturbed up to –2D. However, noticeable interactions between the fluid flow and particles at the bed surface are visible in this region corresponding to the sediment motion under live-bed conditions. From this point further downstream, the flow field is clearly influenced by the presence of the pipeline and sediment particles. As the flow approaches the pipeline, the main flow is separated with acceleration above and beneath the pipeline. The scour hole beneath the pipeline has already formed and the flow acceleration between the bed and the pipeline develops with little restriction. The accelerated flow under the pipeline propagates downstream to around 2D before reattaching with the flow from above. Further downstream, the lee-wake vortices form and propagate on the downstream side (x > 3D). Between the lee-wake vortices and the bed, the flow is hindered by the bed, particularly in the region of the sediment mound at x = 1.8D (1.5D < x < 3D).

In addition, the free surface becomes perturbed due to the complex flow structures as shown in the figure. As the flow accelerates at x = –1D, the surface dips and subsequently

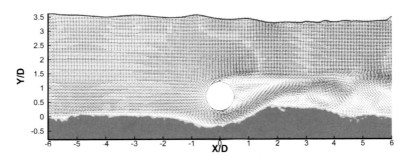

Figure 8: Velocity field in the vicinity of the pipeline at t = 1.5 min.

fluctuates as vortex shedding takes place in the lee-wake region until about x = 5D where the surface starts to recover to the inflow level. In this particular case, the Froude number Fr is computed to be above 0.2 and free surface variation is expected. By resolving the free surface, the internal flow field can adjust more realistically and timely as the bed profile evolves. This is particularly important for the case with large Froude numbers in order to resolve the flow field correctly and hence improve the accuracy of the scour predictions as discussed above.

5.1.3 Wave induced scour under a pipe

To test model performance under waves, a wave induced scour test around a horizontal pipe was reproduced numerically based on the physical experiments conducted by Sumer and Fredsøe (1990). For the test, the water depth was 0.4 m, the wave period (T) was 1.22 s and the bottom orbital velocity was 0.24 m/s. The pipe diameter was 0.03 m resulting in a KC = 11. The sediment particles with median diameter of 0.18 mm were distributed inside the pit initially with the pipeline lying on the surface at the middle of this section. For the simulation the mesh resolution was 0.15 cm, and the time step 1 x 10^{-4} s. The critical solid volume fraction was set at 0.65, with an initial bed fraction of 0.649.

The bed profile at 0.5 min and 1 min is compared to the measurements of Sumer and Fredsøe (1990) in Figure 9. The bed surface is plotted as the isoline of the solid volume fraction at 0.6. The overall scour pattern is well captured as seen in these two figures, where the gentle bed slopes on both sides of the pipe are well reproduced. The scour profile is also in good agreement with the measurements. At t = 0.5 min, small fluctuations are observed in the bed surface between x = −10 cm and x = 10 cm, which illustrates that the bed in the model responds immediately to the flow field fluctuations due to vortex shedding from the wave pipe interaction. As time progresses further (t = 1 min), the scour hole beneath the pipe becomes deeper which reduces the flow acceleration under the pipe alongwith the scour rate. Overall, the model predicts the scour depth under the pipe with good accuracy, even though the maximum scour depth at t = 1 min is slightly underestimated.

5.2 3D scour around a complex foundation

The mesh used in this study was selected to have the same dimensions as the flume in which the experimental measurements of Tavouktsoglou et al. (2017) were made. Thus the numerical flume was 15 m long 1.2 m wide and 0.5 m deep. The mesh was comprised of a total of 1,500,000 cells with an increasing refinement closer to the structure (see Figure 10). This was done to capture as accurately as possible the flow separation off the structure's surface. The sediment box (i.e., Lagrangian field) is confined in a 0.5 x 0.5 m inverted

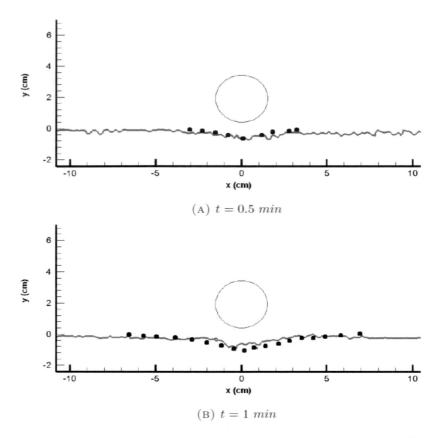

(A) $t = 0.5\ min$

(B) $t = 1\ min$

Figure 9: Comparison of scourFOAM results with Sumer and Fredsøe (1990) for scour due to waves. Red line: model; dots: Sumer and Fredsøe (1990) physical model results.

pyramid surrounding the base of the structure in the centre of the domain (see Figure 10). The inverted pyramid was selected in order to reduce the amount of Lagrangian particles needed to fill the virtual sand pit and thus reduce the computational time.

The equilibrium scour depth induced by a structure is the most important engineering aspect of scour. Numerous empirical methods for the prediction of the equilibrium scour depth exist which can be applied with some success, but their shortcoming is that they cannot provide an estimate of the overall 3D scour features. This section presents the results of the numerical simulation and provides a comparison with experimental measurements for a cylindrical base structure. Figure 11 presents the results of this comparative study.

As can be seen from the figure, the numerical model predicts well the 3D scour around the structure. The discrepancies and variation shown in Figure 11 are mainly attributed to the slight variation of the location of scour features in the x-y plane rather than the underestimation or overestimation of scour. This is expected as the locations of these features are highly effected by the slightest of variations in the starting flow and bed conditions. It can be seen that the maximum scour depth in both cases occurs at an angle of 45 degrees relative to the flow direction which agrees with potential flow theory. There is a small tendency for the model to over predict the equilibrium scour depth in front of the structure. This is mainly attributed to the slightly different flow profile which is used in the numerical model. According to Tavouktsoglou et al. (2017) higher near-bed velocities (as is the case

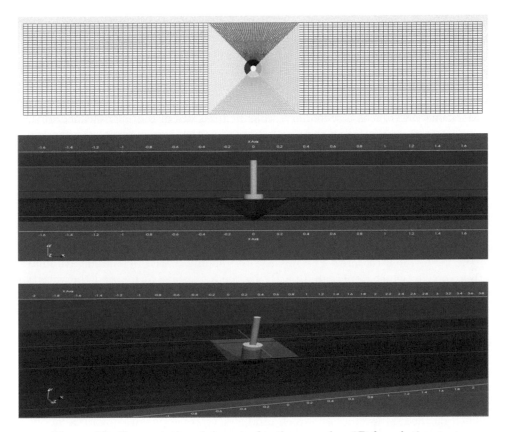

Figure 10: Computational domain for the complex 3D foundation test.

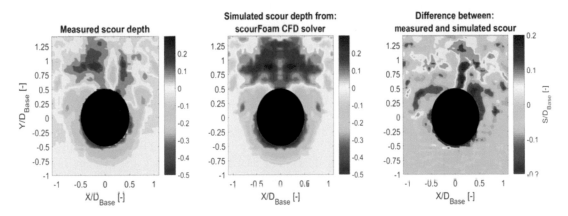

Figure 11: Equilibrium scour depth for the complex 3D foundation test.

in the numerical model) induce a stronger depth-averaged horizontal pressure gradient and thus deeper scour depths. Further observation of the data reveals that the model captures accurately the lateral extent of the scour hole both in the streamwise and cross-flow direction. This aspect is also important for design purposes as it controls the lateral extent of the scour protection. Therefore, the overestimation of the extent of the scour hole may lead to the overdesign of the scour protection system. In the case of the underestimation of the scour extent, the scour protection may be designed to be less extensive than required and thus pose a threat to the stability of the foundation.

The scourFOAM solver also captures accurately the secondary scour holes which radiate away from the structure at an angle ±130 degrees relative to the flow direction at the lee of the structure. This scour feature is associated with the lee-wake vortices which are induced by the flow structure interaction and are a common feature in cylindrical geometries (Olsen and Melaaen, 1993). These secondary scour features are also of significance for the design of scour protection systems as they may destabilize the edge detail of the scour protection and thus make the entire protection susceptible to failure. The simulation of this feature is also important as many existing sediment transport solvers have difficulties reproducing.

A final feature which is picked-up by the scourFOAM model and shows its ability to accurately capture the physical processes associated with scour around complex geometries is the deposition zone at the lee of the structure. This feature is caused by a secondary scour mechanism, the large-scale counter-rotating streamwise phase-averaged vortices (LSCSVs), which effectively creates a strong up-ward pressure gradient which re-suspends sediment from the lee of the structure. The LSCSVs are mainly driven by the longitudinal counter-rotating vortices which are created partly by the horseshoe vortex and the variation of the shedding frequency over the height of the structure (Baykal et al., 2015). In the case of the cylindrical base structure the interaction between the flow and the components of the foundation with different diameters leads to the disruption of the LSCSVs. This in turn, reduces the erosion rate at the lee of the structure and thus more deposition takes place. Figure 11 shows that the sediment mound at the lee of the structure is not in line with the principal flow axis. This is attributed to a small flow asymmetry of around 5% which was present in the flume. A closer examination of this feature shows that the depression in the middle of this deposition zone is deeper in the case of the experimental results. This is also explained by the flow asymmetry which produces a stronger lee wake vortex on one side and thus a stronger upward flow gradient at the lee.

After the pile is installed the scour depth develops rapidly; however, it is important to predict the time evolution of scour for several reasons:

- To determine the scour development during a storm or extreme flood event;

- To calculate the time window available to install scour protection after the foundation is installed; and

- To evaluate the total scour depth in layered and mixed sediments.

Numerous methods for predicting the time evolution of scour around monopiles have been developed (Harris et al., 2010 contains a brief review). The shortcoming of these methods is that they have all been developed based on laboratory data. This means that their accuracy is subject to scale effects and their performance varies from case to case. Therefore, the accurate prediction of scour development using a numerical model can provide an invalauble tool for use in engineering analysis, design and practice. Figure 12 presents the results of the comparison between the experimental and numerical results. In this figure the scour depth is measured adjacent to the structure at an angle of 45 degrees relative to flow direction. It can be seen that the present model predicts well the time development of scour. The model has a tendency to over predict the scour but this is attributed to the stronger pressure gradient induced by the flow in the simulation case. The difference between the experimental and numerical results is small enough to be attributed to the slightly different flow profiles that were present in each case. To verify this, the scour prediction correction factor of Tavouktsoglou et al. (2016) is applied to the simulation data. By applying a 90% correction factor recommended by Tavouktsoglou et al. (2016) to the simulation results

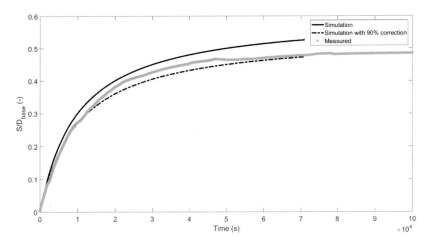

Figure 12: Effect of correction factor on the model prediction.

we can obtain the dashed line in Figure 12 which shows that the agreement between the two data-sets has significantly improved, with the numerical solution following more closely but slightly underestimating the scour depth during the stabilization phase of the measured process. In conclusion, the scourFOAM model appears to represent a valuable tool to aid the design of coastal structures in real world engineering practice.

6   Conclusions

The scourFOAM sediment transport solver represents the state-of-the-art in the numerical modelling of scour. ScourFOAM is a three-phase (air-water-sediment) solver that has been developed over several years begining with the work of Kelly (2013) and culminating in the model presented in this chapter. The model is based on the open CFD source code Open-FOAM, modified to include an additional non-cohesive sediment phase. Implementation of the sediment phase employs a Lagrangian approach based on the multiphase particle in cell (MP-PIC) method. Whilst MP-PIC has been previously employed to simulate gas fluidised beds and batch sedimentation successfully, it has not been used to simulate highly dynamic free surface morphodynamic problems involving bed evolution. Results from numerical experiments employing the model have shown that:

- The air-water-sediment 3-phase approach is effective in resolving the physical processes involved in sediment transport. The model is strictly valid only for a dispersed sediment phase. The main assumption in the model is that the particles can be treated at sub-grid scale, which is typically the case in the scour over sandy bed.

- Results from simulations of scour around horizontal pipelines indicate that the strong turbulence-particle interactions around the structure, affect, the rate of sediment and scour transport substantially.

- In the present model, the effects of particles on the continuum phase are resolved at an Eulerian cell scale. The good agreements between the predicted scour and measurements suggest that the sub-grid particle-turbulence effects can be considered to be less important to the overall scour process.

- The good agreement between simulation results and experiments indicates that the MP-PIC approach is well suited to simulate sediment transport processes. The proposed method has resolved the numerical stability issues, which have hindered the application of similar approaches in scour modelling in the past.

- The Euler-Lagrange multiphase approach has proven to be advantageous over conventional continuum approaches. Particle modelling approaches like the MP-PIC method are extremely powerful in their ability to resolve the scour process at the particle scale, and their potential to reveal more detailed micro scale processes is very encouraging. However, the limited knowledge we have to date, with regards to the flow-particle and particle-particle interactions, poses a significant problem.

- Realisation of improved parallelisation is another aspect for further development. This is required in order to expand the model application to study larger scale engineering problems such as scour around prototype scaled structures and the scour associated with high velocity flows.

Acknowledgements

The authors acknowledge the contributions made by Dr. Justin Finn and Dr. Yaru Li during the early stages of the scourFOAM model development. The authors also wish to thank Dr. Giovanni Cuomo (Research Director at HR Wallingford) for his support of this work over many years.

References

Aghaee, Y. and Hakimzadeh, H. (2010). Three dimensional numerical modelling of flow around bridge piers using LES and RANS. River Flow, Dittrich, Koll, Aberle and Geisenhainer (eds.). P(2010): 211–218.

Alper Oner, A., Salih Kirkgoz, M. and Samil Akoz, M. (2008). Interaction of a current with a circular cylinder near a rigid bed. Ocean Engineering 35: 1492–1504.

Andrews, M. and O'Rourke, P. (1996). The multiphase particle-in-cell (mp-pic) method for dense particulate flows. International Journal of Multiphase Flow 22: 379–402.

Apte, S. V., Mahesh, K. and Lundgren, T. (2003). A Eulerian-Lagriangian model to simulate two-phase/particulate flows. Centre for Turbulence Research Annual Research Brief, 161–171.

Baykal, C., Sumer, B. M., Fuhrman, D. R., Jacobsen, N. G. and Fredsøe, J. (2015). Numerical investigation of flow and scour around a vertical circular cylinder. Philosophical Transactions of the Royal Society A: Mathematical, Physical and Engineering Sciences 373(2033): 20140104.

Chauchat, J., Cheng, Z., Nagel, T., Bonamy, C. and Hsu, T. (2017). SedFoam-2.0: A 3-D two-phase flow numerical model for sediment transport. Geoscientific Model Development Discussions 10(12): 4367–4392.

Chen, X., Li, Y., Niu, X., Li, M., Chen, D. and Yu, X. (2011). A general two-phase turbulent flow model applied to the study of sediment transport in open channels. International Journal of Multiphase Flow 37(9): 1099–1108.

Cundall, P. A. and Strack, O. (1979). A discrete numerical model for granular assemblies. Geotechnique 29: 47–65.

Finn, J. R., Li, M. and Apte, S. V. (2016). Particle based modelling and simulation of natural sand dynamics in the wave bottom boundary layer. Journal of Fluid Mechanics 796: 340–385.

García, M. H. and Parker, G. (1991). Entrainment of bed sediment into suspension. Journal of Hydraulic Engineering 117(4): 414–435.

Harris, J. M. Whitehouse, R. J. S. and Benson, T. (2010). The time evolution of scour around offshore structures. Proceedings of the Institution of Civil Engineers, Maritime Engineering, 163(A1): 3–17.

Higuera, P., Lara, J. L. and Losada, I. J. (2013). Realistic wave generation and active wave absorption for Navier-Stokes models: Application to OpenFOAM. Coastal Engineering 71: 102–118.

Hirt, C. W. and Nichols, B. D. (1981). Volume of fluid (VOF) method for the dynamics of free boundaries. Journal of Computational Physics 39(1): 201–225.

Issa, R. I. (1986). Solution of the implicitly discretised fluid flow euqations by operatorsplitting. Journal of Computional Physics, 40–65.

Jasak, H. (1996). Error analysis and estimation for the finite volume method with applications to fluid flows. Ph.D. thesis, Imperial College, University of London.

Jha, S. K. and Bombardelli, F. A. (2010). Toward two-phase flow modelling of nondilute sediment transport in open channels. Journal of Geophysical Research: Earth Surface 115(F3).

Kelly, D. M. (2013). Development of the scourFOAM model. HR Wallingford Internal Research Report. Wallingford, Oxon, UK.

Launder, B. E. and Spalding, D. B. (1974). The numerical computation of turbulent flows. Computer Methods in Applied Mechanics and Engineering 3(2): 269–289.

Li, M., Pan, S. and O'Connor, B. A. (2008). A two-phase numerical model for sediment transport prediction under oscillatory sheet flows. Coastal Engineering 55(12): 1159–1173.

Li, Y., Kelly, D. M., Li, M. and Harris, J. M. (2014). Development of a new 3D Euler-Lagrange model for the prediction of scour around offshore structures. Coastal Engineering Proceedings 1(34): sediment.31.

Liang, D., Cheng, L. and Li, F. (2005). Numerical modeling of flow and scour below a pipeline in currents: Part II. Scour simulation. Coastal Engineering 52(1): 43–62.

Liu, X. and García, M. H. (2008). Three-dimensional numerical model with free water surface and mesh deformation for local sediment scour. Journal of Waterway, Port, Coastal, and Ocean Engineering 134(4): 203–217.

Macpherson, N. N. G. and Weller, H. (2009). Particle tracking in unstructured, arbitrary polyhedral meshes for use in cfd and molecular dynamics. Commun. Numer. Meth. Engng. 25: 263–273.

Mao, Y. (1986). The interaction between a pipeline and an erodible bed. Ph.D. thesis, Tech. Univ. of Denmark.

Mathieu, A., Chauchat, J., Bonamy, C. and Nagel, T. (2019). Two-phase flow simulation of tunnel and lee-wake erosion of scour below a submarine pipeline. Water 11(8): 1727.

Nordin, N. (2000). Complex chemistry model of diesel spray combustion. Ph.D. thesis, Chalmers University of Technology, Gothenburg.

Olsen, N. R. and Melaaen, M. C. (1993). Three-dimensional calculation of scour around cylinders. Journal of Hydraulic Engineering 119(9).

O'Rourke, P., Zhao, P. and Snider, D. (2009). A model for collisional exchange in gas/liquid/solid fluidized beds. Chemical Engineering Science 64: 1784–1797.

O'Rourke, P. and Snider, D. (2010). An improved collision damping time for MP-PIC calculations of dense particle flows with applications to polydisperse sedimenting beds and cliding particle jets. Chemical Engineering Science 65: 6014–6028.

Orszag, S. A. (1970). Analytical theories of turbulence. Journal of Fluid Mechanics 41(02): 363–386.

Patankar, N. and Joseph, D. (2001). Lagrangian numerical simulation of particulate flows. International Journal of Multiphase Flow 793: 1685–1706.

Penko, A., Calantoni, J. and Slinn, D. (2009). Mixture theory model sensitivity to effective viscosity in simulations of sandy bedform dynamics. OCEANS 2009, MTS/IEEE Biloxi—Marine Technology for Our Future: Global and Local Challenges, 1–9.

Rhie, C. and Chow, W. (1983). A numerical study of the turbulent flow past an isolated airfoil with trailing edge separation. AIAA Journal 1525–1532.

Rijn, L. C. v. (1984). Sediment transport, Part II: suspended load transport. Journal of Hydraulic Engineering 110(11): 1613–1641.

Roulund, A., Sumer, B. M., Fredsøe, J. and Michelsen, J. (2005). Numerical and experimental investigation of flow and scour around a circular pile. Journal of Fluid Mechanics 534: 351–401.

Rusche, H. (2003). Computational fluid dynamics of dispersed two-phase flows at high phase fractions. Ph.D. thesis, Imperial College, University of London.

Snider, D. M. (2001). An incompressible three-dimensional multiphase particle-in-cell method for dense particle flows. Journal of Computational Physics 170: 523–549.

Soulsby, R. (1997). Dynamics of Marine Sands: A Manual for Practical Applications. Thomas Telford.

Sumer, B. and Fredsøe, J. (1990). Scour below pipelines in waves. Journal of Waterway, Port, Coastal, and Ocean Engineering 116(3): 307–323.

Sun, R. and Xiao, H. (2016). Sedifoam: A general-purpose, open-source CFD-DEM solver for particle-laden flow with emphasis on sediment transport. Computers & Geosciences 89: 207–219.

Tavouktsoglou, N. S., Harris, J. M., Simons, R. R. and Whitehouse, R. J. (2016, June). Equilibrium scour prediction for uniform and non-uniform cylindrical structures under clear water conditions. In ASME 2016 35th International Conference on Ocean, Offshore and Arctic Engineering. American Society of Mechanical Engineers.

Tavouktsoglou, N. S., Harris, J. M., Simons, R. R. and Whitehouse, R. J. S. (2017). Equilibrium scour-depth prediction around cylindrical structures. Journal of Waterway, Port, Coastal, and Ocean Engineering, 143(5).

Tavouktsoglou, N. S., Harris, J. M., Whitehouse, R. J. S. and Simons, R. R. (2018). CFD simulation of clearwater scour at complex foundations. In Scour and Erosion IX: Proceedings of the 9th International Conference on Scour and Erosion (ICSE 2018), Taipei, Taiwan.

Van Rijn, L. C. (1993). Principles of sediment transport in rivers, estuaries and coastal seas (Vol. 1006). Amsterdam: Aqua publications.

Yoshizawa, A. (1993). Bridging between eddy-viscosity-type and second-order turbulence models through a two-scale turbulence theory. Phys. Rev. E Stat. Phys. Plasmas Fluids Relat. Interdiscip. Topics 48: 273–281.

Zubeldia, E. H., Fourtakas, G., Rogers, B. D. and Farias, M. M. (2018). Multi-phase SPH model for simulation of erosion and scouring by means of the shields and Drucker-Prager criteria. Advances in Water Resources 117: 98–114.

Chapter 8

A Coupling Strategy for Modelling Dynamics of Moored Floating Structures

Tristan de Lataillade,[1,]* *Aggelos Dimakopoulos,*[2] *Chris Kees*[3] and *Lars Johanning*[4]

1 Introduction

Assessing the performance of a multiphysics system such as moored floating structures subject to environmental loads is a challenging task, which can be performed by means of physical or numerical modelling. The latter can be categorised depending on the inclusion of: nonlinear effects (linear or nonlinear models), dynamic effects (static/quasi-static or dynamic models), and the interaction between multiple physical processes (coupled or uncoupled models). Until recently, relatively simple numerical tools using linear, uncoupled and static/quasi-static models were proven to be relatively reliable and computationally efficient for simulating floating structures that are not expected to significantly respond to environmental conditions, e.g., moored vessels in a port or floating LNG platforms, as the slow or limited motion of the structure makes the use of these tools appropriate. The recent growth of business sectors such as marine renewables has nevertheless been a game changer for the design of floating and moving structures. In order to harvest energy, offshore renewable energy devices are designed to be deployed in sites that exhibit a high energy potential, which usually translates to a hostile environment with harsh conditions. For example, the construction and operation of a floating offshore wind platform and associated components such as mooring cables and turbine must be designed to withstand severe environmental loads that may cause failure due to extreme loading or even fatigue. Other devices, such as some wave energy converters (e.g., attenuators, point absorbers) are designed to actually resonate with the dominant wave frequency of the deployment site, making the use of coupled nonlinear approaches vital for developing successful devices. The performance of these smaller scale devices cannot be reliably assessed using linear, static/quasi-static and uncoupled tools as these either require tuning parameters on a case-by-case basis to perform adequately, or in some cases, are simply inappropriate for assessing the performance of the structure.

[1] Research Engineer at US Army Corps of Engineers ERDC, Vicksburg MS, US and EngD Researcher at the University of Edinburgh, Edinburgh, UK.

[2] Principal Engineer at HR Wallingford, Wallingford, UK.

[3] Research Hydraulic Engineer at US Army Corps of Engineers ERDC, Vicksburg MS, US (now Associate Professor at Louisiana State University, Baton Rouge LA, US).

[4] Professor of Ocean Technology at University of Exeter, Exeter, UK.
   Emails: A.Dimakopoulos@hrwallingford.com; cekees:@gmail.com; L.Johanning@exeter.ac.uk

* Corresponding author: delataillade.tristan@gmail.com

Nonlinear processes in Fluid–Structure Interaction (FSI) applications for floating bodies typically include, but are not limited to: viscous effects, vortex shedding, wave overtopping, wave-wave interactions, and mooring damping. Many of these physical processes are interdependent, i.e., their presence or magnitude affects the other effects, hence the need of a viable coupling approach when they are to be modelled numerically. Prototype testing and large scale physical models, are technically able to reproduce real-world conditions in an exact manner, but are generally considered prohibitively expensive for early development, due to their high cost of initial production and testing, as well as any subsequent iterations in their design. These methods are instead commonly used in later design stages, before deploying devices for commercial use. In early design stages, engineers have the option of using small scale physical testing or coupled numerical models. The former option can indeed offer useful insights in physical processes, but small scale experiments typically suffer from scaling effects that make it difficult for accurate quantitative predictions. High-fidelity coupled models therefore offer a strong alternative to simulate highly nonlinear FSI processes numerically. While promising, they still present several shortcomings (e.g., numerical stability issues, time-consuming simulations) that current and future research is anticipated to address. When compared to simpler models, high-fidelity models such as Computational Fluid Dynamics (CFD) typically require access to significant computational power such as High Performance Computing (HPC) clusters. While this can be prohibitive in certain cases, it is becoming lesser due to recent and continuing growth of available computational power at a reduced cost.

The objective of this chapter is to propose a viable coupling strategy employing a combination of two-phase flow CFD model with a fully dynamic multibody and mooring dynamics model, suitable for high-fidelity simulation of floating moored structures. A key difference between the models mentioned above and the one developed here is that the coupling between moorings and structure is strong, and that fluid velocities from the CFD solver are retrieved to calculate hydrodynamic pressures along the mooring line. The CFD and solid dynamics solver are based the Proteus® (https://proteustoolkit.org) and Chrono® (https://projectchrono.org/) open-source projects. The implementation herein is based on these particular tools, but it is also our ambition to provide a blueprint for interested researchers and engineers aiming to develop similar coupling strategies based on different tools. A decision has therefore been made to make the description of the coupling strategy as generic as possible. It is worth noting that this chapter follows on the work performed in de Lataillade (2019) where a more detailed account of the particulars of the strategy implementation is presented.

This chapter therefore describes the complete development process, implementation, and testing of a high-fidelity numerical framework for FSI of moored floating structures. The main numerical techniques and tools selected for this purpose are:

- CFD using Finite Element Method (FEM) for fluid flow, free surface tracking, and any auxiliary models solving Partial Differential Equations (PDEs) on the fluid mesh;

- Arbitrary Lagrangian–Eulerian (ALE) formulation for the motion of mesh-conforming solids within the fluid mesh, and mesh deformation through the equations of linear elastostatics;

- Discrete Element Method (DEM) coupled with capable Multibody Dynamics (MBD) solver for rigid structures, including collision detection features;

- Statics/quasi-statics model for estimating tensions of catenary mooring lines at equilibrium;

- FEM method for mooring dynamics using beam theory with gradient deficient Absolute Nodal Coordinate Formulation (ANCF) elements.

Following this introduction, the present chapter contains 5 additional main sections. To start with the governing equations of each uncoupled model used in this work are presented in Section 2. Then, an overview of the different fluid-structure coupling schemes is given in Section 3. This is followed in Section 4 by an overview of the coupling strategy employed here for the different coupled interfaces: fluid-structure, fluid-mooring, and mooring-structure. The coupled framework in then validated with several simulations in Section 5. Finally, discussion, conclusions and future work are laid out in Section 6.

2 Uncoupled numerical models

In typical FSI applications, the numerical domain Ω contains two main subdomains: the fluid domain Ω_f and the solid domain Ω_s. The fluid subdomain may contain multiple phases such as air Ω_a and water Ω_w-with $\Omega_f = \Omega_w \cup \Omega_a$-separated by a free surface $\Gamma_\sim$, while the solid domain may contain different types of solids such as rigid bodies and flexible structures. A schematic representation of the different domains described above is presented in Figure 1. The boundary of the numerical domain is denoted as $\partial\Omega$, with partial boundaries denoted as Γ such as the interface between solid and fluid $\Gamma_{f\cap s}$.

In the current work, we consider a two-phase fluid domain with air and water, and a solid domain for rigid bodies representing floating structures, while flexible beams are used to represent mooring lines. The proposed approach is implemented using the CFD toolkit Proteus® for simulating the physics in the fluid domain and the multiphysics simulation engine Chrono® for simulating MBD in the solid domain. The theoretical background and numerical implementation for each model are briefly presented below.

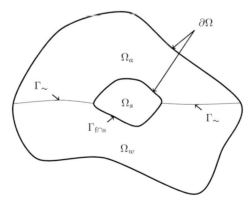

Figure 1: Illustration of arbitrarily shaped domain for typical FSI applications with two-phase flow and floating body.

2.1 Fluid dynamics

Fluid dynamics and all models using the fluid domain are simulated through Proteus®, a computational toolkit for solving PDEs that is highly specialised for fluid mechanics (`https://proteustoolkit.org`). Proteus® uses FEM to numerically solve transport equations within the fluid domain. A brief description of the toolkit, in relation to the methodology used to simulate the case studies in this work, is presented below.

2.1.1 Governing equations

The toolkit solves the Navier-Stokes equations for two-phase incompressible, immiscible flow. The momentum and continuity equations are as follows:

$$\begin{cases} \dfrac{\partial \mathbf{u}_f}{\partial t} + \rho \mathbf{u}_f \cdot \nabla \mathbf{u}_f + \nabla p - \nabla \cdot \bar{\bar{\tau}} = 0 \\[2mm] \nabla \cdot \mathbf{u} = 0 \end{cases} \tag{1}$$

where t is the time, $\nabla = (\frac{\partial}{\partial x}, \frac{\partial}{\partial y}, \frac{\partial}{\partial z})$, x, y and z are the spatial coordinates, $\mathbf{u}$ is the fluid velocity field in vector form, p is the fluid pressure, $\bar{\bar{\tau}}$ is the viscous shear stress tensor and ρ is the fluid density. The shear stress tensor may additionally include turbulent stresses, but these are not considered in the current work.

When the fluid domain contains two incompressible fluid phases (e.g., air and water), the fluid density ρ and dynamic viscosity ν take different (i.e., discrete) values according to the phase, which depends on the position considered in the fluid domain. The two phases are separated by an interface called the free surface. It is well-known that the free surface within the context of two-phase CFD models can be tracked implicitly by setting up transport equations to capture its evolution. This concept was first used by Hirt and Nichols (1981) with the Volume Of Fluid (VOF) method, according to which a transport equation for the water volume fraction was used to capture the evolution of the free surface. Another approach for implicit tracking is the level set method, which treats the distance of any point from the free surface as a field variable and uses relevant transport equations to track its evolution (Sussman et al., 1994).

In this work, the free surface is tracked by using a coupled level set/VOF method (Kees et al., 2011). This approach consists of the following steps:

1. The air volume fraction α_a is calculated according to the following transport equation written in conservative form:

$$\frac{\partial \alpha_a}{\partial t} + \nabla \alpha_a \mathbf{u}_f = 0 \tag{2}$$

2. The distance from the free surface ϕ_{sdf} is calculated according to the non-conservative form of the level set transport relation:

$$\frac{\partial \phi_{\text{sdf}}}{\partial t} + \mathbf{u} \nabla \phi_{\text{sdf}} = 0 \tag{3}$$

where ϕ_{sdf} is a signed distance function taking negative values in the water phase and positive values in the air phase. After being transported by Equation (3), ϕ_{sdf} is not a true signed distance function anymore and must be corrected in order to satisfy the eikonal relation:

$$\|\nabla \phi_{\text{sdf}}\| = 1 \tag{4}$$

which can be recovered by keeping the interface where $\phi_{\text{sdf}} = 0$ fixed and using a pseudo-steady state solution approach (Sussman et al., 1994).

3. The signed distance function is then corrected by the VOF solution from Equation (2) to ensure mass conservation, according to the procedure presented in Kees et al. (2011).

4. The corrected signed distance function ϕ_{sdf} is then used to recalculate the air volume fraction α_a with a smoothed Heaviside function. This step is performed to ensure that α_a remains sharp along the interface.

This approach combines two important elements of the VOF and the level set methods: mass conservation from the VOF equation and smooth and sharp representation of the interface using a signed distance function. The computational cost associated with this approach is nevertheless higher than using a single transport equation for VOF or level set method. Further progress has been made in Quezada de Luna et al. (2019) by developing a faster conservative level set scheme, but it was not implemented in the Proteus toolkit at the time of performing the simulations presented herein.

2.1.2   Boundary conditions and wave generation

Proteus® supports the use of Dirichlet and Neumann boundary conditions within the FEM framework, which are applied to the integral form of the Navier-Stokes equations and other PDE models that use the fluid mesh. The boundary conditions can be classified as Dirichlet, Advective, and Diffusive. The Dirichlet conditions, impose an explicit value of the solution at the boundary based on the weak imposition (Bazilevs and Hughes, 2007). The Advective conditions set the flux through a boundary face. The Diffusive conditions set the gradient flux (equivalent to a Neumann condition). For wall boundaries (e.g., tank walls, or surface of structures), typical free-slip or no-slip conditions are usually imposed. For open boundaries such as the top of a numerical tank, atmospheric conditions are used, imposing a Dirichlet pressure (usually a constant $p = 0$ for flat top) and Dirichlet VOF (usually the air value, $\alpha_a = 1$ in our case).

Waves are introduced in the domain at the generating boundaries and any reflection or transmission of the waves are absorbed using the relaxation zone method. Relaxation zones (generation or absorption zones) correspond to regions of the domain that introduce a source term in the fluid momentum equation:

$$\frac{\partial \mathbf{u}_f}{\partial t} + \rho \mathbf{u}_f \cdot \nabla \mathbf{u}_f + \nabla p - \nabla \bar{\bar{\tau}} = \alpha_b \alpha_0 \left(\mathbf{u}_f - \mathbf{u}_t \right) \tag{5}$$

where α_b is a blending function, α_0 is a constant, and $\mathbf{u}_t$ is the target velocity (fluid velocity from wave theory in generation zones, and zero velocity in absorption zones).

Specifically, the method implemented here is based on the work presented by Dimakopoulos et al. (2019), with $\alpha_0 = \frac{30}{T}$ given the wave period T, with generation zones of length $L_{\text{gen}} = 1\lambda$, where λ is the wavelength, absorption zones of length $L_{\text{abs}} = 2\lambda$, and using the blending function α_b introduced by Jacobsen et al. (2012):

$$\alpha_b = \frac{e^{d^{3.5}} - 1}{e - 1} \tag{6}$$

where $0 < d < 1$ is the scaled distance from the boundary to the end of the relaxation zone. An illustration of relaxation zones in a typical numerical wave tank is presented in Figure 2. This methodology is capable of generating and absorbing waves from various theories, including regular (linear and nonlinear), focused, as well as plane and directional random

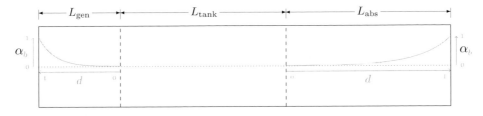

Figure 2: Relaxation zones and blending functions in numerical wave tank.

waves. In the current work, regular waves are generated using the Fenton Fourier theory for steady waves (Fenton, 1988), whilst random waves are generated using superimposition of linear wave components from a known spectrum.

2.1.3   Numerical solution

The Proteus® toolkit is based on the FEM, a residual-based solution method that produces an approximation to the true solution. The discretisation of all transport equations for the fluid domain—which includes free surface tracking—follow the weak formulation of the FEM, based on the introduction of test functions in the integral form of the transport equations.

A triangular (in 2D) or tetrahedral (in 3D) mesh was used to discretise the flow variables in space. P1 Lagrange elements with affine linear nodal basis and simplex Gaussian quadrature were used. Piecewise linear basis functions were used to interpolate the field variables within the element, and these were also used as test functions, following the Galerkin method. A first-order accurate implicit backward Euler scheme was employed for advancing the solution algorithm in time, as it is a simple numerically robust scheme. The timestep is restricted according to the flow velocities, following the CFL criterion. More details on the discretisation scheme can be found in de Lataillade (2019).

2.2   Solid dynamics

For solid dynamics, the open source multiphysics simulation engine Chrono® is used. Chrono® solves the equations of motion of floating structures and allows the definition of flexible mooring lines using beam theory and FEM. Additionally, mooring statics are solved with a catenary equation solver developed for the purpose of this work which can also be used for initial conditions of the mooring dynamics module. A brief description of these approaches is presented in the remainder of the section.

2.2.1   Rigid body dynamics

The equations of translational and rotational motion for a rigid body can be summarised as follows:

$$m\ddot{\mathbf{r}}_b = \Sigma \vec{f} \tag{7}$$

$$\bar{\bar{\mathbf{I}}}_t \ddot{\boldsymbol{\theta}}_b = \Sigma \vec{m} \tag{8}$$

where m is the mass, $\bar{\bar{\mathbf{I}}}_t$ is the 3×3 moment of inertia tensor of the rigid body, $\Sigma \vec{f}$ and $\Sigma \vec{m}$ are the sum of the forces and moments acting on the body, respectively, $\mathbf{r}_b$ and $\boldsymbol{\theta}_b$ are the position and rotational vector of the body, respectively, and $\ddot{\mathbf{r}}_b$ denotes acceleration of $\mathbf{r}_b$ (i.e., second-order time derivative). The position vector $\mathbf{r}_b$ corresponds to the absolute position of the centre of mass of the body (surge, sway, heave) in the global frame of reference, and the rotational vector $\boldsymbol{\theta}_b$ corresponds to the rotation of the body (roll, pitch, yaw) according to the global frame of reference. In case of moored floating structures, forces and moments are typically caused by hydrodynamic forces and the mooring system. For convenience, these equations are often expressed in terms of generalised coordinates or state vector $\mathbf{s} = \begin{pmatrix} \mathbf{r}_b & \boldsymbol{\theta}_b \end{pmatrix}^\mathsf{T}$ and generalised force vector $\mathbf{f} = \begin{pmatrix} \vec{f} & \vec{m} \end{pmatrix}^\mathsf{T}$ with a 6×6 mass matrix $\mathbf{M}$:

$$\mathbf{M}\ddot{\mathbf{s}} = \Sigma \mathbf{f} \tag{9}$$

In the current implementation, Chrono® is used to discretise Equation (9) in time with a backward Euler scheme. The time step of the discretisation is typically lower than the one used for the fluid dynamics model, due to the cheaper computational cost of an MBD problem when compared to a multi-phase flow FEM problem within the context of a typical FSI simulation.

2.2.2   Mooring statics

Mooring statics allow the geometrical and physical description of mooring lines subject to their own weight (or submerged weight) and at equilibrium. To solve for the layout of the mooring cable, the positions of the anchor and fairlead are known, as well as the position of the seabed (if any) upon which part of the line can rest. The following properties of the mooring line are also given: the unstretched line length L necessary to find the shape of the line, the submerged weight w_0 necessary to calculate tensions in the line, and the axial stiffness EA that defines the elasticity of the line (which will affect both the shape and tension in the line). Any catenary line can be analytically represented through the catenary equations:

$$x = a \sinh^{-1}\left(\frac{s}{a}\right) \tag{10}$$

$$y = a \cosh\left(\frac{x}{a}\right) \tag{11}$$

where s is the distance along the catenary, and a a variable defining the shape of the cate- nary. The parameter a is found through different means depending on the four possible layouts of the line: fully stretched, fully lifted, partly lifted, and with no horizontal span. The possible configurations are illustrated in Figure 3. When the line is fully stretched (straight between anchor and fairlead) or has no horizontal span (straight between fairlead and seabed), then the line is technically not a catenary and it is straightforward to find a geometrical description of the line and the tension at the fairlead. When the line is partly or fully lifted, the transcendental equations with respect to a (Equation (11)) are typically solved using bisection or Newton-Raphson methods to find the layout of the catenary. For a single catenary mooring line subject to its own weight, the horizontal tension at any point in the lifted line is $T_H = aw_0$. The vertical tension varies according to the position s within the catenary, and can be found with simple trigonometry as $T_V = T_H \tan(\theta_l)$ where θ_l is the

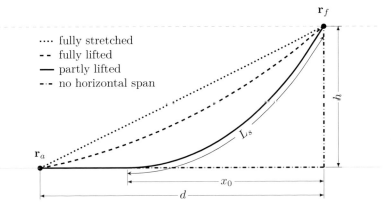

Figure 3: Possible configurations of a single catenary line where $\mathbf{r}_a$ and $\mathbf{r}_f$ are the positions of the anchor and the fairlead, respectively, d and h are the horizontal and vertical distance between the anchor and fairlead, respectively, and x_0 and L_s are the horizontal span and lifted line length for the partly lifted line, respectively.

angle formed by the catenary at the point considered. For a fully lifted line, the following transcendental equation is solved:

$$\sqrt{L_s^2 - h^2} = 2a \sinh\left(\frac{d}{2a}\right) \tag{12}$$

where L_s is the lifted line length, d the horizontal distance between fairlead and anchor, and h the vertical distance between fairlead and anchor. For a partly lifted line, the following transcendental equation is solved:

$$a \cosh\left(\frac{x_0}{a}\right) - 2a = h \tag{13}$$

where x_0 is the horizontal span of the line. Technically, x_0 is also an unknown so Equation (13) is iteratively solved with updated values of x_0 until a is found for the right line length. Furthermore, if elasticity and/or multi-segmented lines (i.e., varying properties w and EA) are used, iterations are necessary for both the fully and partly lifted case before converging to a solution. The algorithm developed for solving the procedure above are described in more detail in de Lataillade (2019).

2.2.3 Mooring dynamics

For dynamic simulations, mooring lines are assumed to be elastic cables with isotropic material and circular cross-section. Each line is discretised in 1D elements that are governed by flexible beam theory. As isotropic materials have symmetric stress and strain tensors, the constitutive equation linking stresses and deformation is the generalised Hooke's law for linear elastic material, expressed as:

$$\boldsymbol{\sigma} = C \cdot \boldsymbol{\epsilon} \tag{14}$$

where $\boldsymbol{\sigma}$ is the 6×1 material stress vector, $\boldsymbol{\epsilon}$ is the 6×1 strain tensor and C is the 6×6 stiffness matrix. The dynamic equilibrium equation for linear elasticity that links stresses, deformation and body forces is:

$$\nabla \cdot \bar{\bar{\boldsymbol{\sigma}}} + \mathbf{f} = \rho_s \ddot{\mathbf{h}} \tag{15}$$

where $\bar{\bar{\boldsymbol{\sigma}}}$ is the 3×3 stress tensor, ρ_s is the material density and $\mathbf{h}$ is the deformation vector such that, mooring lines are discretised. The deformation of the cross-section of the beams can therefore be assumed negligible (i.e., considered rigid in its own plane). With this assumption, shear stresses become negligible as the positive tensile stress along the longitudinal axis of the line becomes dominant. A local coordinate system $\boldsymbol{\xi} = (\xi, \eta, \zeta)$ where ξ is the longitudinal direction and η and ζ are the transverse directions, is defined. Assuming that Poisson's effect is negligible (i.e., $\nu = 0$), Equation (14) reduces to:

$$\sigma_\xi = E\epsilon_\xi \tag{16}$$

with the longitudinal strain being:

$$\epsilon_\xi = \frac{\partial h_\xi}{\partial \xi} + \zeta \kappa_\eta - \eta \kappa_\zeta \tag{17}$$

where $\kappa_\eta = \frac{\partial^2 h_\zeta}{\partial \xi^2}$ and $\kappa_\zeta = \frac{\partial^2 h_\eta}{\partial \xi^2}$ are the curvature of the beam. The axial force f_ξ and bending moments m_η and m_ζ are expressed as:

$$f_\xi = \int_A \sigma_\xi \mathrm{d}A; \quad m_\eta = \int_A \zeta\sigma_\xi \mathrm{d}A; \quad m_\zeta = -\int_A \eta\sigma_\xi \mathrm{d}A \tag{18}$$

Cables are often described in terms of axial stiffness EA_0 where A_0 is the undeformed cross-sectional area and bending stiffness (also known as flexural rigidity) EI, where I is the moment of inertia for flexion of the cable (which is the same in the η and ζ direction for isotropic materials). For a chain, it is common practice to set the bending stiffness to zero (i.e., $I := 0$) and to forbid compression stress (i.e., $\sigma_\xi := 0$ when $\epsilon_\xi < 0$) as the mooring line will deform/buckle before sustaining any compression stress.

In the implementation used here, the beams are defined by the gradient deficient ANCF method provided by Chrono®. According to the formulation proposed by Gerstmayr and Shabana (2006), each beam or cable element is described through the position vector and a single direction gradient for each of its two end nodes. By using the Green strain tensor, the implementation of the ANCF cable elements leads to the following expression for the axial strain:

$$\epsilon_\xi = \frac{1}{2}\left((\nabla_\xi \mathbf{x})^\mathsf{T} \nabla_\xi \mathbf{x} - 1\right) \tag{19}$$

where ∇_ξ is the gradient along the longitudinal axis and $\mathbf{x}$ the spatial coordinates along the cable. This formulation does not take into account torsional effects as only a single direction gradient is used. Note that torsional effects are not expected to be of significant importance for the simulations presented in this work. Other formulations can be used to include them if necessary, such as Euler-Bernoulli beams. The numerical solution is calculated through the FEM module of Chrono® using a Backward Euler scheme to solve the system of equations emerging from the mooring line discretisation into segments (Recuero and Negrut, 2016).

3 An overview of fluid-structure coupling schemes

Coupling schemes used for modelling fluid-structure interaction include two main types of schemes that handle the exchange of information between the fluid and the solid phase: i) monolithic schemes and ii) partitioned schemes, which can be implicit or explicit. These coupling schemes can be combined with any numerical method for representing moving solids in a fluid such as ALE mesh, immersed boundary method or cut-cell method. The main traits of these coupling schemes, as well as their subcategories, are summarised below.

3.1 Monolithic schemes

Monolithic schemes are the optimal choice for numerical stability and accuracy. In these schemes, the calculation of the fluid pressure and solid acceleration are intertwined, leading to an interdependent solution that is found simultaneously for the fluid and the solid. The idea can be described at a fundamental level by the following equation:

$$\nabla p^{n+1} = \rho_f \cdot \ddot{\mathbf{x}}_{\Gamma_{f\cap s}}^{n+1} \quad \text{on } \Gamma_{f\cap s} \tag{20}$$

where $\Gamma_{f\cap s}^{n+1}$ is the position of the fluid–structure interface at timestep $n+1$, ∇p is the fluid pressure gradient representing the local fluid acceleration, ρ_f is the fluid density, and $\ddot{\mathbf{x}}_{\Gamma_{f\cap s}}$ is the acceleration of the boundary $\Gamma_{f\cap s}$. The total acceleration acting on the boundary includes gravity and hydrodynamic forces and any other stresses imposed by additional considerations (e.g., flexible structures) or external restrictions (e.g., moorings and springs). The solid boundary acceleration $\ddot{\mathbf{x}}_{\Gamma_{f\cap s}}$ is a function of the hydrodynamic pressure itself (among other variables), hence Equation (20) results in an implicit system. This equation is eventually fed into the pressure solver which is arguably the most critical part of the numerical solution for incompressible flows, as it is the condition that ensures mass conservation. Finding the numerical solution for the fluid pressure is also considered the most time-consuming part for incompressible solvers.

In the special case of a rigid and free floating body, $\ddot{\mathbf{x}}_{\Gamma_{f \cap s}}$ is solely dependent on the weight and the fluid pressure. It is therefore relatively straightforward to develop an algorithm using Equation (20) as an implicit system of equations for the fluid pressure which can be combined with the overall pressure solution for the fluid and thus result in a truly monolithic scheme for, e.g., floating structures. This approach was followed in Chen et al. (2016) in the context of a Particle-In-Cell (PIC) model and the cut-cell method for modelling floating structures. This work has shown the merits of using monolithic schemes as there were no numerical stability issues reported. The most significant drawback of monolithic schemes is nevertheless its lack of modularity. By including additional forces such as mooring, springs or solid stresses, the global system of equations for the pressure at t^{n+1} must be redesigned to take into account the additional forces. For example, including mooring with linear springs and damping would require to express the forces in Equation (20) at $t = t^{n+1}$, without using data from t^n, as doing otherwise will compromise the monolithic character of the scheme. This task can be proven particularly challenging for implementing a simple spring mooring system. When coupling the CFD model with external and highly sophisticated software libraries for, e.g., introducing cable or chain moorings (or even other types of solid deformable structures), maintaining a true monolithic scheme may prove an impossible task, as an analytical expression to express the forces at t^{n+1} is not available. In this case, some level of partitioning must be introduced, which compromises the monolithic character of the scheme.

3.2   Partitioned schemes

Partitioned schemes solve the fluid and solid kinematics separately, and can be divided in three main categories: explicit, implicit, and semi-implicit schemes. All of them allow for greater modularity as each solver can be treated as a black-box, using available data from previous time-step or iteration to proceed. While this enables the use of highly-specialised software for each of the coupled problems, it is important to note that partitioned schemes can be subject to numerical instabilities or time consuming calculations, depending on the chosen scheme, as discussed further below.

3.2.1   Explicit schemes

In explicit schemes, calculations for the solid and fluid kinematics are separated and only performed once per time step. A representative sequence of calculations can be expressed as:

$$p^{n+1} = \mathcal{F}(\mathbf{u}_f^n, p^n, \Omega_s^{P,n+1}) \tag{21}$$

$$\delta\Omega_s^{n+1} = \mathcal{S}(p^{n+1}) \tag{22}$$

$$\Omega_s^{n+1} = \Omega_s^n + \delta\Omega_s^{n+1} \tag{23}$$

where $\mathbf{u}_f$ is the fluid velocity, $\mathcal{F}$ and $\mathcal{S}$ are the fluid and solid discrete time advancement operators (explicit or implicit), respectively, $\delta\Omega_s$ is the motion increment of the solid domain, and $\Omega_s^{P,n+1}$ is a prediction of the position of the solid domain at t^{n+1}. Equation (21) can be varied by further breaking down steps and by employing different time advancement schemes for $\mathcal{F}$ and $\mathcal{F}$, or using subcycling (e.g., use smaller times steps for the solid within a global time step). If $\Omega_s^{P,n+1} = \Omega_s^n$, then this is known as the Conventional Serial Staggered (CSS) scheme. It is also possible to calculate p^{n+1} from a predicted solid domain position $\Omega_s^{n+1,P} = \Omega_s^n + \delta\Omega_s^{n+1,P}$ with:

$$\delta\Omega_s^{n+1,P} = \alpha_0 \dot{\Omega}_s^n \Delta t + \alpha_1 \left(\dot{\Omega}_s^n - \dot{\Omega}_s^{n-1} \right) \Delta t \tag{24}$$

where $\dot{\Omega}_s$ is the velocity of the solid domain, and α_0 and α_1 are known coefficients controlling the order of time-accuracy of the prediction. Using predictions from Equation (24) has been introduced as the Generalised Serial Staggered (GSS) in Piperno et al. (1995). Second-order time-accuracy of the coupling scheme can be achieved with the Improved Serial Staggered (ISS) scheme (Lesoinne and Farhat, 1998) using the midpoint rule with prediction:

$$\delta\Omega_s^{n+1,P} = \dot{\Omega}_s^n \frac{\Delta t}{2} \tag{25}$$

Note that using Equation (25) actually leads an asynchronous method where the fluid solution is technically found for the predicted position of the solid at $t^{n+\frac{1}{2}}$ and used to advance the solid solution from t^n to t^{n+1}.

Explicit partitioned schemes are the most flexible approach, in terms of implementing multiple physical processes relevant to the solid body motion/response. The calculation step that involves the update of the position of the body is not intertwined with the hydrodynamic equation, as in the monolithic schemes, and this allows for the introduction of external forces or solid stresses without any particular complications. It is also evident that this scheme is at the lowest possible threshold in terms of computational resource requirements. This nevertheless comes with the most severe stability issues among any other alternatives, due to the lag between the solutions of fluid and solid kinematics, which will be discussed further below.

3.2.2 Implicit schemes

Implicit partitioned schemes were developed in order to mitigate some of the issues experienced in explicit schemes, whilst maintaining the flexibility of adding multiple physical processes relevant to the solid body response and motion. The idea is to iterate through Equation (21) until convergence is achieved, up to a user defined tolerance. Common approaches to converge to a solution are Block Gauss-Seidel (BGS), block Jacobi, and block Newton. Using fixed-point formulation, the BGS approach can be compactly defined as minimising the following residual:

$$\mathcal{R}^{k+1} = \mathcal{S} \circ \mathcal{F}(\mathbf{u}_f^n, p^n, \Omega_s^k) - \delta\Omega_s^k = \delta\Omega_s^{k+1} - \delta\Omega_s^k \tag{26}$$

where k is the iteration number, $\mathcal{S}$ represents the solid solver, and $\mathcal{F}$ represents the fluid solver. Because multiple iterations are usually required for convergence of the solution, these schemes are likely to become prohibitively expensive in the context of modelling complex FSI applications. The bottleneck is usually the iteration for the fluid step, which can be very demanding computationally when using highly refined meshes. Note that a single iteration of Equation (26) is equivalent to a fully explicit scheme.

These schemes are less prone to numerical stability issues than explicit schemes as relaxation factors can be used when iterating between the fluid and solid solver (e.g., see Devolder et al. (2015)), but they remain less robust than monolithic schemes.

3.2.3 Semi-implicit schemes

It is worth mentioning semi-implicit schemes, which are more computationally efficient than implicit schemes, and can be made more stable than explicit schemes. The main constraint of these schemes is that they require segregation of the fluid velocity and pressure equations, which can be achieved using a projection scheme such as the Chorin-Temam scheme (Chorin, 1968; Temam, 1969a,b). This allows to first solve once per time step for the fluid velocity and then to iterate between the fluid pressure and solid solutions. It is essentially equivalent to replacing the full fluid solver $\mathcal{F}$ of Equation (26) by a pressure equation. Although these schemes are faster than implicit schemes, they still require significant additional computational cost when compared to explicit schemes due to iterations with the pressure and solid solver, especially for incompressible flows, where the solution of a Poisson equation for the pressure is a bottleneck in terms of computational time. Furthermore, updating the pressure equation but not the velocity field may pose flow continuity issues if large deformations are involved, and may require additional treatment to ensure mass conservation.

3.3   Coupling instabilities

As mentioned in the previous section, while monolithic schemes are inherently stable, partitioned schemes may experience stability issues. Within the framework of moving/floating structures, these stability issues are caused by the added mass effect, which is briefly described below. The added mass of a solid submerged in a fluid—resulting from the moving solid accelerating the fluid in its vicinity—depends on the density of the surrounding fluid and on the acceleration of the solid within this fluid. Introducing the added mass matrix, Equation (9) for a single rigid body can be written as:

$$(\mathbf{M} + \mathbf{A})\ddot{\mathbf{s}} = \widehat{\mathbf{f}}_f + \mathbf{f}_e \tag{27}$$

where $\mathbf{M}$ and $\mathbf{A}$ are the mass and added mass matrix, respectively, $\mathbf{s}$ is the generalised coordinate array, $\mathbf{f}_e$ is the generalised external force array (e.g., gravity), $\mathbf{f}_f$ is the generalised fluid force array, and $\widehat{\mathbf{f}}_f$ is the generalised fluid force array exempt from added mass contributions. Because CFD models directly calculate $\mathbf{f}_f$, the added mass contribution is inherently taken into account. This means $\mathbf{A}$ is unknown, leading to a time discretised formulation for explicit partitioned schemes of the form:

$$\mathbf{M}\ddot{\mathbf{s}}^{n+1} = \mathbf{f}_f^n + \mathbf{f}_e^n = \widehat{\mathbf{f}}_f^n - \mathbf{A}\ddot{\mathbf{s}}^n + \mathbf{f}_e^n \tag{28}$$

where the added mass matrix $\mathbf{A}$ cannot be retrieved separately, leading to an added mass force term calculated in t^n. When using an explicit partitioned scheme, the fluid forces are integrated at a specific position of the solid domain in time—Ω_s^n—but when the solid evolves from t^n to t^{n+1}, the effect of the change in acceleration of the solid ($\ddot{\mathbf{s}}^{n+1} \neq \ddot{\mathbf{s}}^n$) within the fluid is ignored. Ignoring the effect of the added mass while stepping for the solid produces a force term acting opposite to the previous acceleration of the solid. This effect can lead to a diverging solution, due to strong oscillations in the acceleration of the solid. Such instability can clearly be visualised when solving Equation (28) for a few time steps while setting $\widetilde{\mathbf{f}}_f$ and $\mathbf{f}_e$ as constants over time and with $\mathbf{A} > \mathbf{M}$. This instability also applies to implicit or semi-implicit schemes (when no relaxation is used), by replacing the time dependence n of $\ddot{\mathbf{s}}$ in Equation (28) by the iteration number k. Note that if this effect is significant, it renders the coupling unconditionally unstable, i.e., increasing the spatial or temporal refinement does not recover stability, as reducing numerical diffusion errors actually destabilises the system even more.

Stabilising simulations prone to the added mass effect has been an active field of research for the past decades. For implicit schemes, relaxation factors can be used to penalise oscillations as in Söding (2001); Yvin et al. (2013, 2014); Devolder et al. (2015). For explicit schemes, Robin–Neumann boundary conditions can be used (Fernández et al., 2013) or, alternatively, Robin and Dirichlet–Nitsche boundary conditions with a penalty term on the pressure from the fluid solver (Burman and Fernández, 2007, 2009), or prediction-correction steps (Banks et al., 2017a,b).

In the work presented here, Dirichlet–Neumann coupling is used with no prediction-correction steps. Numerical instabilities related to the added mass effect are countered by introducing an accurate estimation of the added mass (following the method presented by Söding (2001)) and using it to take into account the change of acceleration of the solid within the fluid, leading to:

$$\mathbf{M}\ddot{\mathbf{s}}^{n+1} = \mathbf{f}_f^n - \widetilde{\mathbf{A}}\left(\ddot{\mathbf{s}}^{n+1} - \ddot{\mathbf{s}}^n\right) + \mathbf{f}_e^n \tag{29}$$

$$\Rightarrow \left(\mathbf{M} + \widetilde{\mathbf{A}}\right)\ddot{\mathbf{s}}^{n+1} = \widehat{\mathbf{f}}_f^n + \left(\widetilde{\mathbf{A}} - \mathbf{A}\right)\ddot{\mathbf{s}}^n + \mathbf{f}_e^n \tag{30}$$

where $\widetilde{A}$ is the estimation of the added mass. It is clear that using a good estimation of the added mass—i.e., $\widehat{\mathbf{A}} \approx \mathbf{A}$–leads to having Equation (29) equivalent to the time discretised version of Equation (42), which is unconditionally stable. More details on this technique can be found in de Lataillade (2019).

4 Coupling strategy

This section proposes a coupling strategy that combines the fluid dynamics and solid dynamics solvers implemented in the FSI framework developed for this work. The solution sequence for each model is presented in Figure 4, along with the main communication lines between models and communicated variables. The sequence is summarized in the list below:

1. **Mesh motion model:** The fluid mesh is deformed through linear elastostatics, driven by the displacement of the fluid-structure boundaries $\Gamma_{f\cap s}$ which is retrieved from the multibody dynamics solver from the previous time-step. The model returns the mesh velocity $\mathbf{u}_m$ for the ALE formulation to correct advection fluxes in subsequent models that use the deformed fluid mesh.

2. **Navier-Stokes model:** The model calculates fluid velocities and pressures by solving the Navier-Stokes equations, on the updated mesh topology and using corrected fluxes.

3. **Free surface model:** For two-phase flow, the new position of the free surface is obtained, based on the updated mesh topology, mesh velocity, and fluid velocities which were calculated by the previous models within the same time-step. Fluid densities and viscosities are also updated according to the new position of the free surface.

4. **Added mass model:** When instabilities due to added mass effect are present, the added mass is estimated in order to later stabilise the partitioned coupling scheme. The estimation of the added mass increment is based on the method proposed in Söding (2001) and details of the implementation with the context of a CSS coupling scheme are included in de Lataillade (2019).

5. **Multibody dynamics model:** The new position of bodies are calculated with the hydrodynamic forces and, if included, the added mass estimation. Mooring dynamics are also calculated using a strong coupling with rigid bodies linked to the cables, such as fairleads and anchors. Hydrodynamic forces acting on the cable are taken into account using the fluid velocity solution from the Navier-Stokes model, and collision detection is enabled along the cable.

The coupling interfaces are presented in more details as follows: fluid-structure coupling in Section 4.1, fluid-mooring coupling in Section 4.2, and mooring-structure coupling in Section 4.3.

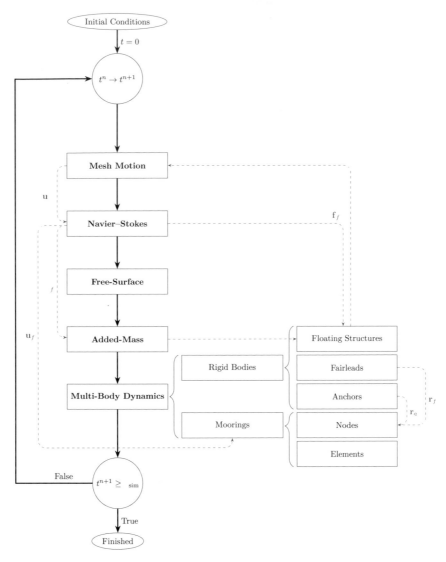

Figure 4: Workflow of numerical models with current coupling approach (adapted from de Lataillade (2019)).

4.1 Fluid-structure coupling

4.1.1 Fluid-structure interface

When coupling a fluid problem with a structure problem, new conditions are applied to both problems in order to ensure a well-defined fluid-structure interface $\Gamma_{\text{f}\cap\text{s}}$: the continuity of velocity and the continuity of stresses. The continuity of velocity is expressed as:

$$\mathbf{u}_f = \mathbf{u}_s \qquad \text{on } \Gamma_{\text{f}\cap\text{s}} \tag{31}$$

where $\mathbf{u}_f$ and $\mathbf{u}_s$ are the velocity of the fluid and the velocity of the solid at the interface, respectively. The continuity of stresses is expressed as:

$$\bar{\bar{\boldsymbol{\sigma}}}_f \cdot \mathbf{n} = \bar{\bar{\boldsymbol{\sigma}}}_s \cdot \mathbf{n} \qquad \text{on } \Gamma_{\text{f}\cap\text{s}} \tag{32}$$

where $\bar{\bar{\boldsymbol{\sigma}}}_f$ and $\bar{\bar{\boldsymbol{\sigma}}}_s$ are the fluid and solid stresses, respectively, and $\mathbf{n}$ is the normal vector to $\Gamma_{f \cap s}$. For the work presented here, and as is commonly the case for FSI problems, the fluid-structure interface is coupled where the Dirichlet-Neumann method where the continuity of velocity, Equation (31), is imposed as a Dirichlet condition on the fluid velocity, while the continuity of stresses, Equation (32), is imposed as a Neumann condition on the solid. Hydrodynamic forces $\vec{f}_f$ and moments $\vec{m}_f$ acting on the structure are integrated along the fluid-structure interface $\Gamma_{f \cap s}$ as:

$$\vec{f}_f = \int_{\Gamma_{f \cap s}} \bar{\bar{\boldsymbol{\sigma}}}_f \cdot \mathbf{n} \mathrm{d}\Gamma \tag{33}$$

$$\vec{m}_f = \int_{\Gamma_{f \cap s}} (\mathbf{x} - \mathbf{r}) \times (\bar{\bar{\boldsymbol{\sigma}}}_f \cdot \mathbf{n}) \mathrm{d}\Gamma \tag{34}$$

where $\mathbf{x}$ is a point on $\Gamma_{f \cap s}$ and $\mathbf{n}$ is the normal vector to $\Gamma_{f \cap s}$.

While a variety of coupling schemes are possible (see Section 3), the partitioned, fully explicit scheme known as the CSS scheme is used for the work presented here. This scheme allows for great modularity, making it possible to use two distinct and highly specialised frameworks: Proteus® for describing the fluid dynamics and Chrono® for solving the structural problem.

4.1.2 Moving boundaries and mesh motion

When mesh-conforming structures are moving within the fluid mesh, the latter must be deformed in order to accommodate this motion while retaining good overall quality and prevent mesh entanglement. This mesh deformation must also be taken into account by all unsteady PDE models that are spatially discretised with the fluid mesh. Internal and external mesh-conforming boundary motion is achieved here by means of the ALE method (Hirt et al., 1974). According to this method, three different domains-spatial, material and reference—are mapped to each other. These domains correspond to the Eulerian (spatial), Langragian (material) and ALE (reference) frames of reference (see Figure 5). The practical differences with respect to, e.g., temporal evolution of processes are summarised in the following equation, showing the rate of change (time derivative) of a flow variable f:

$$\frac{df}{dt} = \begin{cases} \frac{\partial f}{\partial t} & \text{Langragian frame} \\ \frac{\partial f}{\partial t} + \mathbf{u}_f \cdot \nabla f & \text{Eulerian frame} \\ \frac{\partial f}{\partial t} + (\mathbf{u}_f - \mathbf{u}_m) \cdot \nabla f & \text{ALE frame} \end{cases} \tag{35}$$

where $\mathbf{u}_m$ is the mesh velocity, i.e., the rate of mesh node displacement. The mesh velocity is then introduced in the Navier-Stokes equations and other transport models so that advective fluxes take $\mathbf{u}_m$ into account.

The deformation of mesh elements itself is handled here by the method of linear elasto-statics (Dwight and Dwight, 2009):

$$\nabla \cdot \bar{\bar{\boldsymbol{\sigma}}}_m + \mathbf{f} = 0 \tag{36}$$

where $\mathbf{f}$ is the body force, and $\bar{\bar{\boldsymbol{\sigma}}}_m$ is the stress tensor that can be expressed in terms of Lamé parameters as $\bar{\bar{\boldsymbol{\sigma}}}_m = \lambda \operatorname{tr}\left(\bar{\bar{\boldsymbol{\epsilon}}}\right) \bar{\bar{\mathbf{I}}} + 2\mu\bar{\bar{\boldsymbol{\epsilon}}}$ where $\bar{\bar{\mathbf{I}}}$ is the identity tensor and $\bar{\bar{\boldsymbol{\epsilon}}}$ is the strain tensor expressed as:

$$\bar{\bar{\boldsymbol{\epsilon}}} = \frac{1}{2}\left(\nabla \mathbf{h} + \nabla \mathbf{h}^{\mathsf{T}}\right) \tag{37}$$

where $\mathbf{h}$ is the displacement vector, and μ and λ are the Lamé parameters:

$$\lambda = \frac{\nu E}{(1 + \nu)(1 - 2\nu)}; \qquad \mu = \frac{E}{2(1 + \nu)} \tag{38}$$

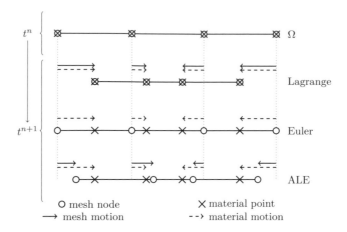

Figure 5: Schematic representation of Lagrangian, Eulerian, and ALE domains on a 1D mesh.

According to this method, the mesh is allowed to deform as a (fictitious) linearly elastic material. Variable stiffness can be used within the mesh according to defined metrics. In the implementation presented here, the Young's modulus is scaled according to the mesh element volume:

$$E = \frac{E_0}{\det J_{\mathcal{E}}} \tag{39}$$

where E_0 is a user-defined constant, and $J_{\mathcal{E}}$ is the Jacobian of the mesh element $\mathcal{E}$ whose determinant is proportional to the volume of the element. Therefore, in highly refined regions, the elements are stiffer whilst in coarser regions the elements are more flexible. This results in a tendency of the method to maintain the size and shape of more refined elements, which are typically more critical for the quality of the solution, whilst allowing coarser elements to absorb most of the mesh deformation.

While this method maintains acceptable mesh quality for typical engineering FSI that are reasonably short, it was found in de Lataillade (2019) that for long simulations that, e.g., may correspond to a full duration of an extreme storm, mesh quality can deteriorate over time due to multiple nonlinear wave cycles that drive the motion of the floating body. The use of a mesh quality monitor process will very likely address these issues. The application of a mesh monitor function model that deforms the mesh according to target element volumes was explored in (de Lataillade, 2019), based on the techniques proposed by Grajewski et al. (2009). In this work, only the linear elastostatics method is used, as the mesh quality was found to remain acceptable for the whole duration of the simulations presented herein.

4.2   Fluid-mooring coupling

Due to their slender geometry, mooring cables cannot realistically be discretised as a mesh-conforming structure within the fluid mesh. Usually, the cross-sectional scale of cables is indeed much smaller than the typical flow scales and other structures such as floating platforms. The Dirichlet-Neumann interface coupling described in Section 4.1 cannot be applied here in a practical manner. The fluid–mooring coupling is therefore one-way: the forces from the fluid are applied to the cable, but the effect of the cable on the fluid is ignored. While the fluid forces can have significant effects on the dynamics of the cable, the cable forces on the fluid are justifiably negligible due to the scale of cables relative to the structure, making it unlikely for the effect of the cable on the fluid to have any significant impact on the response of the structure itself. The total hydrodynamic force $\mathbf{f}_f$ per unit length acting along a mooring cable can be decomposed as follows:

$$\mathbf{f}_f = \mathbf{f}_d + \mathbf{f}_m + \mathbf{f}_b \tag{40}$$

With $\mathbf{f}_d$ the drag force, $\mathbf{f}_m$ the inertia force, and $\mathbf{f}_b$ the buoyancy force. Due to the geometry of the cable, it is possible to use the semi-empirical Morison's equations for loads on slender cylindrical structures. The drag force is therefore expressed as:

$$\mathbf{f}_d = \frac{1}{2}\rho_f \left(d_n C_{d,n} \left\| \mathbf{u}_{r,n} \right\| \mathbf{u}_{r,n} + d_t C_{d,t} \left\| \mathbf{u}_{r,t} \right\| \mathbf{u}_{r,t} \right) \tag{41}$$

where subscripts $_n$ and $_t$ denote for normal and tangential components, respectively, $\mathbf{u}_r$ is the velocity of the fluid relative to the cable, C_d is the drag coefficient, and d is the drag diameter (with usually $d_t = \pi^{-1} d_n$). The inertia force is calculated as:

$$\mathbf{f}_m = \rho_f A_0 \left(C_{m,n} \dot{\mathbf{u}}_{r,n} + C_{m,t} \dot{\mathbf{u}}_{r,t} + \dot{\mathbf{u}} \right) \tag{42}$$

where C_m is the added mass coefficient and $\dot{\mathbf{u}}$ is the absolute velocity of the fluid. Finally, the buoyancy force is calculated as:

$$\mathbf{f}_b = \frac{\rho_c - \rho_f}{\rho_c} \mathbf{g} A_0 \tag{43}$$

The drag and inertia forces are calculated using the actual fluid velocity retrieved from the Navier-Stokes model at each node of the cable. These forces are then applied along the cable through triangular loads as shown in Figure 6. In order to retrieve the solution for the fluid velocity at arbitrary locations within the fluid mesh (i.e., at the cable node position in our case), a particle localisation algorithm is used. The technique described by Haselbacher (2007) has been implemented in the current work, allowing for a computationally efficient retrieval of the solution at arbitrary locations. From an initial position (fluid mesh element and coordinates within this element), the algorithm finds the element containing the target particle by means of boundary intersection between this initial position and the target location.

Flow past mooring lines may additionally cause Vortex-Induced Vibrations (VIV) that are caused by vortex shedding processes and associated with oscillations in drag and lift forces, typically occurring at a much higher frequency than the waves. These effects are not included in the scheme. Estimating VIV effects may be needed to provide reliable predictions of mooring line failure due to fatigue, a case which is not considered in the current work.

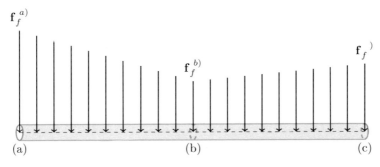

Figure 6: Schematic representation of triangular load applied on cable by interpolation from hydrodynamic forces calculated at nodes (a), (b), (c).

Inclusion of VIV effects may involve adding empirical or semi-empirical techniques to further extend the hydrodynamic forces calculation, which is now performed using the Morisson's equation. Resolving the flow around the mooring line directly is also possible but potentially not practical for engineering applications due to the mesh size required.

4.3   Mooring-structure coupling

The structure and its mooring system are part of the same global system of equations through a strong coupling. The fairlead, which is essentially a node at one end of the cable, has its position relative to the centre of mass of the structure enforced with a spherical mate constraint, allowing for rotation around the fairlead, but preventing any translation relative to the centre of mass. The anchor is simulated as a rigid body that is fixed in space at an absolute position, and the node at the other end of the cable is attached to it through a spherical mate constraint, preventing any translation of this node relative to the anchor position. The centre of mass of the structure that is linked to the fairlead through a spherical mate constraint is essentially added as an additional Degree of Freedom (DOF) to the FEM problem solved by Chrono® hence there is a strong coupling (rest is the same).

Contact of cables with the seabed and other structures or obstacles is enabled with Discrete Element Method (DEM) and achieved with Bullet®, a dependency of Chrono® specialising in collision detection. A cloud of collision-enabled contact spheres is created along mooring cables to allow for collision detection. Each node of the discretised cable has an associated contact sphere with a collision diameter which can be the same or different to the actual cable diameter. The seabed is represented as a contact plane when flat, and is otherwise represented with a triangular surface mesh in a similar manner to contact surfaces of other obstacles such as floating structures. Choosing values for an outward envelope (search range for potential collision) and an inward margin (range in which fast detection algorithms can be used), all collision-enabled entities can detect contact with other entities. While this technique allows for collision detection of mooring lines against structures, it is important to note that it cannot be reliably used for cable-to-cable collision detection, as only nodes of cables are collision-enabled while elements can still travel through each other undetected. Because no cable-to-cable collision is expected in the simulations presented here, a cloud of collision-enabled spheres along cables is sufficient for describing contact with the seabed and other structures. The mooring–structure coupling approach and the collision model used here is represented schematically in Figure 7.

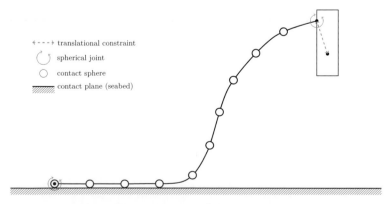

Figure 7: Schematic representation of constraints and contact entities between mooring, structures, and seabed

5   Case studies

The modelling approach discussed in the previous sections is applied to FSI case studies involving waves and moored floating bodies. Numerical simulations of floating bodies are validated against experiments in Section 5.1 with the following cases: free decay of a heaving sphere in Section 5.1.1, wave-induced rolling response of a rolling caisson in Section 5.1.2, and heave and roll response of a floating caisson to an extreme event (focused wave) in Section 5.1.3. The mooring model is validated for catenaries against experimental data in Section 5.2 for: static tensions for different positions of the fairlead relative to the anchor in Section 5.2.1, and dynamic mooring line damping under forced sinusoidal motion of the fairlead in Section 5.2.2. Finally, combining all the different validated aspects, a full-fledged 3D numerical simulation of a floating platform for offshore wind turbines moored with 3 catenary lines is compared to experimental data in Section 5.3. In this last section, simulation of the uncoupled mooring model is first presented in Section 5.3.2, followed by the simulation with all the models presented in this work coupled together in Section 5.3.3.

All parallel simulations presented below ran on an HPC cluster composed of AMD Interlagos Opteron with 2.3GHz core speed, 32 cores per node, and 64GB memory per node. The number of cores used for a simulation is case-dependent, and is therefore mentioned in each relevant subsection.

5.1   Validation of FSI for floating bodies

5.1.1   Free oscillation—heaving cylinder

The case presented here describes the free oscillation of a floating cylinder. The cylinder motion is restricted in all directions with the exception of vertical motion (heave motion). The domain setup is based on the experiment conducted by Ito (1977), where a floating cylinder with length $L = 1.83$m and diameter $D = 15.24$cm is placed horizontally across the width of the tank at the mean water level. The small clearance between the tank walls and the edges of the cylinder (1.27cm) makes it possible to simulate the 3D experimental setup in a 2D numerical domain. The density of the cylinder is half the density of water, making the cylinder draught at equilibrium equal to its radius. The water depth is $h_{\mathrm{mwl}} = 1.22$m. The numerical tank used here has a total length of $L_{\mathrm{tank}} = 20$m and height of $H_{\mathrm{tank}} = 2.44$m, with absorption zones of length $L_{\mathrm{abs}} = 9$m, enough to dissipate outgoing waves generated by the cylinder motion. The cylinder is placed in the middle of the tank with its centre at a distance of 10m from each boundary (see Figure 8).

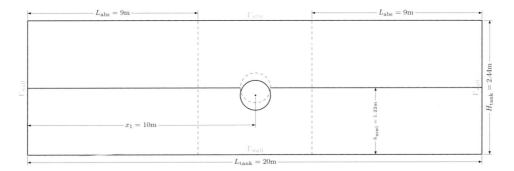

Figure 8: Geometry of numerical domain for the simulation of free oscillation in heave of a floating cylinder. Solid line: initial position, dashed line: equilibrium position.

The numerical simulation is initialised with still water and the cylinder being lowered 2.54cm from its equilibrium position. The boundary conditions are free-slip at the numerical tank walls Γ_{wall}, zero pressure head at the atmosphere (top boundary of the tank Γ_{atm}), and no-slip condition on the cylinder $\Gamma_{\text{f}\cap\text{s}}$. The domain is discretised with triangular elements of minimum characteristic element size h_e^0 constant up to a distance of ± 2.54cm from the still water level and at the boundaries of the cylinder, with gradual coarsening of the mesh applied beyond this area (10% increase in neighbouring characteristic element size). To assess convergence of the coupled fluid and rigid body models, the simulation is repeated with different time-step, CFL numbers and mesh sizes. For all cases, the MBD solver uses a fixed time step of 1×10^{-5}s, which is much smaller than the fluid time-step.

The results are presented in Figure 9a for different values of h_e^0 and Figure 9c for different values of Δt. As shown in these figures, the numerical solution converges, and agreement

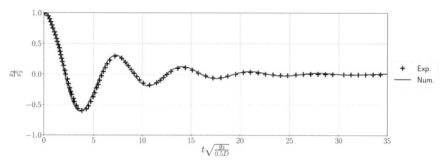

(a) Heave decay of floating cylinder for $\Delta t = 1e^{-3}$, $h_e^0 = 0.02D$. Experimental results from Ito (1977).

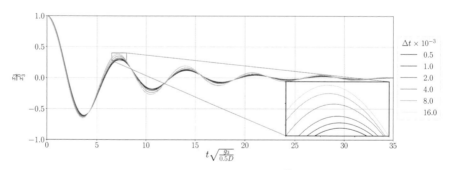

(b) Temporal sensitivity with $h_e^0 = 0.02D$

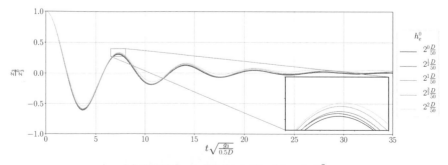

(c) Spatial sensitivity with $\Delta t = 1e^{-3}$s

Figure 9: Time-series and sensitivity analysis of the free oscillation of a heaving cylinder.

with experimental data is already excellent for $\Delta t = 1e^{-3}$ and $h_e^0 = 0.02D$. The case shown in Figure 9a had a mesh composed of $885,433$ elements and took 9.4h running in parallel on 96 cores to produce 4s of data. The Root Mean Square (RMS) variation from the most refined simulations features a temporal order of convergence of 1, and a spatial order of convergence of 2, which is in accordance with the numerical schemes employed here.

5.1.2 Wave-induced motion-rolling Caisson

This case is based on the experiment conducted by Jung et al. (2006), studying the rolling motion of a floating rectangular caisson. In the experiment, the floating caisson has a length of 0.3m, height of 0.1m and width of 0.9m. The caisson is mounted on the tank walls through a pair of bars and hinges sitting at the mean water level, aligned so that the axis of the joint goes through the centre of mass of the caisson. The roll moment of inertia has been measured experimentally as $I = 0.236\text{kg m}^2$. Similar to the previous case, the experimental set-up can be simulated using a 2D numerical domain due to the small clearance from the flume walls.

Free oscillation and decay tests were first performed with the fluid initially at rest, and introducing a roll displacement of 15° from the equilibrium position of the caisson, with no initial velocity. For the numerical set-up tank length, $L_{\text{tank}} = 5\text{m}$ with absorption zones of length, $L_{\text{abs}} = 2\text{m}$ on either end of the tank are used. Boundary conditions are similar to the previous case, with free-slip condition on tank walls Γ_{wall}, atmospheric condition along the top boundary of the tank Γ_{atm}, and no-slip condition on the boundaries of the caisson $\Gamma_{\text{f} \cap \text{s}}$. The numerical simulation runs for at least $\Delta T_{\text{sim}} = 4\text{s}$ in order to record the oscillating signal for the same length of time as Jung et al. (2006). The fluid domain is spatially discretised with $h_e^0 = 0.005\text{m}$ around the mean water level and up to a distance of 0.45m from the barycentre of the floating body, with a gradually coarsened mesh used in the same manner as described in the previous case. The two-phase flow solver uses a timestep of $\Delta t = 5 \cdot 10^{-3}\text{s}$ while the rigid body solver uses a fixed time step of $1 \cdot 10^{-5}\text{s}$. Note that although friction from the hinges acting on the rolling motion of the caisson is not mentioned in Jung et al. (2006), it is argued in the literature that friction energy losses due to this support system were probably present during the experiment, as raised by Calderer et al. (2014); Bihs and Kamath (2017); Chen et al. (2016) and confirmed herein. A linear dissipation term C_ω proportional to the angular velocity is therefore added to the equation of motion of the caisson in order to represent these losses from the experimental results, and was set to $C_\omega = 0.275$ after trial and error, similarly to Bihs and Kamath (2017). Note that this test case was also presented in de Lataillade et al. (2017) on an older implementation of the present FSI framework and without taking the friction into account.

Results of free-oscillations tests are presented in Figure 10, along with experimental data and the results from the Particle-In-Cell (PIC) model from Chen et al. (2016). It is observed that the natural period of the rolling motion is very well predicted numerically, regardless of the usage of a dissipation term. For the amplitude decay, however, a large numerical underestimation is observed when $C_\omega = 0$ but this disagreement with the experiment exists for both Chen et al. (2016) and the current model. Very good agreement with the experiment and the current model can be recovered with $C_\omega = 0.275$, confirming that it is likely that friction occurred during the experiment.

The response of the rolling motion against regular waves is subsequently investigated. Using the same numerical setup, the leftmost boundary and relaxation zone are adapted for regular wave generation boundary, and the length of the numerical domain is adjusted according to the wavelength λ of the generated wave, as shown in the schematic representation of the numerical domain in Figure 11. The RAO of the caisson is calculated using different wave conditions, as shown in Table 1. Each case runs for a time equivalent to 30 generated wave periods, and the amplitude of the response is averaged over the last 10 periods. Run-

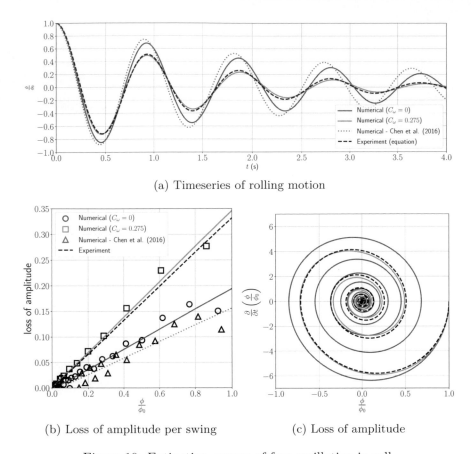

(a) Timeseries of rolling motion

(b) Loss of amplitude per swing

(c) Loss of amplitude

Figure 10: Extinction curves of free oscillation in roll.

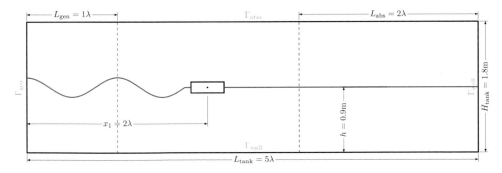

Figure 11: Geometry of numerical domain for the simulation of free roll oscillation of a floating caisson.

ning in parallel on 160 cores, it took 4h to complete the case with the sortest wave period ($T = 0.6$s), and 48h for the case with the longest wave period ($T = 1.4$s). The response of the rolling caisson is shown in Figure 12, and compared with the PIC model from Chen et al. (2016), experimental data from Jung et al. (2006) and linear potential flow theory. Results with the calibrated friction coefficient $C_\omega = 0.275$ from the free-decay tests are also plotted in Figure 12, while the PIC model does not consider friction losses. In general terms,

Table 1: Wave characteristics and resulting roll response for computing Response Amplitude Operator (RAO) of Figure 12.

#	T	ω	λ	H	ka
—	s	rad s^{-1}	m	m	—
1	0.60	10.47	0.56	0.017	0.0950
2	0.70	8.98	0.77	0.015	0.0616
3	0.70	8.98	0.77	0.023	0.0944
4	0.70	8.98	0.77	0.029	0.1191
5	0.80	7.85	1.00	0.029	0.0912
6	0.85	7.39	1.13	0.033	0.0919
7	0.93	6.76	1.35	0.016	0.0372
8	0.93	6.76	1.35	0.027	0.0628
9	0.93	6.76	1.35	0.032	0.0745
10	0.93	6.76	1.35	0.040	0.0931
11	1.00	6.28	1.56	0.044	0.0887
12	1.10	5.71	1.88	0.057	0.0953
13	1.20	5.24	2.22	0.032	0.0453
14	1.20	5.24	2.22	0.060	0.0849
15	1.20	5.24	2.22	0.067	0.0948
16	1.30	4.83	2.57	0.060	0.0732
17	1.40	4.49	2.93	0.061	0.0653

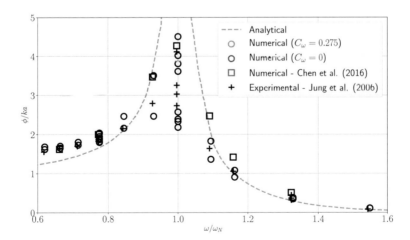

Figure 12: RAO of rolling caisson under regular wave loads. Analytical (linear theory) curve and experimental points digitised from Jung et al. (2006); monolithic model results (PICIN) digitised from Chen et al. (2016).

results are in good agreement with experimental data along the tested frequency range. It is worth noting that the two CFD models (Proteus and PIC) present better agreement than the potential flow theory in the longest wave regime ($\omega \leqq \omega_N$), as nonlinear effects are taken into consideration in both models. Close to resonance, results without the dampening coefficient are overestimating the measured response, while the inclusion of the friction terms dampens the rolling oscillations noticeably. This suggests that the calibrated value

$C_\omega = 0.275$ obtained from free oscillations may not be valid for the forced oscillation case, as the dissipation term may be nonlinear. The results presented here show the capabilities of the model to reliably predict the response of a floating body to different wave loads.

5.1.3 Response to extreme event—floating structure

This validation case is based on the experiment of Zhao and Hu (2012), featuring a 2-DOFs (heave and roll) inverted T-shaped floating structure under extreme wave loads. The experiments were performed in a numerical tank with a flat bottom. The lower platform of the structure has length and height of 0.500m and 0.123m, respectively. In order to prevent any overtopping water from being transmitted behind the structure, a superstructure of length 0.200m and height 0.250m was mounted on the lower platform. The clearance between the side walls and the floating structure is 5mm on either side, such that the wave processes can be considered two-dimensional, as in the previous cases. The water depth at rest was $h_\mathrm{mwl} = 0.4$m and the structure was allowed to heave and roll by attaching the structure to a support system consisting of a cylindrical joint and a heaving rod. A detailed account of the physical model including body properties (mass, moment of intertia etc.) is given in Zhao and Hu (2012). The domain is represented schematically in Figure 13.

A focused wave is generated to represent interaction of an extreme event with the floating body, with an amplitude of $a_f = 0.06$m, focus time $t_f = 20$s and the focus point of the wave coinciding with the location of the barycentre of the body ($x = 7$m). The domain is discretised temporally with CFL = 0.1 and spatially with $h_e^0 = 0.005$m, the latter being kept constant up to a distance of $\pm a_f$ around the still water level and, similarly to previous cases, the mesh gradually coarsened beyond that. The boundary conditions are set similarly to the cases in the previous section and the focused wave is generated at the leftmost boundary, with the assistance of a generation zone of length 1λ (the wavelength corresponding to peak frequency). Dirichlet boundary conditions for the free surface elevation and the fluid velocity profile were calculated using the JONSWAP spectrum. Due to nonlinear wave-wave interaction, the actual focus point and time might be different from the analytical values. Using a trial-and-error approach, the wave components must therefore be assigned appropriately to chosen phases so that the wave focuses at the intended location. The following phase shift ϕ_num is therefore applied numerically on all frequency components of the JONSWAP spectrum, and differs with the analytical phase shift ϕ_ana as follows:

$$\phi_\mathrm{ana} = -kx_f + 2\pi f t_f \tag{44}$$

$$\phi_\mathrm{num} = -k(x_f - 0.5) + 2\pi f t_f \tag{45}$$

with k and f are the frequency and wavenumber of each component, respectively, x_f the focus point and t_f the focus time of the wave. The focused wave is first generated in the

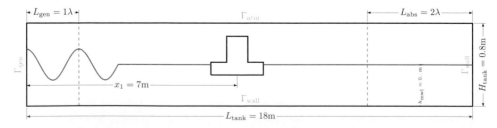

Figure 13: Geometry of numerical domain for the simulation of 2 DOFs floating caisson under extreme wave loads.

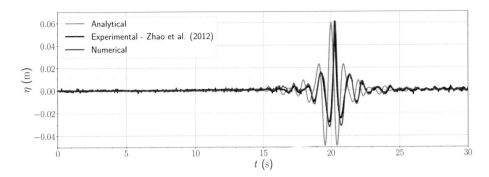

Figure 14: free surface elevation of focused wave over time at $x = x_f$.

numerical flume without the structure in order to validate it against the experimental wave. The free surface elevation obtained numerically over time at $x = x_f$ is shown in Figure 14, along with the experimental data and the analytical solution, with the latter derived from linear wave theory. It is observed that the model is in good agreement with the experimental data. Note that the analytical solution presents deeper wave troughs and appears to focus at $t_f = 20.0$s, while both the experimental and numerical results focus at $t_f = 20.3$s, probably due to nonlinear interaction of wave components.

The structure is now introduced in the numerical domain and its interaction with the focus wave is simulated. With a total of $490,543$ mesh elements, running in parallel on 196 cores, the simulation produced 30s of data in 57h. Snapshots of the numerical results around the focus time of the extreme wave are shown in Figure 15, and compared to photos from the experiment. These show a clear qualitative similitude of free surface profiles as well as floating body position and rotation between the numerical and experimental results. Furthermore, the overtopping events are well represented numerically with water hitting the superstructure at $t = 20.3$s (which corresponds to the actual numerical and experimental focus time), and retreating from the structure at $t = 20.7$s.

Quantitative results in terms of response of the floating body are shown in Figure 16 for heave and roll. The results are also compared to the numerical results from the PIC model of Chen et al. (2016) where a completely different approach is used for modelling the fluid and, most importantly, a monolithic scheme is used for fluid–structure coupling. Because the monolithic approach ensures an inherently stable simulation with no added mass effect, the response obtained by Chen et al. (2016) can be used for code-to-code validation against the added mass stabilised CSS scheme that is used here. In terms of heave, the experiment and the two numerical models predict the peak frequency accurately at $f_p^Z = 0.97$Hz which is the one obtained experimentally, and both underestimate the amplitude of the heave response of the body (16.8% for the explicit scheme and 10.9% for the monolithic scheme). In terms of roll, the peak response frequency is underestimated with the explicit scheme by 6.67%, while the monolithic scheme and experimental results are in agreement with $f_p^\phi = 1.00$Hz. The amplitude in roll is however better estimated with the explicit scheme with an overestimation of the response of 5.6%, against an overestimation of 22.5% for the monolithic scheme. Furthermore, it appears clearly on the time-series of Figure 16c that the response frequency of the explicit scheme is initially in phase with the experimental response but that phase shift occurs after the peak of the focused wave has passed ($t > 21$s).

It is worth noting that running this simulaton without an added mass stabilisaton scheme for fully explicit partitioned schemes leads to an unconditionally unstable simulation. It is therefore demonstrated that both the explicit coupling scheme with the added mass stabilization is a viable approach, as results are in good agreement with experimental data

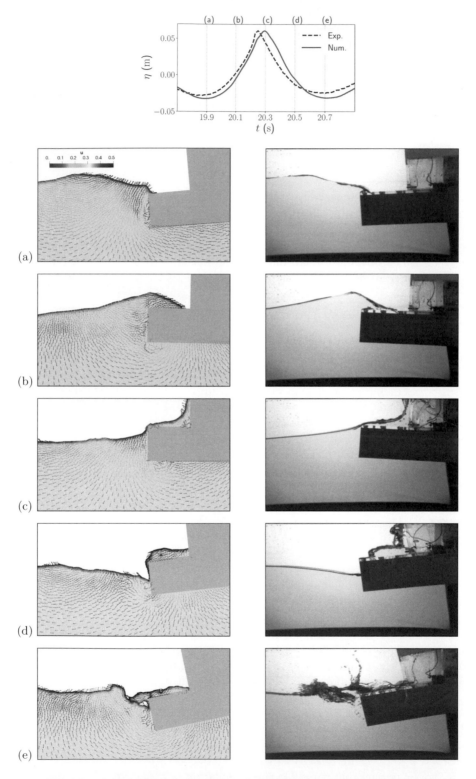

Figure 15: Snapshots of floating body hit by focused wave ($t_f = 20$s) at 19.9s, 20.1s, 20.3s, 20.5s, and 20.7s as shown on the graph at the top. Left: snapshots from numerical model, right: photographs from the experiment of Zhao and Hu (2012).

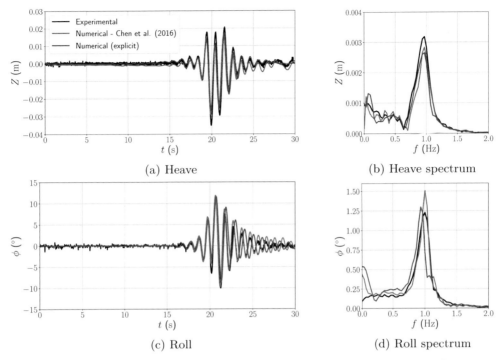

(a) Heave

(b) Heave spectrum

(c) Roll

(d) Roll spectrum

Figure 16: Response in heave and roll of floating body to focused wave loads. Experimental results from Zhao and Hu (2012), numerical (PIC) results from Chen et al. (2016).

and the ones obtained by the inherently stable monolithic numerical scheme of Chen et al. (2016).

5.2 Validation of mooring model

In this section, simulations are undertaken with the aim to validate the static/quasi-static and dynamic mooring models developed here. The validation cases presented here follow the experimental setup of Johanning and Smith (2006); Johanning et al. (2007), provides measurements of mooring tensions under static and dynamic conditions. A catenary mooring line is attached to a floating buoy that is driven horizontally through forced motion, with its fairlead kept at a height of $h = 2.651$m from the seabed. The line has a length $L = 6.98$m, diameter $d_0 = 2.5e^{-3}$m, submerged weight $w_0 = 1.036$N m^{-1}, and axial stiffness $EA_0 = 560$kN. The characteristics of the cable used numerically here are the same as the experimental ones (see Table 2), with drag and added mass coefficients taken from typical studless chain values. As the water level is $h_{mwl} = 2.8$m, the cable is fully submerged at all times. A schematic representation of the setup is shown in Figure 17. Note that all numerical simulations presented in this section ran in serial mode on a laptop, due to the fact that the computationally intensive fluid solver is not involved.

5.2.1 Mooring statics

For the static analysis, the mooring line is held at 16 different surge positions and the resulting static cable tensions at the fairlead are recorded by Johanning et al. (2007). Numerically, the dynamic model uses 100 ANCF elements for the mooring line, and is driven from a fully stretched position ($X = \sqrt{L^2 - h^2} = 6.457$m relative to the anchor) back to $X = 5.5$m at a

Table 2: Characteristics of mooring lines for Johanning et al. (2007) test case.

Seg. #	Seg. Type	L m	d_0 mm	w_0 $\mathrm{N\,m^{-1}}$	EA_0 kN	$C_{d,n}$	$C_{d,t}$	$C_{m,n}$	$C_{m,t}$
1	Chain	6.980	2.500	1.036	560	2.4	$1.15\pi^{-1}$	1.00	0.5

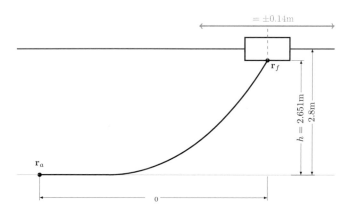

Figure 17: Setup for mooring dynamic validation case.

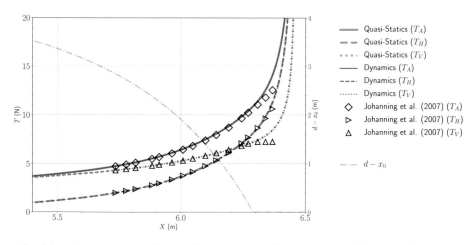

Figure 18: Mooring static analysis: Comparison of experimental data with static and dynamic models comparison. T_A: axial tension; T_H: horizontal tension; T_V: vertical tension; $d - x_0$: length of cable on seabed. Experimental data from Johanning et al. (2007).

rate of $\dot{X} = -5\mathrm{mm\,s^{-1}}$ with time steps of $\Delta t = 1e^{-4}\mathrm{s}$. The motion is slow enough in order to minimise any dynamic effects. The static catenary model calculates tensions analytically for the same surge range as the dynamic model.

The results are presented in Figure 18, with tensions recorded every 0.01m for the dynamic and static models and compared to the 16 experimental points. The length of the mooring line on the seabed for the different surge positions according to the static model is also plotted for reference. The static equations and mooring dynamics model present a total agreement for the tensions at the fairlead across the whole surge range. This result is meaningful in terms of cross-validation because the static/quasi-static and dynamic models use two very distinct approaches: analytical catenary equations and FEM respectively. Both approaches present an excellent agreement for the partly lifted regime. Analytical and numerical results start diverging from recorded experimental tensions once the line is fully lifted ($X > 6.3$m). Note that this might be an inconsistency in the experiment as the vertical tension at the fairlead recorded from Johanning et al. (2007) is constant once the line is fully lifted, while it is expected to further increase. The reasons for this remain unknown, but this might be due to slight tilting of the device during the experiment when the line becomes fully lifted.

5.2.2 Mooring dynamics

After validation tests for static tensions, a surge oscillation is now prescribed at the fairlead, following a sinusoidal evolution in time for a total excursion of $0.1h_{mwl}$. The tests were carried out for different initial positions X^0 (with corresponding static pretension T_{A0}) and frequencies, as shown in Table 3. The experimentally recorded surge displacement is used as input for the prescribed motion of the fairlead. From this sinusoidal motion in surge, the dissipated energy of the horizontal tension T_H is calculated from the area within the closed-loop (hysteresis) of the X-T_H curve. Note that a total of sixteen cases were run experimentally, but only nine cases are presented here due to the experimental data that was made available. The case with most significant nonlinear effects (number 16 in Table 3) is chosen to perform a cable mesh sensitivity analysis by keeping a constant timestep of $\Delta t = 1 \times 10^{-4}$s. Results for the error in energy dissipation compared to the most refined case are shown in Figure 19 where spatial convergence of order 1.9 is observed as the number of elements n_ε increases. Following these results, a discretisation of $n_\varepsilon = 100$, and $\Delta t = 1 \times 10^{-4}$s is selected.

Indicator diagrams for each case are presented in Figure 20, showing the variation in horizontal tension over a single cycle of the surge excursion for the numerical simulation and the experiment. The horizontal tension calculated by the quasi statics model for the same surge excursion is also included. Results show that the quasi-statics model is in good agreement with the mooring dynamics model regarding minimum and maximum horizontal tensions. Minimum experimental tensions are also in good agreement with the numerical results across all cases. However, for case 12 and any other subsequent cases with a higher

Table 3: Parameters and results for mooring line damping test cases (numbered following Johanning et al. (2007)).

Case	X^0	$T_{A0}{}^{exp}$	$T_{A0}{}^{num}$	$\frac{L_s}{L}$	ω	$\frac{E_L{}^{exp}}{w_0 h u}$	$\frac{E_L{}^{num}}{w_0 h a}$
#	m	N	N	—	rad/s	—	—
2	5.778	4.890	4.929	0.611	0.341	0.009	0.003
6	5.964	6.040	6.106	0.704	0.429	0.004	0.011
8	6.049	6.800	6.888	0.761	0.483	0.007	0.022
10	6.143	7.940	8.053	0.837	0.561	0.081	0.049
12	6.243	9.650	9.818	0.942	0.644	0.196	0.112
13	6.269	10.220	10.371	0.974	0.779	0.157	0.154
14	6.307	11.050	11.351	1.000	0.722	0.197	0.253
15	6.339	11.780	12.530	1.000	0.787	0.452	0.428
16	6.367	12.540	14.051	1.000	0.939	0.684	0.843

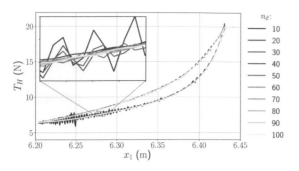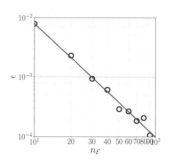

Figure 19: Spatial sensitivity on cable dynamics with $\Delta t = 1 \times 10^{-4}$s on case 16.

pretension stress, numerical and quasi-static results show higher maximum tensions than the experiment. This discrepancy is probably associated with potential problems in the experimental layout when the mooring line becomes fully lifted, as in the static tension cases.

The hysteresis of the curves formed in the indicator diagrams are due to nonlinear effects induced by the cyclic motion of the fairlead and can be used to calculate the damping of the line. The energy dissipation caused by drag forces on the mooring line can be calculated as follows:

$$E_L = \int_t^{t+T} T_H \dot{X} \mathrm{d}t \tag{46}$$

where T_H is the horizontal tension, T the period of oscillation, and X and $\dot{X}$ are surge displacement and velocity, respectively. The non-dimensional energy damping $\frac{E_L}{w_0 ha}$ (with a being the amplitude of oscillation) is plotted for each case as a function of non-dimensional pre-tension in Figure 21. Relatively good agreement between experimental and numerical results show that nonlinear effects are well simulated, with low damping when pretensions are low, and an exponential increase in damping as pretensions become higher and the line is fully lifted. In case 2, 6, and 8, the damping is low both experimentally and numerically, meaning that there is barely any dissipated energy. This can be verified in the indicator diagrams where all curves for these cases are in relatively good agreement with the quasi-statics model (which does not simulate any nonlinear effect). For case 10 and above, energy damping becomes significant. There is a slight inconsistency in the experimental results, again around the point where the line becomes fully lifted, with the damping of case 12 higher than the damping of cases 13 and 14, both of which have a higher pretension that should lead to higher damping under forced oscillation. This inconsistency is not present in the numerical model where any case with a higher pretension consequently yields a higher damping. The good agreement between numerical and experimental damping for different pretensions, has proven that the gradient deficient ANCF mooring model can successfully capture nonlinear behaviour that occurs in real-world applications.

5.3 Moored floating bodies: the OC4-DeepCwind validation case

5.3.1 Setup

This validation case follows the physical model tests of a 1:50 Froude scale model of the OC4-DeepCwind semi-submersible platform for offshore wind turbines (Robertson et al., 2014; Koo et al., 2012). The experimental tests were performed in the wind/wave basin at the Maritime Research Institute Netherlands (MARIN) by the University of Maine DeepCwind program in 2011. The semi-submersible platform consists of four cylinders connected to

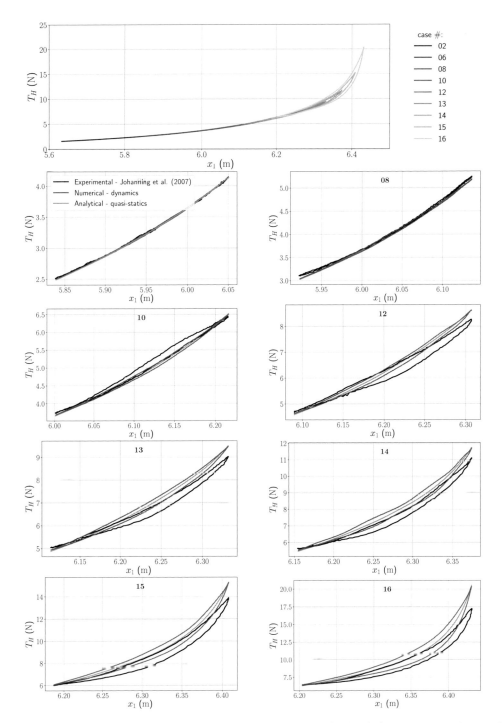

Figure 20: Indicator diagrams of horizontal tension at fairlead (case 02 not shown due to similarity with case 06).

each other by a set of trusses as shown in Figure 22, with the wind turbine mounted on the central cylindrical component. The physical properties of the semi-submersible are shown in Table 4 as retrieved from the references cited above. The mooring system of the structure

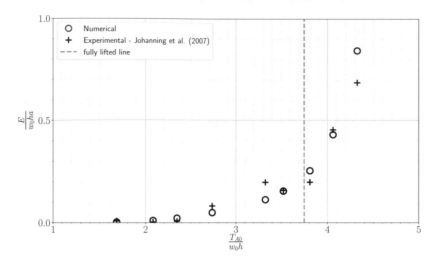

Figure 21: Energy loss as a function of pretension.

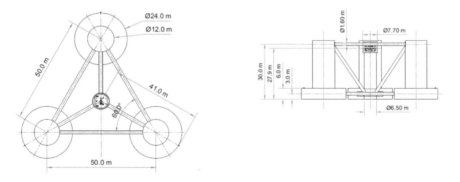

Figure 22: Geometry of OC4-DeepCwind semi-submersible (Koo et al., 2012)—lengths in prototype scale.

Table 4: OC4-DeepCwind semi-submersible platform characteristics.

	Symbol	Prototype		1:50 scale		
		no turbine	turbine	no turbine	turbine	unit
Mass	m	13,444,000	13,841,160	107.552	110.729	kg
Barycentre position above keel	—	5.600	10.110	0.112	0.202	m
Roll radius of gyration	$r_{I_{xx}}$	23.910	0.478	31.610	0.623	m
Pitch radius of gyration	$r_{I_{yy}}$	24.900	0.498	32.340	0.647	m
Yaw radius of gyration	$r_{I_{zz}}$	32.170	0.643	32.170	0.643	m

consists of three catenary chains with identical characteristics and lengths, arranged at an angle of 120° between them. The main properties of the mooring system are shown in Table 5, with several hydrodynamic properties such as drag and added mass coefficients calculated following Hall and Goupee (2015). The experimental setup and data from this model were also used to validate a lumped-mass mooring model in Hall and Goupee (2015) and a floating wind turbine simulator in Coulling et al. (2013). Note that all dimensions and

Table 5: Mooring system characteristics for the OC4-DeepCwind semi-submersible platform.

Property	Symbol	Prototype	1:50 scale	Unit
Length	L	835.500	16.710	m
Angle between lines	—	120	120	°
Fairlead radius	—	40.868	0.817	m
Fairlead depth	—	14.000	0.280	m
Anchor radius	—	837.600	16.752	m
Anchor depth	—	200.000	4.000	m
Equivalent diametre	d	133.760	2.675	mm
Nominal diametre	d_0	79.900	1.598	mm
Linear density	ρ_c	116.6	0.466	$\mathrm{kg\,m^{-1}}$
Linear displacement	$\delta\mathbf{x}$	14.05×10^{-3}	5.62×10^{-6}	$\mathrm{m^2}$
Axial Stiffness	EA_0	753.600×10^6	6.029×10^3	N
Drag coefficient (normal)	$C_{d,n}$	1.080	1.080	—
Drag coefficient (tangential)	$C_{d,t}$	0.213	0.213	—
Added mass coefficient (normal)	$C_{m,n}$	0.865	0.865	—
Added mass coefficient (tangential)	$C_{m,t}$	0.269	0.269	—

measurements hereafter are presented in 1:50 model scale to reflect actual dimensions that were used experimentally and are reproduced here numerically.

According to the experimental setup, the still water depth was $h_{\mathrm{mwl}} = 4$m, with the draught of the semi-submersible equal to 0.4m and the turbine and mooring system mounted on the structure. Regular and irregular sea-states were tested with and without wind forcing. In the numerical simulation presented below, the response of the platform is investigated, but not the response of the turbine to wind loads, as only the mass and inertia characteristics of turbines are incorporated in the model. As opposed to FSI simulations presented previously, this test case must be modelled numerically in 3D, due to the geometry of the platform. In this section, interaction of the structure with monochromatic waves with height $H = 206.08$mm, period $T = 1.71$s, and wavelength $\lambda = 4.66$m was simulated. This case presented the most significant nonlinear effects in the physical modelling tests, making it arguably the most interesting case to investigate numerically. The numerical domain is shown in Figure 23 and has a total length of $L_{\mathrm{tank}} = 5\lambda$, width of $W_{\mathrm{tank}} = 1\lambda$, and height of $H_{\mathrm{tank}} = 1.5h_{\mathrm{mwl}}$. Waves are generated at the left boundary of the tank (placed at $x = 0$) using the Fenton Fourier Transform method (Fenton, 1988) with a generation zone spanning over 1λ from the generating boundary, while an absorption zone of 2λ is placed at the other end of the tank. The platform is positioned at $x = 2\lambda$ away from the generating boundary, and it is centred across the width of the tank. Due to the cylindrical shape of the structure and the relatively slender features, with respect to the wavelength, wave reflection at the side walls is not expected to be significant. The upstream mooring line (Line 1) is placed parallel to the direction of wave propagation while the remaining two lines (Line 2 and 3) form an angle of 60° from the direction of wave propagation. Collision detection between mooring lines and the seabed is enabled. The material properties of the seabed are: friction coefficient of 0.3, normal damping coefficient of 1, and normal stiffness of $3 \times 10^6 \mathrm{Pa\,m^{-1}}$. These properties ensure that collision-enabled nodes of the cable do not sink or bounce when entering into contact with the seabed.

Assuming an experimentally recorded length of L = 16.71m for all mooring lines, the numerical pretension calculated from the static model is 8.33N. This however does not match the experimental pretensions that vary according to the line considered. The line lengths are therefore adjusted numerically in order to match the recorded experimental pretensions:

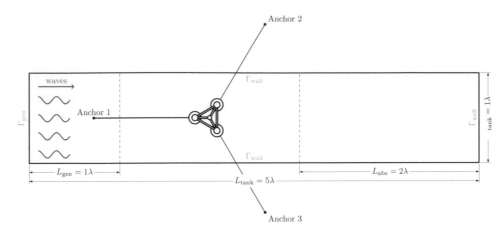

Figure 23: Schematic top-down view of numerical tank for OC4-DeepCwind test case (not to scale).

Figure 24: Snapshot of numerical domain (zoomed) for OC4-DeepCwind case at $t = 21$s.

Line 1 with a target experimental tension of 8.992N is adjusted to $L_1 = 16.674$m (yielding 8.993N), Line 2 and 3 with target pretensions of 8.540N and 8.520N are both adjusted to a length of $L_2 = L_3 = 16.699$m (yielding 8.530N), noting that the resulting variations to the length from the initial estimate are negligible (<1%).

Boundary conditions are set as free-slip on the bottom and solid walls Γ_{wall} of the tank, atmospheric at the top boundary Γ_{atm} of the tank, unsteady two-phase flow inlet at the left boundary Γ_{gen} of the tank, and no-slip conditions at the semi-submersible structure $\Gamma_{\text{f}\cap\text{s}}$. A rendered illustration of the layout is shown in Figure 24. An unstructured tetrahedral mesh is used with a constant $h_e^0 = 0.02$m up to a distance of $\pm\frac{H}{2}$ from the still water level and at the semi-submersible structure. This gives relatively good discretisation of the wave around the free surface with $h_e^0 < \frac{H}{10}$ and $h_e^0 < \frac{\lambda}{200}$. The mesh is coarsened gradually with a maximum increase of 10% in characteristic element size between adjacent elements. Mooring lines are discretised with 100 gradient deficient ANCF elements along each line and the multibody dynamics solver uses a fixed time step of $\Delta t = 1 \times 10^{-4}$s.

The simulation is run twice: first as an uncoupled model with only the multibody dynamics solver (no fluid dynamics) and using the experimentally recorded position of the semi-submersible to drive the motion of the fairlead in the numerical model, and a fully coupled simulation using the fluid dynamics solver to simulate waves and interaction with the semi-submersible structure. The uncoupled simulation can be resolved in less than an hour on a single processor, while fully coupled case leads to an expensive numerical setup with a relatively large fluid mesh: $n_{\mathcal{N}} = 6,551,460$ nodes and $n_{\mathcal{E}} = 40,509,745$ elements. Running on $4,800$ processors for 168 hours (i.e., a week), the fully coupled simulation produced 44.5s of data.

5.3.2 Uncoupled model results

As mentioned above, validation of the mooring system is performed first by using the displacement of the semi-submersible recorded at the experiment. The displacement signal is smoothed with a low-pass cut-off filter at 1Hz, in a similar manner to Hall and Goupee (2015). The simulation was run using both the mooring quasi-statics and dynamics models for comparison.

A selected time frame of the tensions at the three different fairleads over time is presented in Figure 25a, while Figure 25b presents the averaged tension over the averaged heave range for 40 periods. The power spectra of each line from data for these 40 wave periods are shown in Figure 26. It is evident that the quasi-statics model significantly underestimates the response in tension on the mooring lines, with an averaged underestimation of 71.6% for Line 1, 60.0% for Line 2, and 57.8% for Line 3. It is also out of phase by approximately $60°$ when compared to the experimental measurements on every line. On the contrary, tensions from the dynamics mooring model show excellent agreement in terms of phase with the experimental tensions. However, the uncoupled mooring dynamics model is more conservative than the experiment in terms of response with an overestimation of 14.9% for Line 1, 12.9% for Line 2, and 25.4% for Line 3. This difference can be justified by

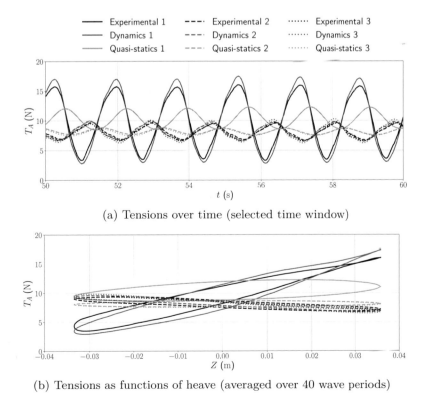

(a) Tensions over time (selected time window)

(b) Tensions as functions of heave (averaged over 40 wave periods)

Figure 25: Tensions at fairleads for uncoupled OC4-DeepCwind semi-submersible simulation.

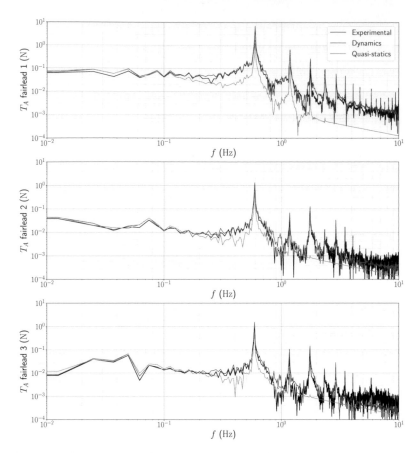

Figure 26: Spectra of tensions at fairleads for uncoupled semi-submersible validation case (spectral analysis was carried out for 40 wave periods).

uncertainties such as the empirical value of the drag coefficients, especially since the geometry of the mooring chain reported in Hall and Goupee (2015) is different from conventional commercial chains. Note however that Hall and Goupee (2015) obtained better agreement with experimental data on this particular case using the drag coefficients of Table 5.

5.3.3 Coupled model results

This full-fledged 3D case uses all the models and tools that have been described and developed before: two-phase flow module, ALE mesh with elastostatics, rigid body motion, mooring statics/quasi-statics and dynamics models, added mass estimator model, and tools such as the particle localisation algorithm for retrieving the fluid velocity at mooring cable nodes. The simulation ran on $4,800$ cores for 168 hours produced $\Delta T_{\text{sim}} = 44.5$s of data. The response of the OC4-DeepCwind platform is plotted in Figure 27 for both the numerical and the physical model. The time series includes response in all DOFs, with the experimental signal shifted by -16.2s to ensure consistent timing of the arrival of the first wave between the experiment and the numerical simulation. Figure 28 shows the tensions at the fairleads from the numerical model against experimental data.

It clearly appears that surge, heave, and pitch of the platform are in phase and in an overall reasonable agreement between the experiment and the simulations. Note that the magnitude of the pitch is also very small, with $|\theta| < 2°$ at any time. Sway, yaw, and roll

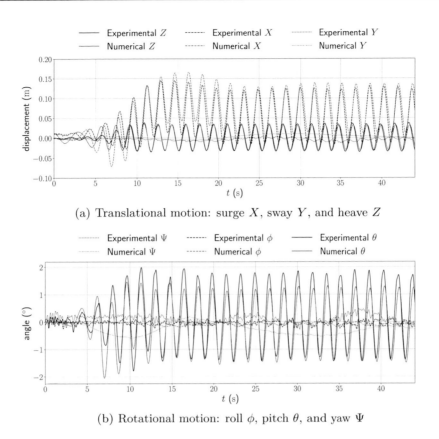

(a) Translational motion: surge X, sway Y, and heave Z

(b) Rotational motion: roll ϕ, pitch θ, and yaw Ψ

Figure 27: 6-DOFs Response of OC4-DeepCwind semi-submersible platform to wave loads in fully coupled simulation.

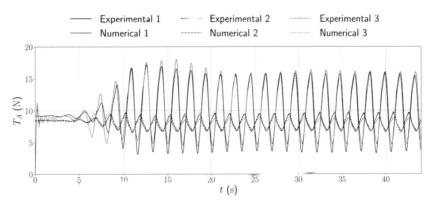

Figure 28: Tensions at fairleads of OC4-DeepCwind semi-submersible platform in fully coupled simulation.

are negligible in both cases, as can be expected from the direction of wave incidence and the mooring arrangement. The surge motion follows a similar trend numerically and experimentally, with the structure displaced from its static equilibrium position before restoring forces in the mooring system bring it back to a new mean position along the main wave propagation axis. The mean surge offset remains constant after approximately $t = 25$s in both cases with values of $\bar{X}^{\text{num}} = 0.073\text{m}$ and $\bar{X}^{\text{exp}} = 0.062\text{m}$ by the numerical model and the experiment, respectively. Note that in Figure 27a, the heave signal has been translated by $+0.012\text{m}$, as the structure in the numerical model is not at equilibrium at $t = 0$. This error of 3% in draught can be due to slight differences in mass provided by Koo et al. (2012) and the actual mass of the whole system (including moorings) to reach the theoretical draft of 0.4m, but also due to discretisation errors resulting from meshing the surface geometry of the structure that could result in a slightly different volume.

The responses of the platform and mooring system to wave loads are discussed below by considering the last ten wave periods of the simulation and the corresponding time frame of the experiment, where the flow and mooring processes can be considered fully developed. In terms of hydrodynamic response of the structure itself, the heave and surge amplitude are underestimated numerically by 7.55% and 10.27% respectively. For the mooring system, it appears that fairlead tension signals in all mooring lines are in phase with the experimental data. Mean tensions acting on all lines have a difference of less than 4% from the experimental data. Tension at the upstream fairlead is significantly higher than at the other fairleads both in the numerical and the physical models. The amplitude of the response for the numerical model is underestimated by 7.44%. Note that these differences are relatively close to the ones observed in hydrodynamic responses (heave, surge).

At the fairleads attached to Line 2 and 3, numerical predictions of tensions at the fairlead are underestimated by 30.70% and 34.49%. These seem to be relatively large differences, but it is noted that both the experimental and the numerical results for Line 2 and 3 are approximately one order of magnitude lower than the upstream line (Line 1) and comparable to the absolute difference between the experiment and the numerical model in Line 1. It is therefore more important to resolve accurately the stresses in Line 1, as it has the largest effect in restraining the motion of the floating structure.

Results from the coupled model compare better to the experiment relative to the coupled results of Hall and Goupee (2015), where an underestimation of 26% in heave and a leading phase of 40° were observed. Errors of the same order of magnitude were present in mooring line tensions in Hall and Goupee (2015), particularly on Line 1 (upstream). While it is not clear why this happened in Hall and Goupee (2015), this relatively large difference could be partly due to the fact that potential flow theory was used to describe the hydrodynamics of the structure while a CFD model for producing the results presented in this work, as the latter approach is considered to be a higher fidelity hydrodynamic model. In any case, this further highlights the suitability of the framework developed here to simulate coupled problems of moored floating bodies, both in terms of hydrodynamic response of the floating structure and of tensions in the mooring system.

The fluid velocity magnitude and direction are calculated along the upstream line using the particle localisation algorithm and are shown in Figure 29a over a full period of oscillation. The fluid velocity magnitude is relatively significant in the vicinity of the free surface and the fairlead, where wave kinematics are predominant, especially for deep water waves. The relative velocity of the cable with respect to the fluid velocity is shown in Figure 29b for the same time interval (magnitude and direction). It appears that the line is never fully lifted during a full cycle, with at least approximately $0.25L$ remaining on the seabed at all times (zero velocity area). It also appears that the largest relative velocities are found between the centre of the line and a distance of $0.25L$ away from the fairlead (at $0.5L < s < 0.75L$, with $s = 0$ at the anchor), which consequently corresponds to the part of the line with the strongest drag forces. This implies that, in this particular case, the anchor does not experience any significant force induced by the mooring line as one quarter of it lies on the seabed at all times, and that the largest contributions to mooring line damping come from the middle section of the lifted line.

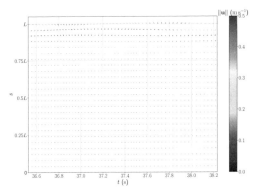

(a) Fluid velocity for one period of oscillation

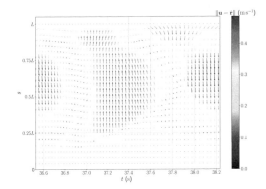

(b) Velocity of cable relative to fluid for one period of oscillation.

Figure 29: Velocity of fluid and cable along Line 1, where s is the position along cable (from anchor to fairlead) and velocity vectors are in X-Z plane.

Results obtained here clearly show the capability of the FSI framework developed in de Lataillade (2019) for simulating complex moored floating structure problems in a satisfactory manner. Results from the numerical model are realistic in terms of both hydrodynamic response and mooring line tensions experienced by the semi-submersible, even in the presence of highly nonlinear dynamic effects. It is also proven that the coupled model can provide valuable information to inform the design of the structure, that would be difficult to obtain otherwise (e.g., motion of mooring lines).

6   Conclusions

The simulations in Section 5 present verification and validation for each submodel and for the coupling strategy as a whole, highlighting the capabilities and several limitations of the FSI framework developed here. The fluid dynamics model encompassing the Navier-Stokes, free surface tracking models and ALE moving mesh approach, is capable of dealing with nonlinearities and complex processes such as nonlinear wave propagation, wave-wave interaction, wave overtopping, and 3 DOFs modelling of floating structures. Good agreement with experimental data was obtained over all the simulations presented for floating structures. It is also important to note that the partitioned and explicit ALE coupling was numerically stable and in good agreement with results from fully implicit, monolithic models.

The mooring dynamics model based on Chrono® and using gradient deficient ANCF elements has also shown to be robustly implemented. Very good agreement with experimental data was obtained, with some discrepancies observed when the mooring line becomes fully lifted. It is argued that such differences are a result of physical modelling limitations, as results from the numerical simulations are found to be consistent with exponential growth in tensile stresses once the line becomes fully lifted. For static/quasi-static analysis, the analytical model returned virtually the same results as the dynamic model, and both were close to experimental data. The statics/quasi-statics model can be therefore used to set initial conditions for the nodal positions and direction gradients of the gradient deficient ANCF mooring model for all dynamic simulations.

The moored floating semi-submersible case of Section 5.3 demonstrated that the FSI framework works robustly as a whole when all models are coupled together for modelling 6 DOF response, and on a large scale (i.e., parallelised with several thousands of processors).

By achieving a good agreement between experimental and numerical data, the framework also showed its viability in realistically simulating the response of complex multiphysics systems in 3D. The results obtained with the uncoupled approach (mooring dynamics model with prescribed motion from experimental data) for the particular wave conditions studied were more conservative than the experimental data and the numerical results from a lumped mass model developed and used by Hall and Goupee (2015).

However, results from the fully coupled model using the framework developed here showed significantly better agreement with experiments than the results of the coupled model in Hall and Goupee (2015), concerning the main responses of the system. Regarding the dynamic mooring model performance, it is clearly shown that it is able to capture nonlinear effects in terms of mooring damping. These nonlinear effects could be represented with the dynamic model, while the statics/quasi-statics model clearly underestimated the response in tension (e.g., see Figure 25a) as the latter is unable to model damping of the mooring line. This further confirmed that a fully dynamic mooring model should be used for estimating tension in dynamic mooring systems in a more accurate manner. It was also shown that the velocity of the fluid could be retrieved successfully by tracking the nodes of the cable in the fluid mesh with the particle localisation algorithm adapted from Haselbacher (2007) and developed in de Lataillade (2019), allowing for more accurate application of drag and inertia forces along mooring cables.

Overall, the various numerical simulations and results show that the numerical framework developed here has the ability to simulate highly nonlinear processes of fluid flow, floating structure dynamics, mooring dynamics, and all of them coupled together, as the framework provided predictions of realistic responses that compare well to real-world experiments, and to inherently stable numerical models. It can be therefore argued that the partitioned scheme with non-iterative added mass effect stabilization proposed by de Lataillade (2019) is a novel and successful contribution that addresses a long-standing issue of numerical stability of fully explicit schemes, as it has proven to produce stable and satisfactory results for all engineering applications presented here. The quest for ever greater efficiency and accuracy for high-fidelity FSI models is however boundless, as many aspects can always be improved upon. Additional information on current limitations of the framework and potential improvements are provided in Appendix A, along with some more details on the particulars of the implementation, from the software engineering aspect. The main limitation identified is the computational cost and while this is currently the price to pay for producing reliable high-fidelity FSI results, it is expected that, with the continuing trend in increase of computational power availability and the rise of alternative technologies in computers (e.g., quantum computing), such simulations will become more affordable in the near future. While several routes are open for further work, it is clear that the models and coupling strategy established has already shown its relevance and great potential for high-fidelity simulations of moored floating structures subject to various environmental loads.

Acknowledgements

This project was funded by Engineer Research and Development Centre and HR Wallingford through the collaborative agreement No W911NF-15-2-0110 and the EPSRC Industrial Centre of Doctoral Training in Offshore Renewable Energy (IDCORE). All authors would like to acknowledge Dr Xizeng Zhao for providing the experimental snapshots of Figure 15, and Dr Andrew Goupee and Mr Christopher Allen for providing experimental data for Section 5.3. All authors would like to thank Prof David Ingram and Dr Tahsin Tezdogan for the support during early stages of the project. Permission was granted by chief of Engineers to publish these data.

Appendices

A   On limitations and way forward

One of the main limitations of the framework presented here (which is common to all high-fidelity FSI models) is the relatively high computational cost of large simulations. This was particularly apparent in Section 5.3.3, where the simulation used more than $800,000$ core hours before completion. This limitation is still prominent despite the careful selection of the techniques presented here according to their computational efficiency when compared to alternative tools. For example, the added mass stabilised non-iterative coupling scheme between the fluid and the solid solvers avoids any additional iteration of either solver per time step, and this induces significant computational savings when compared to implicit schemes. Computational efficiency is also the reason for using gradual mesh refinement on most simulations presented here. Resolving these issues essentially comes down to simply optimizing aspects of the code, sometimes simply by converting code written in high-level interpreted programming languages (Python) to lower-level compiled languages (e.g., C++), or by employing more efficient approaches or algorithms for the numerical solution. The computational efficiency for the fluid problem can be further improved by allowing gradual coarsening of the mesh along the free surface in areas where the description of the fluid does not need to be accurate, such as in absorption zones. This requires further work on the free surface description model as such coarsening along the air-water interface is not possible without introducing spurious fluid velocities due to the dependence on the mesh element size to smooth the VOF field in the current. Furthermore, mesh motion approaches based on monitor function model, on top of allowing for quality control of the mesh, can be used to dynamically refine around the free surface instead of using a large refined band covering all areas where the free surface is expected to evolve, thus improving computational efficiency.

Another limitation concerns the added mass stabilisation, that is limited to single rigid bodies, and cannot technically be applied as it stands to flexible bodies or multiple bodies that are in close proximity. Inclusion of these effects may require a re-design of the added mass stabilisation model, which may prove to be complicated and impractical. In de Lataillade (2019) it is nevertheless shown that using conservative (overestimated) values for the added mass matrix may result in practically identical results as using the actual added mass values, as long as time discretisation is sufficiently fine. It is therefore very possible that a careful implementation of the current model with appropriate multipliers will be suitable for stabilising flexible bodies and multiple structures.

The simulations presented herein do not include turbulent effects for the CFD model, despite Proteus® having built-in the k-epsilon and k-omega models. The authors made a conscious choice not to include them, as the turbulence models had been designed for single-phase flow and this might have caused spurious results for two-phase flows (Devolder et al., 2017). In addition, resolving turbulent processes is not as significant for this work as resolving the coupling and stability issues discussed herein. It can also be argued that the results from the simulation presented here, are in good agreement with the experiment and other numerical models, and are not significantly affected by the absence of turbulence modelling. Turbulent effects may nevertheless be of increasing importance in case of, e.g., multiple bodies with interacting wakes. Implementing specialised turbulence models for two-phase flow such as the one in Devolder et al. (2017) may further improve the framework.

In terms of mooring modelling, various additional aspects can be implemented such as nonlinear stiffness along the cable (varying with tension), line breaking mechanisms, and anchor movement or dislodgement mechanisms. Note that, as the CFD and MBD solvers are fully partitioned, it is possible to extend the capabilities of the existing framework by modifying or replacing one of the solvers without affecting the coupling strategy. Future work may be performed to include deformable bodies, collision for multiple bodies or failure mechanisms for mooring lines or other solid elements, thus further extending the capabilities of the framework towards a realisting representation of FSI processes.

B   On software development

In this section, some general thoughts on software development aspects of the FSI framework applied in this work are discussed briefly. The framework is based on the Proteus® toolkit for solving fluid transport equations with FEM, and the Chrono® MBD library for introducing rigid body motion, dynamic mooring and collision detection capabilities. While both Proteus® and Chrono® were extensively modified to fit the needs of the FSI framework, the framework itself is built upon the Proteus® toolkit where Chrono® is built as a dependency when compiling the source code. At the time of writing, the code repository of Proteus® is available online at `https://github.com/erdc/proteus`, where all contributions for building the FSI framework presented here are recorded. Therefore, the majority of the software development work that has been done for this research is available online and can be used to reproduce any of the simulations from Section 5.

When building this FSI framework, the Chrono® library was mostly written in C++ (with relatively limited high-level Python access), while Proteus® was written in Python for high-level access and C++ for low-level access and efficiency. Both Python and C++ have therefore been extensively used for the implementation of the different models and tools presented here. Interfacing between these two languages and libraries was a major aspect of software development, as can be seen conceptually in the workflow diagram of Figure 4 where the main necessary communication steps between Proteus® (CFD) and Chrono® (MBD and mooring dynamics) are represented. This communication was mostly achieved with Cython, an interfacing language between Python and C++ that allows for sharing objects or pointers between the two languages. Cython also allows to convert automatically Python-like code to C/C++, and was therefore also used to increase the computational efficiency of several parts of the code in Proteus® that were originally designed in Python, but were proven to be time-consuming (e.g., relaxation zones).

Another major recurring aspect that had to be considered for the implementation of the methods presented here was parallelisation of the code. Most of the cases presented in Section 5 were typically medium or large FSI problems, in terms of domain and mesh size. These are recommended to be run in parallel, as decomposing the simulation into parallel tasks reduces both the power and the memory required by each task. This was achieved on a large scale by using HPC and the Messaging Passing Interface (MPI) for the CFD model. At the time of performing the simulations in de Lataillade (2019), the MBD solver was not parallelised when using FEM and collision detection, but this was not a major issue as the MBD solver-typically much faster than the CFD solver—did not pose a significant time overhead. Note however that for more demanding MBD problems (e.g., collision detection with mutliple bodies, deformable structures, very highly refined cables, etc.), it is recommended to include this parallel capability once ready.

It is also worth noting that, on top of the FSI framework presented here and at the time of writing, various additions were being added to the open-source Proteus® project: Immersed Boundary Method (IBM), mesh adaptivity, sediment solver, turbulence models, shallow water equations, and improvement over free surface tracking methods. Due to its open-source nature, the freely-accessible high-fidelity FSI framework that has been developed and presented here is also likely to undergo further development and improvement within the Proteus® toolkit.

References

Banks, J. W., Henshaw, W. D., Schwendeman, D. W. and Tang, Q. (2017a). A stable partitioned FSI algorithm for rigid bodies and incompressible flow. Part I: Model problem analysis. Journal of Computational Physics 343: 432–468.

Banks, J. W., Henshaw, W. D., Schwendeman, D. W. and Tang, Q. (2017b). A stable partitioned FSI algorithm for rigid bodies and incompressible flow. Part II: General formulation. Journal of Computational Physics 343: 469–500.

Bazilevs, Y. and Hughes, T. J. (2007). Weak imposition of Dirichlet boundary conditions in fluid mechanics. Computers and Fluids 36(1): 12–26.

Bihs, H. and Kamath, A. (2017). A combined level set/ghost cell immersed boundary representation for floating body simulations. International Journal for Numerical Methods in Fluids 83: 905–916.

Burman, E. and Fernández, M. A. (2007). Stabilized explicit coupling for fluid-structure interaction using Nitsche's method. Comptes Rendus Mathematique 345(8): 467–472.

Burman, E. and Fernández, M. A. (2009). Stabilization of explicit coupling in fluid-structure interaction involving fluid incompressibility. Computer Methods in Applied Mechanics and Engineering 198(5-8): 766–784.

Calderer, A., Kang, S. and Sotiropoulos, F. (2014). Level set immersed boundary method for coupled simulation of air/water interaction with complex floating structures. Journal of Computational Physics 277: 201–227.

Chen, Q., Zang, J., Dimakopoulos, A. S., Kelly, D. M. and Williams, C. J. K. (2016). A Cartesian cut cell based two-way strong fluid-solid coupling algorithm for 2D floating bodies. Journal of Fluids and Structures 62: 252–271.

Chorin, A. J. (1968). Numerical solution of the Navier-Stokes equations. Mathematics of Computation 22(104): 745–762.

Chrono, P. (2019). Project chrono—an open source multi-physics simulation engine. Accessed on: 2019-10-12.

Coulling, A. J., Goupee, A. J., Robertson, A. N., Jonkman, J. M. and Dagher, H. J. (2013). Validation of a FAST semi-submersible floating wind turbine numerical model with DeepCwind test data. Journal of Renewable and Sustainable Energy 5(2).

de Lataillade, T., Dimakopoulos, A., Kees, C., Johanning, L., Ingram, D. and Tezdogan, T. (2017). CFD modelling coupled with floating structures and mooring dynamics for offshore renewable energy devices using the proteus simulation toolkit. In European Wave and Tidal Energy Conference (EWTEC).

de Lataillade, T. (2019). High Fidelity Computational Modelling of Fluid-Structure Interaction for Moored Floating Bodies. PhD thesis, IDCORE—The Univerisity of Edinburgh, The Univeristy of Exeter, The University of Strathclyde.

Devolder, B., Schmitt, P., Rauwoens, P., Elsaesser, B. and Troch, P. (2015). A review of the implicit motion solver algorithm in OpenFOAM to simulate a heaving Buoy. NUTTS Conference 2015: 18th Numerical Towing Tank Symposium (2015): 1–6.

Devolder, B., Rauwoens, P. and Troch, P. (2017). Application of a buoyancy-modified k-ω SST turbulence model to simulate wave run up around a monopile subjected to regular waves using OpenFOAM®. Coastal Engineering 125(July 2017): 81–94.

Dimakopoulos, A. S., Kees, C. E. and de Lataillade, T. (2019). Fast random wave generation in numerical tanks. Engineering and Computational Mechanics 172(1): 1–11.

Dwight, R. P. (2009). Robust mesh deformation using the linear elasticity equations. In Computational Fluid Dynamics, pp. 401–406.

Fenton, J. D. (1988). The numerical solution of steady water wave problems. Computers and Geosciences 14(3): 357–368.

Fernández, M. A., Mullaert, J. and Vidrascu, M. (2013). Explicit Robin-Neumann schemes for the coupling of incompressible fluids with thin-walled structures. Comput. Methods Appl. Mesh. Engrg. 267: 566–593.

Gerstmayr, J. and Shabana, A. A. (2006). Analysis of thin beams and cables using the absolute nodal co-ordinate formulation. Nonlinear Dynamics 45: 109–130.

Grajewski, M., Köster, M. and Turek, S. (2009). Mathematical and numerical analysis of a robust and efficient grid deformation method in the finite element context. Journal 31(2): 1539–1557.

Hall, M. and Goupee, A. (2015). Validation of a lumped-mass mooring line model with DeepCwind semisubmersible model test data. Ocean Engineering 104: 590–603.

Haselbacher, A. (2007). An efficient and robust particle-localization algorithm for unstructured grids. Journal of Computational Physics 225: 2198–2213.

Hirt, C. and Nichols, B. (1981). Volume of fluid (VOF) method for the dynamics of free boundaries. Journal of Computational Physics 39(1): 201–225.

Hirt, C. W., Amsden, A. A. and Cook, J. L. (1974). An arbitrary Lagrangian-Eulerian computing method for all flow speeds. Journal of Computational Physics 14: 227–253.

Ito, S. (1977). Study of the Transient Heave Oscillation of a Floating Cylinder. Technical report, Massachusetts Institute of Technology.

Jacobsen, N. G., Fuhrman, D. R. and Fredsøe, J. (2012). A wave generation toolbox for the open-source CFD library: OpenFoam. International Journal for Numerical Methods in Fluids 70: 1073–1088.

Johanning, L. and Smith, G. H. (2006). Interaction between mooring line damping and response frequency as a result of stiffness alteration in surge. In International Conference on Offshore Mechanics and Arctic Engineering (OMAE), pp. 1–10.

Johanning, L., Smith, G. H. and Wolfram, J. (2007). Measurements of static and dynamic mooring line damping and their importance for floating WEC devices. Ocean Engineering 34: 1918–1934.

Jung, K. H., Chang, K.-a. and Jo, H. J. (2006). Viscous effect on the roll motion of a rectangular structure. Journal of Engineering Mechanics 132(2): 190–200.

Kees, C. E., Akkerman, I., Farthing, M. W. and Bazilevs, Y. (2011). A conservative level set method suitable for variable-order approximations and unstructured meshes. Journal of Computational Physics 230(12): 4536–4558.

Koo, B., Goupee, A. J., Lambrakos, K. and Kimball, R. W. (2012). Model tests for a floating wind turbine on three different floaters. Proceedings of the International Conference on Offshore Mechanics and Arctic Engineering—OMAE 7: 455–466.

Lesoinne, M. and Farhat, C. (1998). Higher-order subiteration-free staggered algorithm for nonlinear transient aeroelastic problems. AIAA Journal 36(9): 1754–1757.

Piperno, S., Farhat, C. and Larrouturou, B. (1995). Partitioned procedures for the transient solution of coupled aroelastic problems Part I: Model problem, theory and two-dimensional application. Computer Methods in Applied Mechanics and Engineering 124(1-2): 79–112.

Proteus (2019). Proteus—a python toolkit for computational methods and simulation. Accessed on: 2019-10-12.

Quezada de Luna, M., Kuzmin, D. and Kees, C. E. (2019). A monolithic conservative level set method with built-in redistancing. Journal of Computational Physics 379: 262–278.

Recuero, A. and Negrut, D. (2016). Chrono Support for ANCF Finite Elements: Formulation and Validation Aspects. Technical report, University of Wisconsin–Madison.

Robertson, A., Jonkman, J. and Masciola, M. (2014). Definition of the Semisubmersible Floating System for Phase II of OC4. Technical Report September, National Renewable Energy Laboratory.

Söding, H. (2001). How to integrate free motions of solids in fluids. In 4th Numerical Towing Tank Symposium, pp. 23–25.

Sussman, M., Smereka, P. and Osher, S. (1994). A level set approach for computing solutions to incompressible two-phase flow. Journal of Computational Physics 114(1): 146–159.

Temam, R. (1969a). Sur l'approximation de la solution des équations de Navier-Stokes par la méthode des pas fractionnaires (I). Archive for Rational Mechanics and Analysis 32: 135–153.

Temam, R. (1969b). Sur l'approximation de la solution des équations de Navier-Stokes par la méthode des pas fractionnaires (II). Archive for Rational Mechanics and Analysis 33: 377–385.

Yvin, C., Leroyer, A. and Visonneau, M. (2013). Co-simulation in fluid-structure interaction problem with rigid bodies. In 16th Numerical Towing Tank Symposium, 7.

Yvin, C., Leroyer, A., Guilmineau, E., Queutey, P. and Visonneau, M. (2014). Couplage de Codes pour l'Etude d'Interactions Fluide-Structure de Corps Rigides dans le Domain de l'Hydrodynamique Navale. In 14èmes Journèes de l'Hydrodynamique.

Zhao, X. and Hu, C. (2012). Numerical and experimental study on a 2-D floating body under extreme wave conditions. Applied Ocean Research 35: 1–13.

Future Prospects

Numerical modelling of nearshore (coastal) wave processes with the Navier-Stokes (NS) equations is a scientific area that has been studied for several decades. The research work in this field, reviewed in this book, has resulted in state-of-the-art numerical methods and techniques to solve diverse and complex processes such as wave transformation, nonlinear wave-wave and wave-structure interactions or sediment transport processes. There have been a large number of recent developments, most of which have been made possible by easy access to powerful and cheap computational CPU resources, and GPUs, which possess thousands of cores (unlike CPUs, which can only have few tens of them), and enable users to have access to an affordable and very compact computational cluster. Computational fluid dynamics (CFD), however, still presents challenges that need to be overcome in order to progress yet further. This chapter aims to present, and summarize, a broad series of topics that will likely draw significant interest in the future.

Undoubtedly, despite the apparent demise of Moore's law, the future will bring wider access to more powerful computational resources, which in turn will open new simulation possibilities. First, this fact will enable the simulation of larger or higher resolution cases, and generalize the use of more detailed turbulence modelling/simulation approaches. Although the Reynolds-Averaged Navier-Stokes (RANS) approach is currently dominant, other methods presently used but still not fully developed, such as Large Eddy Simulation (LES) or Direct Numerical Simulation (DNS), will become more commonly applied. Perhaps DNS, which is computationally far too demanding to apply in practical simulations presently, may help provide further insight into turbulence in the future.

Another approach that will benefit from the increase in computational power is multiphysics modelling, e.g., Latham et al. (2013); Ulrich et al. (2013). This type of modelling usually requires a lot of computational resource, because such models are able to simulate wave and structure interactions, *including* the solid mechanics within the structures. Currently, however, these models are still in an early stage of development.

Finally, in the longer term, quantum computing is knocking at the door; this radically new computational framework will need to be explored in depth in the context of CFD model development, which will doubtlessly result in new computational methods and simulation approaches.

In a shorter time frame, there are some existing approaches that, although not extensively used at this time, possess significant advantages and encouraging results, and therefore posses the potential for benefitting the field of CFD for coastal engineering significantly.

1   The lattice Boltzmann method

The lattice Boltzmann method (LBM) (Succi, 2001; Frandsen, 2008; Janssen and Krafczyk, 2011) is a promising approach which is derived from the Lattice Gas Automata (LGA) method, a cellular automaton molecular dynamics model developed in the late 1980s. Despite being a discrete method, the traditional macroscopic NS equations can be derived with LGA, in which the particles in LBM move on a fixed lattice, and at every time step they undergo two processes: propagation and collision.

LBM offers significant advantages with respect to other CFD approaches, such as Eulerian and Lagrangian (Smooth Particle Hydrodynamics, SPH) methods. For example, since the particle flow is accounted for, it can deal with complex boundaries (Yin and Zhang, 2012) and boundary conditions (Chang et al., 2009) in a more consistent way than SPH. Also, this approach is suitable for performing massive simulations very efficiently, as it can be parallelized in GPUs (Janssen and Krafczyk, 2011). Finally, LBM has been successfully coupled with the Discrete Element Method (DEM) (Liu and Wu, 2019), to solve flow within porous media generated by solid mechanical interactions, which is an example of a multiphysics model.

LBM presents some disadvantages too. Perhaps the best-known is the limitation to model high-speed (i.e., high Mach number) flows, although this is not likely to present an issue in coastal engineering applications, except perhaps with the exception of modelling impulsive wave breaking on structures. Another disadvantage, which is shared with other simulation approaches, is the diffusivity of the interface in multiphase cases, which impacts the numerical stability of such models, and the difficulty of dealing with the large density gradient (e.g., air-water, 1:1000) in that region (Yang and Boek, 2013).

2   Arbitrary and hybrid Lagrangian-Eulerian models

The Arbitrary Lagrangian-Eulerian (ALE) technique is another advantageous approach, that is extremely promising for coastal CFD application. The approach merges the benefits of Eulerian and Lagrangian flow descriptions. The vertices of the computational mesh in ALE models are arbitrary in that they are either allowed to move (in a Lagrangian fashion) or remain fixed in a manner that aids rezoning. In ALE, the computational mesh thus evolves *selectively* and independently of the fluid motion. There are typically three options for moving the cell vertices:

- Move with the fluid for Lagrangian computing

- Stay fixed for Eulerian computing

- Move in an arbitrarily prescribed way allowing a continuous rezoning capability.

The ALE cycle for hydrodynamic calculations is divided into three phases. Phase I is a standard Lagrangian calculation. Phase II is the rezoning phase, in which rezone velocities are specified to reduce distortions in the mesh. Phase III performs all the advective flux calculations; these are necessary if the mesh is not purely Lagrangian. The purpose of using three phases is to know the Lagrangian motion before a choice is made for rezoning. Through rezoning, a simulation can be run beyond what a pure Lagrangian mesh-based method, such as that described in Chapter 2, will allow. This enables tracking of the interface between fluids, including localised high resolution when it is required and the ability to maintain higher resolution than a Eulerian method for the same number of computational nodes (Donea et al., 2017).

Moreover, the ALE method can be a particularly powerful tool when combined with techniques for actively controlling mesh quality and mesh refinement. Such methods prevent mesh quality deterioration due to mesh entanglement, the development of skewed cells, and at the same time, they allow for maintaining sufficient refinement around elements such as wall boundaries, fluid interfaces or small scale features. These can be simple techniques, such as the linear elastostatic approach for mesh motion (Dwight, 2009; de Lataillade, 2019), which allows smaller cells to maintain their shape by making them more "rigid" and accommodating mesh deformations by making larger cells more "flexible". More sophisticated approaches such as mesh monitor functions (Grajewski et al., 2010) or mesh adaptivity (Bathe and Zhang, 2009) can yield an even higher degree of control to the mesh refinement but these come with a more complicated implementation and higher computational cost. To-date the ALE method has not been widely used for fluid structure interaction (FSI) problems within a coastal engineering setting, but a successful example of application is shown in chapter 8 of this book.

Other hybrid Lagrangian-Eulerian models have been employed within a coastal CFD framework; perhaps the most well-known of which is the Marker and Cell (MAC) method originally developed by Harlow and co-workers (Harlow and Welch, 1965) which uses passive marker particles to track the free surface. In MAC the particles serve the sole purpose of surface tracking. More recently, the Particle in Cell (PIC) approach, also originally due to Harlow and co-workers (Harlow, 1964), has been revived and applied to various Coastal Engineering problems (Kelly, 2012; Chen et al., 2016; Maljaars et al., 2017). The PIC method attempts to get the best from two worlds using a Lagrangian approach efficiently for both the non-linear velocity advection and interface tracking whilst employing an optimized Eulerian approach to enforce solid boundary conditions and also compute the pressure. With recent innovations the PIC approach seems particularly well suited as a CFD method for fluid structure interaction (FSI) problems, including problems that involve floating structures, within a coastal engineering setting (e.g., Chen et al. (2018, 2019)). In common with all particle, or semi-particle, based methods particle clustering, due to numerical error, is a problem with the PIC method. Recent work by Maljaars et al. (2018) uses a hybridized discontinuous Galerkin (HDG) framework for the Eulerian phase of the computations. The HDG approach appears to allow for strict vanishing of the velocity divergence in a local, cell-by-cell, manner. This, combined with the diffusion free advection offered by the Lagrangian phase, retains a uniform particle distribution over time. This removes the need for particle re-distribution schemes which themselves add additional, unwanted, numerical diffusion. The work of Maljaars et al. (2018) appears very promising in this regard; however, the problem of the free surface remains unresolved.

3 Direct pressure and pressure-marching methods

Direct methods are those in which pressure is solved directly (sometimes exactly). This is in contradistinction to the projection method where the value for the pressure is solved approximately by the iterative solution of an elliptic Poisson equation (PPE). Because the PPE is elliptic it necessitates the use of an implicit solution technique, thus introducing another layer of complexity to parallel codes when compared with simple explicit techniques. The explicit time-marching (ETM) of pressure falls into the direct methods class as defined here. This is because it is not an iterative method. The ETM approach, which includes both the penalty approach of Temam (1984) and the artificial compressiblity approach devised independly by Yanenko (1971) and Chorin (1967), is particularly attractive for free surface flows that involve (potentially) violent fluid structure interactions. The equations in ETM approaches, such as the penalty approach (Temam, 1984), the artificial compressibility approach (Chorin, 1967; Yanenko, 1971), or combinations of these two approaches, are necessarily stiff and thus necessitate the use of a small time-step to ensure numerical stability. This means that such approaches are not competitive with implicit iterative-type approaches for flows without free surfaces, relatively slow moving free surface flows or flows that tend toward a steady state (Dukowicz, 1994). For highly unsteady free surface flows, however, rapid change (which can only be captured by a small time-step) is an inherent feature of the flow motion; thus, the main disadvantage of ETM approaches, namely the requirement of a small time-step, is no longer a disadvantage in this context. Moreover, large benefits can be gained from combining the penalty and artificial compressibility approaches together; this was first understood by Yanenko et al. (1984) who combine the two aforementioned methods by adding a linear combination of the pressure and pressure evolution terms to the velocity divergence. Later this technique was refined and improved by Ramshaw and Mesina (1991) who added the pressure and pressure evolution contributions to the pressure itself. Very little work has been done on combined penalty-artificial compressibility approaches applied to free surface flows. It is noted here that this avenue appears to offer a lot of promise as purely explicit codes are relatively simple to write and straightforward to parallelize.

4   Machine learning

Machine learning (ML) is a series of techniques that allow computers to perform tasks without being explicitly programmed to do so, by "learning" insights from input datasets. ML has a vast potential and is currently a hot topic in multiple scientific fields; CFD and fluid mechanics in general are not an exception. Although the variety of approaches that machine learning encompasses is very wide, the number of works of machine learning applied in coastal engineering is not very large yet. Two review papers, one focussed in general ML and another focussed in Deep Reinforcement Learning applied to fluid mechanics can be found in Brunton et al. (2019) and Garnier et al. (2019). Other examples of particular applications are provided as follows.

ML has proven to be a very useful tool to rationalise the use of computational resources in CFD. In this sense, Stefanakis et al. (2014) explored the maximum runup behind a conical island, a system that depends on multiple inputs, and ML identified the interdependence between them and allowed reducing the total number of simulations required to gain valuable insights. Hennigh (2017) used ML to reduce the computational time and memory required to run Lattice Boltzmann CFD simulations and, most recently, Bar-Sinai et al. (2019) presented an approach to obtain accurate representations of PDEs on coarse grids, which in the future could be applied to NS equations.

One of the most recent trends and promising approaches within ML is Deep Learning (DL), which is expected to play a critical enabling role in the future of modelling complex flows (Kutz, 2017). This technique mimics how brains work, representing the connections between neurons; therefore, the input data gets assimilated with so-called neural networks, which can have different architecture depending on their purpose. The shape of the neural network determines its suitability to perform certain functions, and the internal weights of such neurons produce the output.

Due to its complexity, and the fact that it is still not fully understood, turbulence analysis and modelling is the perfect target to be attacked with DL. Most of the relevant numerical modelling studies that have been performed, unfortunately outside the coastal engineering field, aim at improving turbulence modelling with experimental or numerical (i.e., DNS) turbulence data. Examples applied to RANS turbulence modelling include Ling et al. (2016); Moghaddam and Sadaghiyani (2018); Zhu et al. (2019), whereas LES subgrid closure is studied in Beck et al. (2018).

In view of the results of these studies, it can be concluded that sooner rather than later, ML will have a noticeable impact in CFD models for coastal flows.

5 Coupled models

Navier-Stokes (NS) modelling is the most complete and accurate procedure to simulate fluids, but this approach is extremely demanding computationally and NS equations cannot represent additional physics, e.g., solid mechanics problems. These two limitations can be overcome by means of coupling additional models with the CFD model in an approach known as hybrid modelling (HM).

As a first example, since running very large domains in RANS is still not feasible, some types of "simple" hydrodynamics do not require solving the full NS equations and can be simulated with less complex sets of equations instead. For instance, wave propagation and transformation nearshore can be approximated very efficiently with the Nonlinear Shallow Water Equations (NLSWE) (Stoker, 1957; Brocchini and Dodd, 2008), the Green-Naghdi equations (Green et al., 1974), the fully nonlinear potential flow (FNPF) equations based on the Boundary Element Method (BEM) (Grilli et al., 1989), the Boussinesq equations (Wei et al., 1995), etc. Alternatively, waves in deep waters can be approximated with the FNPF equations (Grilli et al., 2010), the High Order Spectral (HOS) methods Dommermuth and Yue (1987); Ducrozet et al. (2007, 2012), etc.

Such approximations are very efficient but cannot deal with very complex processes, such as wave breaking, in a detailed way. Therefore, producing a hybrid model that couples both equations/models will reduce the computational costs associated to solving NS equations in areas in which the flow conditions allow using simpler equations, while offering NS accuracy where required.

As a second example, waves often interact with structures in what has been called wave-structure interaction simulations (Chen et al., 2014), a special case of the best-known fluid-structure interaction (FSI) field. When structures respond to hydrodynamic forcing, accounting for the interaction between both requires solving additional sets of equations, e.g., Newton's second law or solid mechanics equations, which yield the structure movements and deformations.

There are two ways to establish a coupling between models. First is the one-way coupling (i.e., →), in which information is only transferred from one model to the other, e.g., from the propagation model into the CFD code. This is the simplest approach; it is easy to implement and allows running the complete simulation in the simplified model first. However, reflected waves in the CFD simulation cannot propagate back into the propagation model and interact with the incoming waves, hence, this constitutes a simplified option.

The other approach is two-way coupling (i.e., ↔), in which both models run concurrently and exchange information at the coupling interface (area/volume). This option is more complex to implement, but allows for a seamless simulation framework. There are two methods of solving two-way coupling problems. On the one hand, the interaction between the fluid and the structure is reduced to a single iteration every time step in loosely-coupled interfaces; thus, it does not solve for the equilibrium state between both. On the other hand, in strongly-coupled interfaces, the system is solved iteratively, ensuring that the fluid and structure have reached an instantaneous equilibrium state.

Regardless of the approach chosen, the coupling should be implemented in such a way that the wave kinematics are transferred accurately, to prevent the introduction of additional errors into the simulation.

Hybrid models are not new, but have just started gaining momentum and will most likely continue to do so in the future. Some examples of HM involving wave propagation-

CFD models include the work of Altomare et al. (2014) in which the authors couple a NLSWE model with SPH; Narayanaswamy et al. (2010), in which an SPH model is coupled with a Boussinesq model; or Paulsen et al. (2014), who couple a fully nonlinear potential flow with OpenFOAM®. Another example of hydrodynamic coupling between a Lagrangian model and CFD was introduced in Higuera et al. (2018) and has already been discussed in Chapter 2.

FSI is another field in which HM is currently a "hot topic", mainly due to the interest in renewable energies. Examples of FSI include wave energy converters (WECs) of all kinds (Brito et al., 2016; Ransley et al., 2017; Wei et al., 2019; Windt et al., 2020); floating offshore platforms and wind turbines (Leble and Barakos, 2016; Chivaee et al., 2018; Wang et al., 2019) see also Chapter 8 of this book. Further applications can be observed in Canelas et al. (2018) or Chen and Zou (2019), in which waves interact with vegetation and the coupling is performed via the immersed boundary method (IBM).

Finally, a most promising approach directly related to coastal engineering is the coupling of wave action with the concrete units in breakwaters (Xiang et al., 2012). The ultimate goal would be to couple the hydrodynamics with discrete element method (DEM) to study the displacements caused by extreme events (Xiang et al., 2019).

References

Altomare, C., Suzuki, T., Domínguez, J. M., Crespo, A., Gomez-Gesteira, M. and Caceres, I. (2014). A hybrid numerical model for coastal engineering problems. In Proceedings of the 34th International Conference on Coastal Engineering (ICCE), Seoul, South Korea.

Bar-Sinai, Y., Hoyer, S., Hickey, J. and Brenner, M. P. (2019). Learning data-driven discretizations for partial differential equations. Proceedings of the National Academy of Sciences 116(31): 15344–15349.

Bathe, K. J. and Zhang, H. (2009). A mesh adaptivity procedure for cfd and fluid-structure interactions. Computers & Structures 87(11-12): 604–617.

Beck, A. D., Flad, D. G. and Munz, C. -D. (2018). Deep neural networks for data-driven turbulence models. arXiv preprint arXiv:1806.04482.

Brito, M., Canelas, R. B., Ferreira, R. M. L., García-Feal, O., Domínguez, J. M., Crespo, A. J. C. and Neves, M. G. (2016). Coupling between DualSPHysics and Chrono-Engine: towards large scale HPC multiphysics simulations. In 11th International SPHERIC Workshop, Munich, Germany.

Brocchini, M. and Dodd, N. (2008). Nonlinear shallow water equation modeling for coastal engineering. Journal of Waterway, Port, Coastal and Ocean Engineering 134(2): 104–120.

Brunton, S. L., Noack, B. R. and Koumoutsakos, P. (2019). Machine learning for fluid mechanics. Annual Review of Fluid Mechanics 52.

Canelas, R. B., Brito, M., García-Feal, O., Domínguez, J. M. and Crespo, A. J. C. (2018). Extending DualSPHysics with a differential variational inequality: modeling fluid-mechanism interaction. Applied Ocean Research 76: 88–97.

Chang, C., Liu, C. -H. and Lin, C. -A. (2009). Boundary conditions for lattice Boltzmann simulations with complex geometry flows. Computers & Mathematics with Applications 58(5): 940–949.

Chen, H. and Zou, Q. -P. (2019). Eulerian-Lagrangian flow-vegetation interaction model using immersed boundary method and OpenFOAM. Advances in Water Resources 126: 176–192.

Chen, L. F., Zang, J., Hillis, A. J., Morgan, G. C. J. and Plummer, A. R. (2014). Numerical investigation of wave-structure interaction using openfoam. Ocean Engineering 88: 91–109.

Chen, Q., Kelly, D. M., Dimakopoulos, A. S. and Zang, J. (2016). Validation of the PICIN solver for 2D coastal flows. Coastal Engineering 112: 87–98.

Chen, Q., Zang, J., Kelly, D. M. and Dimakopoulos, A. S. (2018). A 3D parallel particle-in-cell solver for wave interaction with vertical cylinders. Ocean Engineering 147: 165–180.

Chen, Q., Zang, J., Ning, D., Blenkinsopp, C. and Gao, J. (2019). A 3D parallel particle-in-cell solver for extreme wave interaction with floating bodies. Ocean Engineering 179: 1–12.

Chivaee, H. S., Jurado, A. M. P. and Bredmose, H. (2018). CFD simulations of a newly developed floating offshore wind turbine platform using OpenFOAM. In 21st Australasian Fluid Mechanics Conference.

Chorin, A. J. (1967). A numerical method for solving incompressible viscous flow problems. Journal of Computational Physics 2(1): 12–26.

de Lataillade, T. d. (2019). High-Fidelity Computational Modelling of Fluid-Structure Interaction for Moored Floating Bodies. Ph.D. thesis, IDCORE Doctoral Training Centre, Edinburg (submitted).

Dommermuth, D. G. and Yue, D. K. P. (1987). A high-order spectral method for the study of nonlinear gravity waves. Journal of Fluid Mechanics 184: 267–288.

Donea, J., Huerta, A., Ponthot, J. -P. and Rodríguez-Ferran, A. (2017). Arbitrary Lagrangian-Eulerian Methods. Encyclopedia of Computational Mechanics Second Edition, pp. 1–23.

Ducrozet, G., Bonnefoy, F., Le Touzé, D. and Ferrant, P. (2007). 3-D HOS simulations of extreme waves in open seas. Natural Hazards and Earth System Science 7(1): 109–122.

Ducrozet, G., Bonnefoy, F., Le Touzé, D. and Ferrant, P. (2012). A modified high-order spectral method for wavemaker modeling in a numerical wave tank. European Journal of Mechanics-B/ Fluids 34: 19–34.

Dukowicz, J. K. (1994). Computational efficiency of the hybrid penalty-pseudocompressibility method for incompressible flow. Computers & Fluids 23(2): 479–486.

Dwight, R. P. (2009). Computational Fluid Dynamics 2006, Chapter Robust Mesh Deformation using the Linear Elasticity Equations. Springer, Berlin, Heidelberg.

Frandsen, J. (2008). A simple LBE wave runup model. Progress in Computational Fluid Dynamics, An International Journal 8(1-4): 222–232.

Garnier, P., Viquerat, J., Rabault, J., Larcher, A., Kuhnle, A. and Hachem, E. (2019). A review on deep reinforcement learning for fluid mechanics. arXiv preprint arXiv: 1908.04127.

Grajewski, M., Köster, M. and Turek, S. (2010). Numerical analysis and implementational aspects of a new multilevel grid deformation method. Applied Numerical Mathematics 60(8): 767–781.

Green, A. E., Laws, N. and Naghdi, P. M. (1974). On the theory of water waves. Proceedings of the Royal Society of London. A. Mathematical and Physical Sciences 338(1612): 43–55.

Grilli, E. T., Dias, F., Guyenne, P., Fochesato, C. and Enet, F. (2010). Progress in fully onlinear potential flow modelling of 3D extreme ocean waves, pp. 75–128.

Grilli, S. T., Skourup, J. and Svendsen, I. A. (1989). An efficient boundary element method for nonlinear water waves. Engineering Analysis with Boundary Elements 6(2): 97–107.

Harlow, F. and Welch, J. E. (1965). Numerical calculation of time-dependent viscous incompressible flow of fluid with a free surface. Physics of Fluids 8: 191–203.

Harlow, F. H. (1964). The particle-in-cell computing method for fluid dynamics. pp. 319–343. *In*: Alder, B. (ed.). Methods in Computational Physics, Academic Press, New York.

Hennigh, O. (2017). Lat-Net: compressing lattice Boltzmann flow simulations using deep neural networks. arXiv preprint arXiv: 1705.09036.

Higuera, P., Buldakov, E. and Stagonas, D. (2018). Numerical modelling of wave interaction with an FPSO using a combination of OpenFOAM and Lagrangian models. In Proceedings of the 28th International Ocean and Polar Engineering Conference, Sapporo, Japan, June 10–15, 2018, volume ISOPE-I-18-014.

Janssen, C. and Krafczyk, M. (2011). Free surface flow simulations on GPGPUs using the LBM. Computers & Mathematics with Applications 61(12): 3549–3563.

Kelly, D. M. (2012). Full particle PIC modelling of the surf and swash zones. In Proceedings of the 33rd International Conference on Coastal Engineering (ICCE), Santander, Spain 3: 1982–1990.

Kutz, J. N. (2017). Deep learning in fluid dynamics. Journal of Fluid Mechanics 814: 1–4.

Latham, J. -P., Anastasaki, E. and Xiang, J. (2013). New modelling and analysis methods for concrete armour unit systems using FEMDEM. Coastal Engineering 77: 151–166.

Leble, V. and Barakos, G. N. (2016). CFD investigation of a complete floating offshore wind turbine. In MARE-WINT, Springer, pp. 277–308.

Ling, J., Kurzawski, A. and Templeton, J. (2016). Reynolds averaged turbulence modelling using deep neural networks with embedded invariance. Journal of Fluid Mechanics 807: 155–166.

Liu, W. and Wu, C. -Y. (2019). A hybrid LBM-DEM numerical approach with an improved immersed moving boundary method for complex particle-liquid flows involving adhesive particles. arXiv preprint arXiv: 1901.09745.

Maljaars, J. M., Labeur, R. J., Möller, M. and Uijttewaal, W. (2017). Development of a hybrid particle-mesh method for simulating free–surface flows. Journal of Hydrodynamics, Ser. B 29(3): 413–422.

Maljaars, J. M., Labeur, R. J. and Möller, M. (2018). A hybridized discontinuous Galerkin framework for high-order particle-mesh operator splitting of the incompressible Navier-Stokes equations. Journal of Computational Physics 358: 150–172.

Moghaddam, A. A. and Sadaghiyani, A. (2018). A deep learning framework for turbulence modeling using data assimilation and feature extraction. arXiv preprint arXiv: 1802.06106.

Narayanaswamy, M., Cabrera Crespo, A. J., Gómez-Gesteira, M. and Dalrymple, R. A. (2010). SPHysics-FUNWAVE hybrid model for coastal wave propagation. Journal of Hydraulic Research 48: 85–93.

Paulsen, B. T., Bredmose, H. and Bingham, H. B. (2014). An efficient domain decomposition strategy for wave loads on surface piercing circular cylinders. Coastal Engineering 86: 57–76.

Ramshaw, J. D. and Mesina, G. L. (1991). A hybrid penalty-pseudocompressibility method for transient incompressible fluid flow. Computers & Fluids 20(2): 165–175.

Ransley, E. J., Greaves, D. M., Raby, A., Simmonds, D., Jakobsen, M. M. and Kramer, M. (2017). RANS-VOF modelling of the Wavestar point absorber. Renewable Energy 109: 49–65.

Stefanakis, T. S., Contal, E., Vayatis, N., Dias, F. and Synolakis, C. E. (2014). Can small islands protect nearby coasts from tsunamis? An active experimental design approach. Proceedings of the Royal Society A: Mathematical, Physical and Engineering Sciences 470(2172).

Stoker, J. J. (1957). Water Waves, the Mathematical Theory with Applications. Interscience Publishers Inc., New York.

Succi, S. (2001). The Lattice Boltzmann Equation: For Fluid Dynamics and Beyond. Oxford University Press.

Temam, R. (1984). Navier-Stokes Equations: Theory and Numerical Analysis. Studies in Mathematics and its Applications. North-Holland.

Ulrich, C., Leonardi, M. and Rung, T. (2013). Multi-physics SPH simulation of complex marine-engineering hydrodynamic problems. Ocean Engineering 64: 109–121.

Wang, J. -H., Zhao, W. -W. and Wan, D. -C. (2019). Development of naoe-FOAM-SJTU solver based on OpenFOAM for marine hydrodynamics. Journal of Hydrodynamics 31(1): 1–20.

Wei, G., Kirby, J. T., Grilli, S. T. and Subramanya, R. (1995). A fully nonlinear Boussinesq model for surface waves. Part 1. Highly nonlinear unsteady waves. Journal of Fluid Mechanics 294: 71–92.

Wei, Z., Edge, B. L., Dalrymple, R. A. and Hérault, A. (2019). Modeling of wave energy converters by GPUSPH and Project Chrono. Ocean Engineering 183: 332–349.

Windt, C., Davidson, J., Ransley, E. J., Greaves, D., Jakobsen, M., Kramer, M. and Ringwood, J. V. (2020). Validation of a CFD-based numerical wave tank model for the power production assessment of the wavestar ocean wave energy converter. Renewable Energy 146: 2499–2516.

Xiang, J., Latham, J. -P., Higuera, P., Via-Estrem, L., Eden, D., Douglas, S., Simplean, A., Nistor, I. and Cornett, A. (2019). A fast and effective wave proxy approach for wavestructure interaction in rubble mound structures. In Proceedings of Coastal Structures 2019, Hannover, Germany.

Xiang, J., Latham, J. -P., Vire, A., Anastasaki, E. and Pain, C. C. (2012). Coupled Fluidity/Y3D technology and simulation tools for numerical breakwater modelling. In Proceedings of the 33rd International Conference on Coastal Engineering (ICCE), Santander, Spain.

Yanenko, N. N. (1971). The Method of Fractional Steps: The Solution of Problems of Mathematical Physics in Several Variables. Springer-Verlag.

Yanenko, N. N., Shokurov, V. and Shokin, Y. I. (1984). Numerical Methods in Fluid Dynamics. Mir Publishers.

Yang, J. and Boek, E. S. (2013). A comparison study of multi-component Lattice Boltzmann models for flow in porous media applications. Computers & Mathematics with Applications 65(6): 882–890.

Yin, X. and Zhang, J. (2012). An improved bounce-back scheme for complex boundary conditions in lattice Boltzmann method. Journal of Computational Physics 231(11): 4295–4303.

Zhu, L., Zhang, W., Kou, J. and Liu, Y. (2019). Machine learning methods for turbulence modeling in subsonic flows around airfoils. Physics of Fluids 31(1).

Index

Other Useful Reading

Carmo, J. S. A. D. (2020). Physical Modelling vs. Numerical Modelling: Complementarity and Learning. Preprints 2020, 2020070753 (doi: 10.20944/preprints202007.0753.v1).

Frostick, L. E., McLelland, S. J. and Mercer, T. G. (2011). Users guide to Physical Modelling and Experimentation–experience of the HYDRALAB Network, CRC Press, pp 272, ISBN 9780429108211.

Hughes, S. A. (1993). Physical Models and Laboratory Techniques in Coastal Engineering, ISBN 981-02-1540-1, World Scientific, Singapore.

Russell, J. S. (1847). On the practical forms of breakwaters, sea walls and other engineering works exposed to the action of waves. Proc. ICE, Vol VI pp. 135–148.

Wolters, G., van Gent, M., Allsop, W., Hamm, L. and Mühlestein, D. (2009). HYDRALAB III: Guidelines for physical model testing of rubble mound breakwaters. Proc ICE Conf. on Coasts, Marine Structures & Breakwaters, publn. Thomas Telford, London.

T - #0269 - 111024 - C68 - 254/178/12 - PB - 9780367619381 - Gloss Lamination